Constitutive Equations for Polymer Melts and Solutions

BUTTERWORTHS SERIES IN CHEMICAL ENGINEERING

SERIES EDITOR

HOWARD BRENNER
Massachusetts Institute of Technology

ADVISORY EDITORS

ANDREAS ACRIVOS
Stanford University

JAMES E. BAILEY
California Institute of Technology

MANFRED MORARI
California Institute of Technology

E. BRUCE NAUMAN
Rensselaer Polytechnic Institute

ROBERT K. PRUD'HOMME
Princeton University

SERIES TITLES

Chemical Process Equipment: Selection and Design *Stanley M. Walas*

Constitutive Equations for Polymer Melts and Solutions *Ronald G. Larson*

Gas Separation by Adsorption Processes *Ralph T. Yang*

Heterogeneous Reactor Design *Hong H. Lee*

Molecular Thermodynamics of Nonideal Fluids *Lloyd L. Lee*

Phase Equilibria in Chemical Engineering *Stanley M. Walas*

Transport Processes in Chemically Reacting Flow Systems *Daniel E. Rosner*

RELATED TITLES

Catalyst Supports and Supported Catalysts *Alvin B. Stiles*

Enlargement and Compaction of Particulate Solids *Nayland Stanley-Wood*

Fundamentals of Fluidized Beds *John G. Yates*

Liquid and Liquid Mixtures *J.S. Rowlinson and F.L. Swinton*

Mixing in the Process Industries *N. Harnby, M. Edwards, and A.W. Nienow*

Solid Liquid Separation *Ladislav Svarovsky*

Supercritical Fluid Extraction *Mark A. McHugh and Val J. Krukonis*

Constitutive Equations for Polymer Melts and Solutions

Ronald G. Larson
Member of Technical Staff
AT&T Bell Laboratories

Butterworths
Boston London Singapore Sydney Toronto Wellington

Library of Congress Cataloging in Publication Data
Larson, Ronald G.
Constitutive Equations for Polymer Melts and Solutions.
(Butterworths series in chemical engineering).
Includes index.
1. Polymers and polymerization — Rheology
2. Polymer melting. I. Title. II. Series.
QD381.9.R48L37 1988 660.2'8448 87-21807
ISBN 0-409-90119-9

British Library Cataloguing in Publication Data
Larson, Ronald G.
Constitutive equations for polymer melts and Solutions.
(Butterworths series in chemical engineering).
1. Polymers and polymerization — Mathematical Models
2. Rheology — Mathematical models
I. Title
547.8'4'0724 QD381.8
ISBN 0-409-90119-9

Butterworth Publishers
80 Montvale Avenue
Stoneham, MA 02180

10 9 8 7 6 5 4 3 2 1

Printed in the United States of America

To Bebe

CONTENTS

PREFACE

This book presents a description of the most important constitutive equations for stress and birefringence in polymer melts, as well as in dilute and concentrated solutions of flexible and rigid polymers, and in liquid crystalline materials. Presented also are the molecular, structural, or continuum-mechanical models from which the equations are derived, and a comparison of the predictions of the equations with each other and with experimental data. *Constitutive Equations for Polymer Melts and Solutions* is an introduction and guide to constitutive equations, and to molecular and phenomenological theories of polymer motion and flow.

Constitutive equations are mathematical relationships that allow one to calculate the stresses in a liquid, given the flow history. They are for non-Newtonian fluids what Newton's law of viscosity is for Newtonian fluids. With a reliable constitutive equation, one can describe the important polymer flows and shaping processes such as occur in basic and applied research and engineering, molding, extrusion, encapsulation, film blowing, cable jacketing, and fiber spinning.

Constitutive *equations* often are derived from constitutive *models*. A constitutive model is a set of assumptions and idealizations about the molecular or structural forces and motions that produce stress. By comparing the constitutive equation derived from a given model with measured stresses, one can test the model. Thus a constitutive equation is not only useful for prediction of gross motions or flow, but is also useful for testing ideas of polymer molecular physics. Constitutive equations are the link between the physics of polymer motion and macroscopic flow phenomena.

Chapter 1 introduces the flow phenomena commonly associated with viscoelasticity. Since constitutive equations are mathematical objects and require the mathematical tools of tensor calculus for their understanding, the necessary mathematical background is introduced in Chapter 1. At the end of Chapter 1, a simple constitutive equation, the upper-convected Maxwell equation, is introduced in its differential and integral forms.

Chapter 2 contains the elementary models that have become foundational for understanding the rheology of melts and solutions of flexible polymers, and for understanding more advanced constitutive equations. These are the temporary network model, the Hookean dumbbell model, and the bead-spring model of Rouse and Zimm. After a fairly detailed presentation of these models, foundational concepts of linear viscoelasticity and time-temperature superposition are discussed.

A continuum-mechanical approach to constitutive equations is taken in Chapter 3, with emphasis on linear viscoelasticity, the Oldroyd differential constitutive equations, and the Kaye-BKZ class of integral equations. The advantages and limitations of these equations are discussed, preparing the reader for a discussion of molecular constitutive models presented in Chapters 4-8.

Chapters 4-7 deal mainly with melts and concentrated solutions of flexible polymers. Reptation models are covered in Chapter 4, with an emphasis on useful simplified forms of these models, and a comparison of the predictions of reptation models with experimental data for both ideal and commercial materials.

Birefringence and the stress-optic law are also discussed in Chapter 4. Popular semi-empirical models are considered in Chapters 5-7. Models with nonaffine motion are presented in Chapter 5, and Chapter 6 covers the Giesekus and Leonov models as well as network models. Chapter 7 contrasts and compares the simpler constitutive equations with each other and with a variety of polymer melt data. Chapter 7 also points out connections between simple integral and differential constitutive equations and gives the reader some tools with which whole classes of constitutive equations can be evaluated.

Dilute solutions of both flexible and stiff macromolecules are taken up again in much more detail in Chapter 8, which can be read after the elementary bead-spring models of Chapter 2 are mastered. The important effects of hydrodynamic interaction, solvent quality, and chain stiffness and extensibility are addressed for both shear and extensional flows.

Chapter 9 contains constitutive relationships for special flows such as steady shearing, steady extension, slow flows, or deformations of small amplitude. These relationships are relatively free of molecular or ad hoc continuum assumptions and so are valid for most *isotropic* viscoelastic liquids — that is, liquids that are unoriented at equilibrium. Chapter 9 can be read out of sequence if the reader has a familiarity with constitutive equations, gained either from the earlier chapters or elsewhere.

Because of the rapidly expanding interest in *liquid crystalline* polymers, Chapter 10 has been added to present a background for these fascinating materials. The chapter does not advocate any constitutive equations for these materials, since no reliable ones have yet been formulated, but instead seeks to provide the reader with a theoretical background that will help him to understand their basic properties. Much of this theoretical background will almost certainly be contained (in some form) in the constitutive equations that eventually emerge. Thus Chapter 10 presents theories of hindered rotary diffusion in nondilute solutions of rodlike molecules, theories of isotropic to nematic transitions, the statics of liquid crystals, the Ericksen and Leslie-Ericksen theories of flow of nonpolymeric liquid crystals, and the Doi theory of idealistic nontextured nematic polymers. Finally the rheological properties of real liquid crystal polymers are summarized and some directions for future research are mentioned.

ACKNOWLEDGEMENT

I owe an immeasurable debt of thanks to my many colleagues through whom I have been educated in this field. Several individuals have been particularly helpful to me in that they gave of their valuable time to critique portions of the manuscript: Henk Booij, John Dealy, Glenn Fredrickson, Gerald Fuller, Hanswalter Giesekus, Hermann Janeschitz-Kriegl, Saad Khan, Jules Magda, Roger Tanner, Larry Thompson, and Matt Tirrell. In addition, I am indebted to AT&T Bell Laboratories and its management; without their wholehearted support of basic research in general and my work in particular, this volume would certainly not exist.

October 28, 1987 Ronald Larson

CHAPTER 1

Introduction to Constitutive Equations for Viscoelastic Fluids

"The mountains melted from before the Lord"

Prophetess Deborah, Judges 5:5

1.1 INTRODUCTION

The earliest constitutive equations[1] were not rheological and, in general, were so simple and so reliable that they often came to be called "laws." For example, Fourier's "law" of heat conduction,

$$\underset{\sim}{q} = -K\underset{\sim}{\nabla} T \tag{1-1}$$

states that the heat flux, $\underset{\sim}{q}$, is proportional to the gradient of temperature, T. There are many other such linear relationships, such as Fick's "law" of diffusion, Darcy's "law" of permeation through porous media, and Newton's "law" of viscous flow.

These simple "laws" have several traits in common with the rheological constitutive equations that are the subject of this book. They describe transport in materials. They are needed because the conservation equations by themselves are insufficient to determine the field variables, such as T above. Constitutive equations provide needed material-dependent relationships between the field variables that appear in material-independent conservation equations. For example, Fourier's law of heat conduction supplies a relationship between temperature and heat flux that supplements the relationship provided by the energy balance equation. The two equations, together with material constants and appropriate boundary conditions, allow the determination of heat flow in a material. Constitutive equations are continuum or macroscopic descriptions, not valid at molecular or microstructural length scales, yet in principle are derivable from statistical theories that account for the microscale. Constitutive equations are economical; they contain only the material-dependent properties needed to determine macroscopic transport.

Rheological constitutive equations deal with fluid flow. Except for Newtonian fluids, which obey a linear law, rheological constitutive equations are almost always nonlinear. In that sense they are distinct from and more complex than the simple constitutive laws cited above. Viscoelastic constitutive equations describe

Fig. 1-1 Rod-climbing of a viscoelastic polyacrylamide solution (right) up a rotating shaft. The Newtonian fluid (left) doesn't climb (from Bird, Armstrong, and Hassager,[3] reprinted from *Dynamics of Polymeric Liquids*, Vol. 2 ©1977 by permission of John Wiley & Sons, Inc.).

polymeric fluids that, though liquid, have some of the properties of elastic solids.

1.2 VISCOELASTIC FLOW PHENOMENA

Polymeric fluids, owing to their viscoelastic character, can exhibit behavior in flow that differs from that of Newtonian fluids in ways sometimes subtle, sometimes striking. If the constitutive equation is to describe faithfully the polymeric fluid, it should make possible an analysis of peculiar flow phenomena, some of which are described below.

1.2.1 Rod-Climbing

In 1947, Weissenberg[2] reported and interpreted the phenomenon of rod-climbing pictured in Figure 1-1. Macromolecular fluids climb a rotating shaft! He deduced that the phenomenon was caused by elastic forces or hoop stresses directed along the streamlines. He wrote, "the pull along [lines of flow] strangulates the liquid and forces it inwards against the centrifugal forces and upwards against gravity." A prosaic example of this effect occurs when egg batter climbs the shaft of a beater. The magnitude of these hoop stresses, nowadays called normal stresses, are often larger than the shear stresses and produce pronounced effects such as rod-climbing.

1.2.2 Extrudate Swell

Another phenomenon that arises at least in part from normal stresses is extrudate swell. Figure 1-2 shows the swelling of a jet of polymeric liquid as it emerges

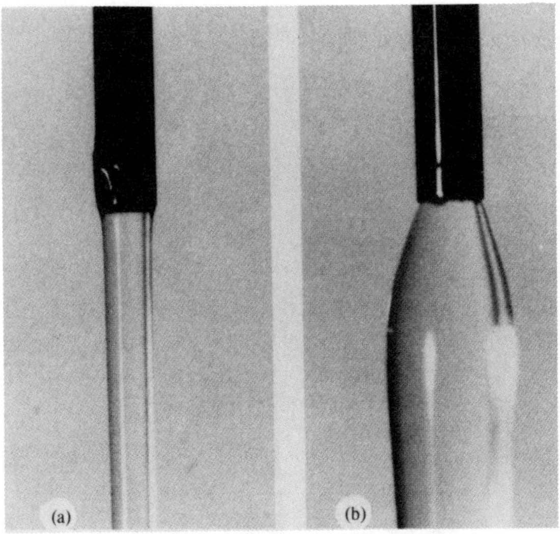

Fig. 1-2 Extrudate swell (b) of a viscoelastic liquid emerging from a die. The Newtonian fluid (a) doesn't swell (from Tanner,[4] reprinted with permission from *Engineering Rheology*, Oxford University Press, 1985).

from a tube. The diameter of the viscoelastic jet can be more than three times the tube diameter, while a Newtonian jet swells only slightly. Extrudate swell plays an important role in industrial extrusion processes.

1.2.3 Tubeless Siphon

Fano[5] in 1908 reported a remarkable phenomenon he observed in studies of biological macromolecular fluids. He immersed the tip of a capillary tube into a beaker of the fluid and began withdrawing fluid from the beaker through the tube. When the fluid level in the beaker dropped below the tip of the capillary tube, fluid, supported in part by its elasticity, continued to flow upwards to the tip of the no-longer-submerged capillary, as shown in Figure 1-3. This is known as the *tubeless siphon*. Fano recognized that the maximum height attainable by the unsupported column of liquid is related to the fluid *spinnability* —i.e., the ability to be drawn into a fluid thread. Spinnability is an important criterion in the manufacture of synthetic fibers.

1.2.4 Vortex Formation in Contraction Flows

When a polymeric fluid flows from a tube of large diameter to one of small diameter, a large doughnut-shaped vortex often forms near the mouth of the small tube; see Figure 1-4. Velocities around the vortex are often unsteady.[8] These effects are unrelated to inertia, since they occur at low Reynolds number. There is no generally accepted explanation of the vortex formation or its unsteadiness, although they are presumably caused by fluid elasticity.

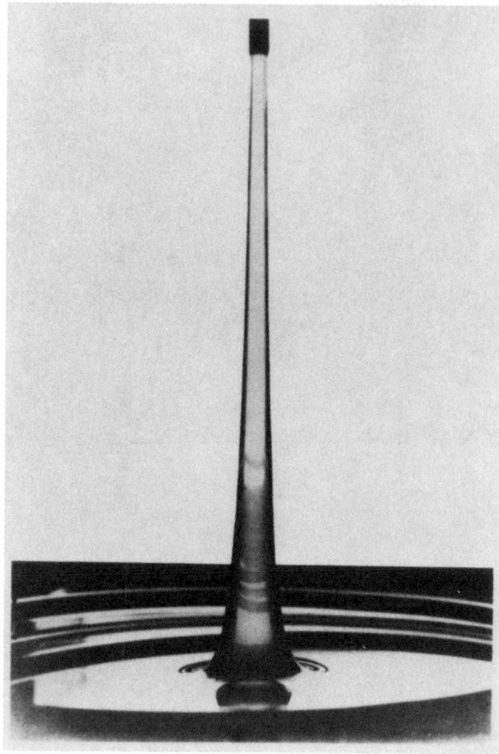

Fig. 1-3 The tubeless siphon formed by an ascending column of polymeric solution (from Peng and Landel,[6] reprinted with permission from J. Appl. Phys.).

1.2.5 Other Examples

Many other unusual phenomena are observed in polymeric flows. The interested reader can find further examples in *Dynamics of Polymeric Liquids*, Vol. 1, by Bird, Armstrong, and Hassager.[3]

1.3 VISCOELASTIC MEASUREMENTS

Quantitative explanation and prediction of viscoelastic flow phenomena such as these require the development of adequate rheological constitutive equations. Before complex phenomena can be tackled confidently, however, a proposed constitutive equation must be shown to be capable of describing the simplest flows.

1.3.1 Shear Thinning

A uniform shearing flow is produced in a fluid that is held between two parallel plates, when one plate is translated in a direction parallel to itself; see Figure 1-5a. Uniform shearing is also produced in the gap between a rotating cone and a stationary plate. The shear rate, $\dot{\gamma}$, is the velocity, V, of the moving surface divided by the gap, h; i.e.,

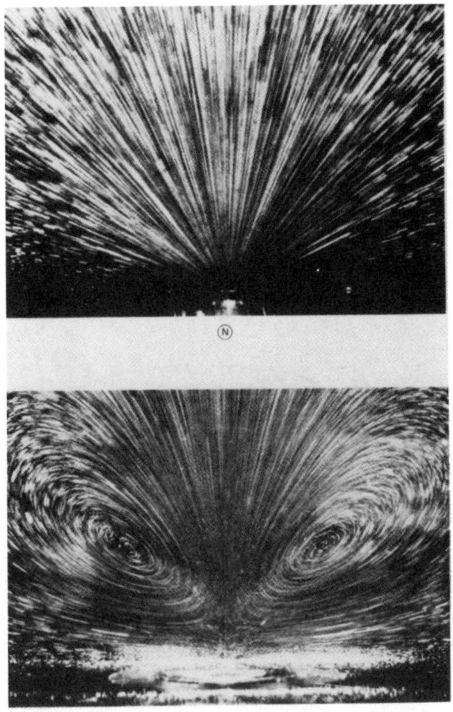

Fig. 1-4 A large toroidal eddy forms in a region of contraction of flow from a reservoir into a small tube when the fluid is viscoelastic (lower photo). The Newtonian fluid (upper photo) forms no such eddy (Photographs by H. Giesekus published in ref. 7, reproduced by permission of the American Institute of Chemical Engineers).

$$\dot{\gamma} = \frac{V}{h} \tag{1-2}$$

In the cone and plate geometry, Figure 1-5b, both the velocity on the surface of the rotating cone and the gap are proportional to the radial distance from the cone tip. This produces a uniform shear rate, $\dot{\gamma} = \omega/\tan\alpha$, where ω is the angular velocity of the cone and α is the angle between cone and plate. The *shear viscosity*, η, measures the resistance that the material offers to this deformation. Specifically, it is the shear stress, σ, divided by the rate of shear, $\dot{\gamma}$:

$$\eta = \frac{\sigma}{\dot{\gamma}} \tag{1-3}$$

σ is the force, per unit area, tangent to the lower plate, exerted by the fluid on the lower plate. For a Newtonian fluid, the viscosity, η, is independent of shear rate, but for a viscoelastic fluid, η decreases with increasing $\dot{\gamma}$; that is, viscoelastic fluids are *shear thinning*. Figure 1-6, which shows η as a function of $\dot{\gamma}$ for a molten low-density polyethylene, "Melt I," is a typical example of shear thinning.[9] At

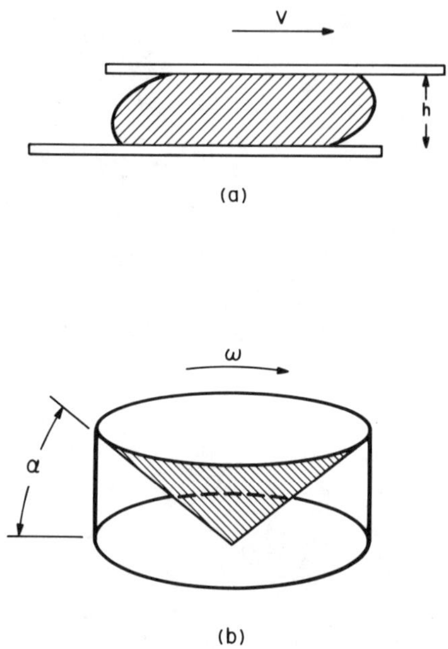

Fig. 1-5 Shearing flows: (a) shear sandwich, (b) cone and plate geometry.

some lower limit of $\dot{\gamma}$, η reaches a constant value, the Newtonian plateau, or *zero shear viscosity*. Thus at small enough shear rates, even a polymer melt acts as a Newtonian liquid.

1.3.2 Normal Stresses in Shear

Shear thinning does not guarantee that a fluid is viscoelastic. A fluid possesses viscoelasticity if it is capable of storing elastic energy. A stronger evidence of viscoelasticity is the existence of normal stress differences in shearing deformations. A normal stress is a force per unit area acting normal or perpendicular to a surface. The primary or first normal stress difference in shear, N_1, is the difference between the normal stress acting in the direction of flow and that acting perpendicular to the shearing surfaces; i.e., perpendicular to the plates in Figure 1-5. In Figure 1-5, the normal stress in the streamwise direction is just the atmospheric pressure, if the capillary force arising from the meniscus is neglected. Thus the magnitude of the primary normal stress difference is just the stress acting perpendicular to the plates minus atmospheric pressure. Since atmospheric pressure also acts on the upper and lower plates, N_1 is the force per unit area tending to separate the plates or pull them together. The sign convention we choose for stress is one in which a positive normal stress is compressive. By this convention, N_1 is always positive. Thus the compressive stress parallel to the plates in the streamwise direction is greater than the compressive stress perpendicular to the plates; that is, *the plates tend to be pushed apart.* The primary normal stress difference is responsible for the rod-climbing phenomenon of Figure 1-1.

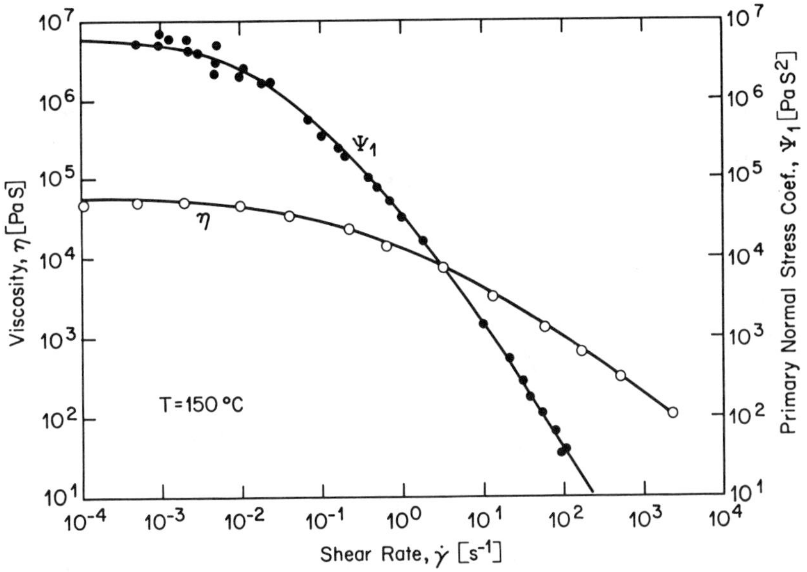

Fig. 1-6 Steady-state viscosity, η , and first normal stress coefficient, Ψ_1 , versus shear rate (from Laun,[9] reprinted with permission from Rheol. Acta).

The shear stress is an odd function of shear rate; that is, it changes sign when $\dot{\gamma}$ changes sign. The primary normal stress difference is an even function of shear rate and does not change sign when the velocity changes direction; whether the flow is to the right or to the left, the force tends to separate the plates rather than pull them together. Therefore at small shear rates, N_1 is not proportional to $\dot{\gamma}$ but to $\dot{\gamma}^2$, which remains positive when $\dot{\gamma}$ changes sign. The ratio, $N_1/\dot{\gamma}^2$ is called the ***primary normal stress coefficient***; like the shear viscosity, it approaches a constant value at low shear rates. The dependence of the primary normal stress coefficient on shear rate for a polyethylene melt, Melt I, is shown in Figure 1-6. The existence of a nonzero N_1 for Melt I indicates that this material is viscoelastic.

Even the presence of normal stress differences, however, does not guarantee that a fluid is viscoelastic. Suspensions of non-Brownian nonspherical particles can have normal stress differences without exhibiting viscoelasticity.

1.3.3 Time-dependent Viscosity

When a viscoelastic fluid is suddenly subjected to steady shearing, the shear stress one measures does not immediately attain a steady value. If this time-dependent stress is divided by the shear rate, a time-dependent viscosity is obtained. Figure 1-7 shows the time-dependent shear stress of Melt I plotted against the time, t, after inception of steady shearing. Also shown is the time-dependent primary normal stress difference.[9] The shear and the normal stress difference rise and may pass through maxima before approaching steady-state values. The viscosity and normal stress difference of viscoelastic materials are

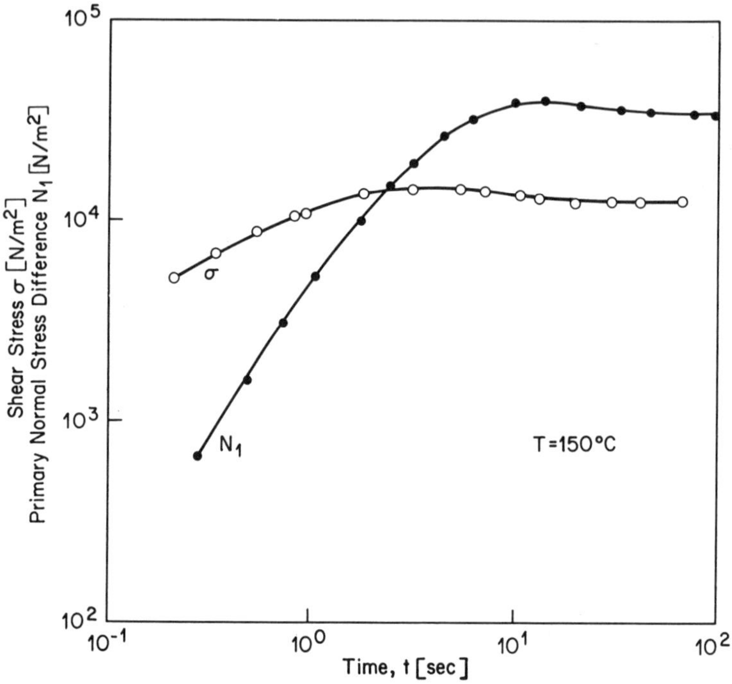

Fig. 1-7 Time-dependent shear stress, σ , and primary normal stress difference, N_1 , after start-up of steady shearing (from Laun,[9] reprinted with permission from Rheol. Acta).

thus both time and shear-rate dependent. As was the case with shear-rate dependence, the existence of time dependence in the viscosity does not imply that a material is viscoelastic. Time-dependence of viscosity can also arise as a result of flow-induced breakdown and buildup of nonelastic structures in the fluid. Such behavior is called *thixotropy*.

1.3.4 Stress Relaxation

A fluid is positively identified as viscoelastic if the stresses in it persist after the deformation has ceased. Figure 1-8 shows the decay of the shear stress and the primary normal stress difference in Melt I after cessation of a steady shearing in which a steady shear viscosity had been attained.[9] The duration of time over which appreciable stresses persist after cessation of deformation gives one estimate of what is called the *relaxation time*, λ, of the material. Molecular theory tells us that for melts λ is also approximated by the ratio $\Psi_1/2\eta$ in the low $\dot{\gamma}$ limit where Ψ_1 and η each approach a constant value.

1.3.5 Recoil

Another sure sign that a fluid possesses viscoelasticity is the occurrence of *recoil*, or reverse deformation, when the fluid is suddenly relieved of an externally imposed stress. This phenomenon is comparable to the snapping back of a

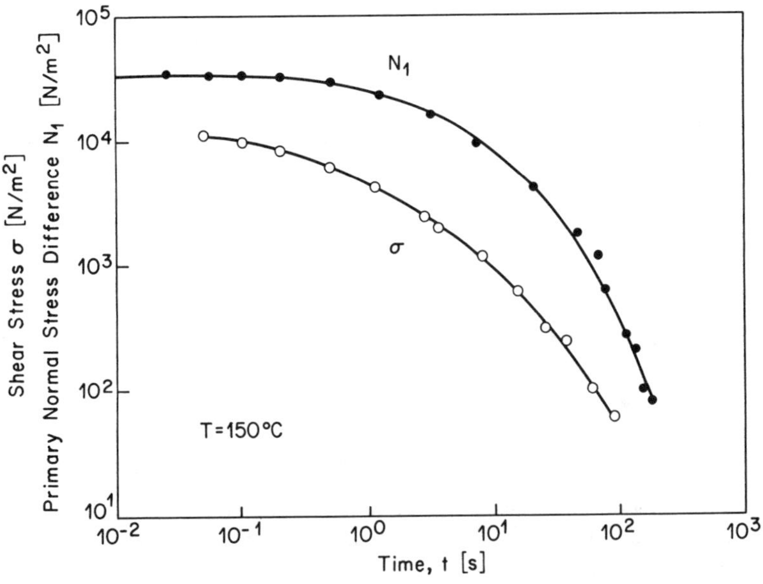

Fig. 1-8 Relaxation of shear stress, σ , and primary normal stress difference, N_1 , after cessation of steady-state shearing (from Laun,[9] reprinted with permission from Rheol. Acta).

stretched rubber band when it is released. Recoil or *elastic recovery* is proof that the material "remembers" its previous shape. Unlike the rubber band, however, viscoelastic materials remember imperfectly, and the recovery process itself takes a longer time to come to completion than can be accounted for by inertia. Suppose a sample is sheared at a rate $\dot{\gamma}$, for time t, and then the shear stress is released. The elastic recovery that then occurs is quantified by γ_r, the amount of shear strain that is finally recovered. Obviously $\gamma_r \leq \dot{\gamma}t$, where $\dot{\gamma}t$ is the amount of strain imposed on the sample just before recovery.[10] In Figure 1-9, γ_r for Melt I is plotted as a function of $\dot{\gamma}t$, the shear imposed before recovery. Note the maximum in recoverable shear.

If the prior shearing is carried out for an interval of time long enough to reach a steady-state stress, then the recoverable shear strain becomes independent of $\dot{\gamma}t$, as seen in Figure 1-9. This steady-state recoverable shear strain, $\dot{\gamma}_{r,s}$, is plotted as a function of the prior shear rate $\dot{\gamma}$ in Figure 1-10. Note that the recoverable strain reaches a value of nearly 10 shear units for this melt.

1.3.6 Sensitivity to Deformation Type

So far we have only discussed shear measurements, which are the easiest to perform. In other types of deformation, the behavior of viscoelastic fluids may differ from that in shear, often in seemingly unpredictable ways. As an example, we consider the behavior of Melt I in a simple elongational deformation. One device for the measurement of steady elongational viscosity is depicted in Figure

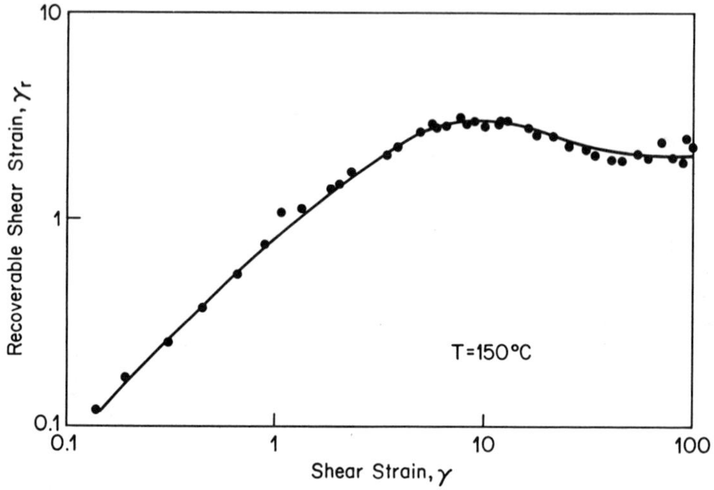

Fig. 1-9 Recoverable shear strain γ_r after imposition of a total shear strain, $\gamma = \dot{\gamma}t$, applied at a steady rate $\dot{\gamma} = 2\ \text{sec}^{-1}$, for a time t (from Wagner and Laun,[10] reprinted with permission from Rheol. Acta).

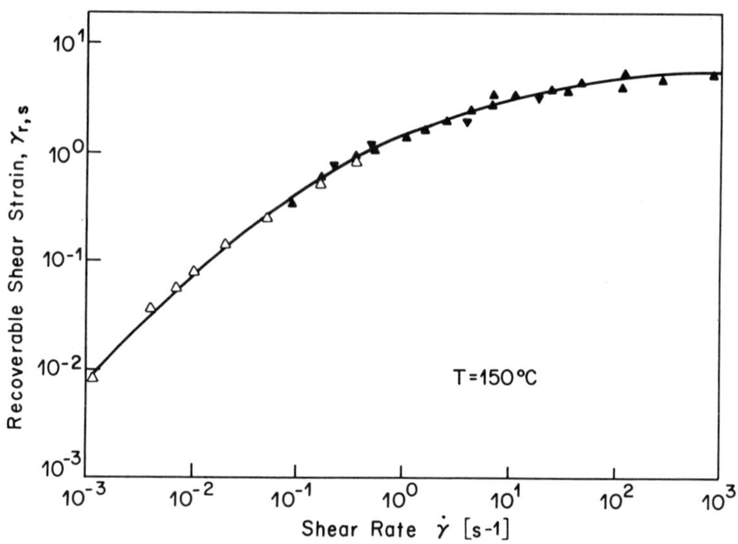

Fig. 1-10 Recoverable shear strain, $\gamma_{r,s}$, after attainment of a steady-state stress in prior steady shearing (from Laun,[10] reprinted with permission from Rheol. Acta).

1-11. The ends of a cylinder of molten polymer floating on an oil bath are gripped by clamps, which rotate at a constant rate, thereby imposing a velocity, V, at one end of the device and $-$V at the other. The velocity gradient is

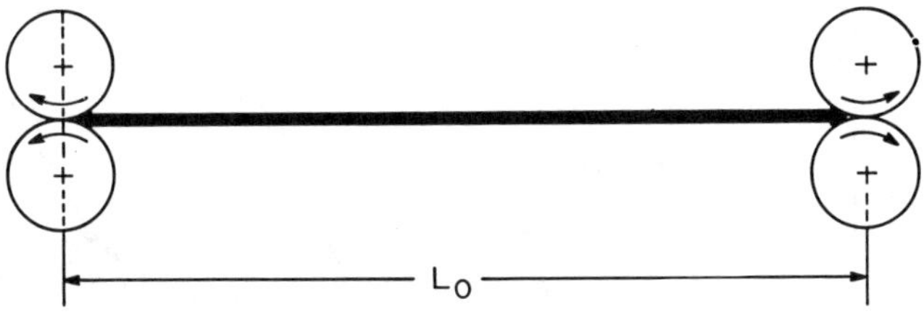

Fig. 1-11 Rotating clamp device used by Meissner to measure elongational viscosity (from Meissner,[12] reprinted with permission from Rheol. Acta).

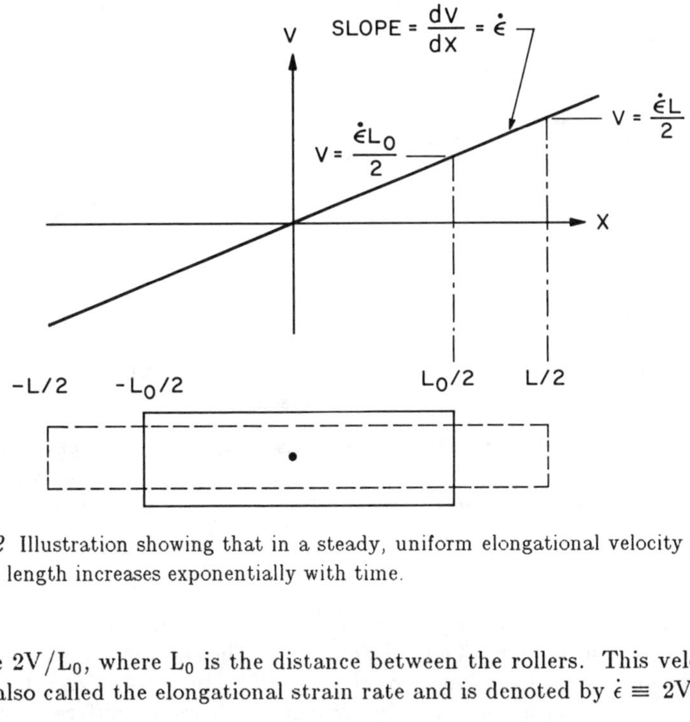

Fig. 1-12 Illustration showing that in a steady, uniform elongational velocity gradient a cylinder's length increases exponentially with time.

therefore $2V/L_0$, where L_0 is the distance between the rollers. This velocity gradient is also called the elongational strain rate and is denoted by $\dot{\epsilon} \equiv 2V/L_0$.

The same deformation can be achieved by gripping the ends of the cylinder and pulling them away from each other such that the length of the cylinder grows exponentially with time; i.e.,

$$L(t) = L_0 \exp(\dot{\epsilon}t) \tag{1-4}$$

To see that this same deformation is obtained in the rotating clamp device, consider Figure 1-12 which depicts a cylinder of fluid undergoing a *uniform* elongational flow with velocity gradient $\dot{\epsilon}$. The velocity at the right end of the cylinder at time zero is $\frac{1}{2}L_0\dot{\epsilon}$. At a later time, t, this end of the cylinder has moved farther downstream and so its velocity must be larger, if the uniform velocity gradient is maintained. Its velocity, V, at time t is $\frac{1}{2}L\dot{\epsilon}$, where L is the cylinder

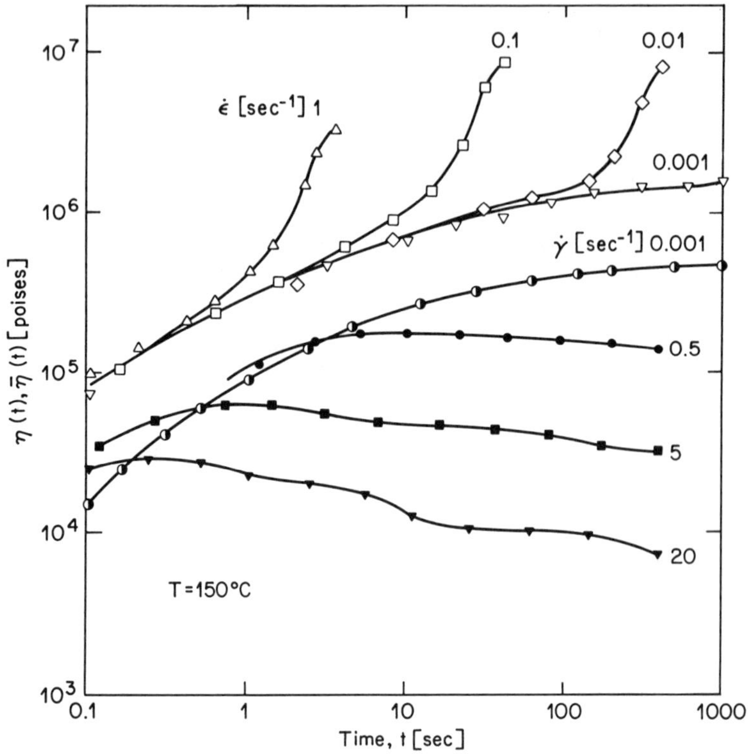

Fig. 1-13 Elongational viscosity, $\bar{\eta}$, and shear viscosity, η , as functions of time after inception of steady straining. The open symbols are elongational viscosities; the closed and half-closed symbols are shear viscosities. (from Meissner,[13] reprinted from J. Appl. Polym. Sci., ©1972 by permission of John Wiley & Sons, Inc.).

length at time t. But V also equals ½(dL/dt), the rate at which the end of the cylinder moves downstream. Thus $dL/dt = L\dot{\epsilon}$. The solution to this differential equation is an exponential function of time (Equation 1-4). Apart from the geometry of the deformation, steady elongation therefore differs from steady shear in that some material lines grow exponentially in time; in shear they grow linearly in time. Hence, steady elongational flows are called *strong flows*; shear is a *weak flow*.

The elongational viscosity, $\bar{\eta}$, is the force along the cylinder's axis per unit cross-sectional area of the deformed cylinder (i.e., stress), divided by the strain rate $\dot{\epsilon}$. Like the shear viscosity, the elongational viscosity depends on both the strain rate and the time following onset of straining. Figure 1-13 shows the elongational viscosity as a function of time for various strain rates, $\dot{\epsilon}$, for Melt I.[13] Contrasted with these curves are the corresponding results for steady shearing. Note the large increases in $\bar{\eta}$ over relatively small intervals of time. This is referred to as *strain hardening*. No such phenomenon is seen in steady shear.

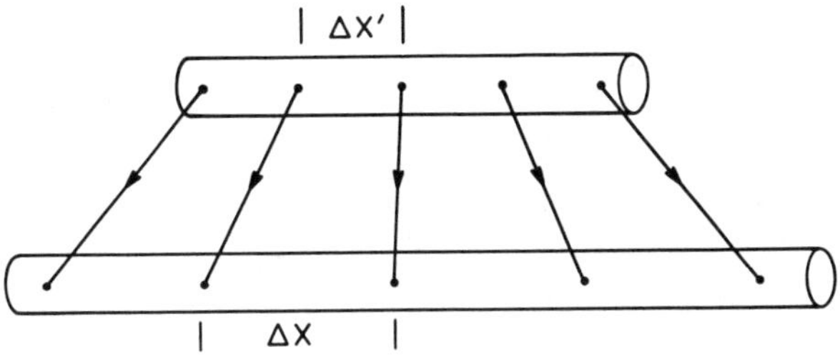

$$F^{-1} = \frac{\Delta X}{\Delta X'}$$

Fig. 1-14 The deformation gradient for stretching of an elastic cord.

Strain hardening doesn't continue indefinitely, but eventually a plateau in the viscosity is reached. In Figure 1-13 one can see the elongational viscosity begin to bend over towards a plateau at long times, t. Measuring elongational viscosities at long times is difficult because the cylindrical sample becomes exceedingly slender. Raible et al.[14] found that when samples of Melt I are stretched to an extreme degree, the elongational viscosity passes through a maximum and decreases as a function of time at long times.

1.4 DEFORMATION GRADIENT, VELOCITY GRADIENT, AND STRESS

1.4.1 The Deformation Gradient

To develop constitutive equations not just for shearing and elongation, but for any velocity field, a mathematical description of arbitrary deformations is needed. This is done with the velocity gradient and the deformation gradient tensors.

We illustrate with a one-dimensional example of the deformation gradient. Consider a filament of material, say an elastic cord, Figure 1-14. First we label each material element in the cord by the ***position***, x, ***that the element occupies at the present time***, t. Suppose t' is a time before the cord has been stretched. We define $x'(t';x,t)$ to be the position at time t' of the element that resides at x at the present time, t. Then

$$F \equiv \frac{dx'}{dx} \tag{1-5}$$

is the deformation gradient. Of more interest to us is its inverse, $F^{-1} = dx/dx'$. F^{-1} is the ratio of the length of the particle in the deformed state to its length in the undeformed state. F^{-1}, in this one-dimensional example, is thus the local stretching to which the element has been subjected. Suppose, for example, that the stretching is uniform; each material element is stretched to the same extent. Then $F^{-1} = L(t)/L(t')$ is the ratio of the length of the cord after the deformation to the length before and is independent of x and x'.

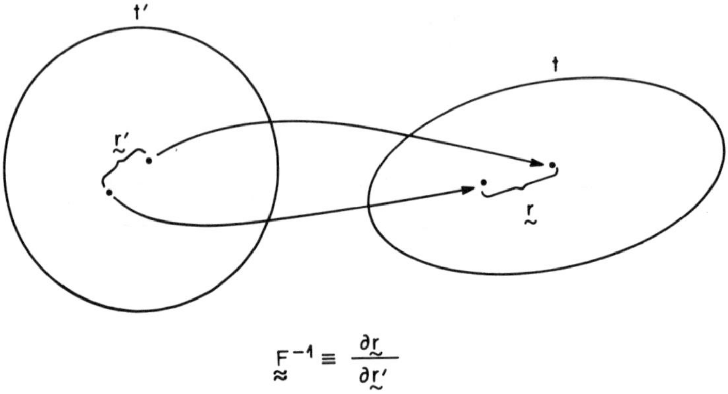

Fig. 1-15 The deformation gradient, $\underset{\approx}{F}^{-1}$, for an arbitrary three-dimensional deformation.

Next consider a three-dimensional material body before and after an arbitrary deformation, Figure 1-15. Consider a vector embedded in the material. Suppose that before the deformation this vector had coordinates, $\underset{\sim}{r}'$, and as a result of the deformation the vector is rotated and stretched so that its new coordinates are given by $\underset{\sim}{r}$. By analogy with the one-dimensional deformation, the three-dimensional tensor, $\underset{\approx}{F}^{-1}$, is defined by

$$\underset{\approx}{F}^{-1}(t,t') \equiv \frac{\partial \underset{\sim}{r}}{\partial \underset{\sim}{r}'} = \begin{pmatrix} \frac{\partial r_1}{\partial r'_1} & \frac{\partial r_2}{\partial r'_1} & \frac{\partial r_3}{\partial r'_1} \\ \frac{\partial r_1}{\partial r'_2} & \frac{\partial r_2}{\partial r'_2} & \frac{\partial r_3}{\partial r'_2} \\ \frac{\partial r_1}{\partial r'_3} & \frac{\partial r_2}{\partial r'_3} & \frac{\partial r_3}{\partial r'_3} \end{pmatrix} \tag{1-6}$$

$\underset{\approx}{F}$, which is the inverse of $\underset{\approx}{F}^{-1}$, is the *deformation gradient tensor.* Note that Equation 1-6 assumes the index convention,* $(\underset{\approx}{F}^{-1})_{ij} = \partial \underset{\sim}{r}_j / \partial \underset{\sim}{r}'_i$. If the deformation is uniform (i.e. $\underset{\approx}{F}^{-1}$ is independent of position) or if the vector $\underset{\sim}{r}'$ is chosen to be much smaller than the distance over which $\underset{\approx}{F}^{-1}$ changes, then $\underset{\sim}{r}$, the

* This convention is the transpose of that used in much of the rheological literature.

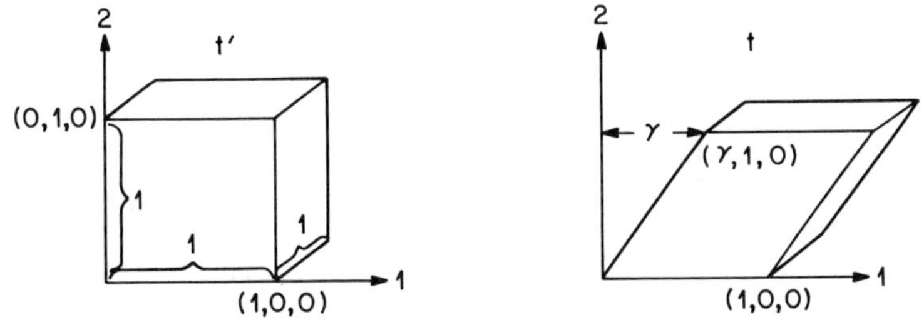

$$\underset{\approx}{F}^{-1} = \begin{bmatrix} 1 & 0 & 0 \\ \gamma & 1 & 0 \\ 0 & 0 & 1 \end{bmatrix}$$

Fig. 1-16 The deformation gradient in shear.

embedded vector at time t, is related to $\underset{\sim}{\mathrm{r}}'$, the same vector at time t′, by

$$\underset{\sim}{\mathrm{r}} = \underset{\sim}{\mathrm{r}}' \cdot \underset{\approx}{\mathrm{F}}^{-1} \tag{1-7}$$

The inverse deformation gradient, $\underset{\approx}{\mathrm{F}}^{-1}$, thus describes the ***cumulative change in orientation and shape of a small material element*** between times t′ and t.

1.4.1.1 Shear Figure 1-16 shows how the components of $\underset{\approx}{\mathrm{F}}^{-1}$ are constructed for simple shear. Take an embedded vector oriented in direction 1, the flow direction, and of unit length at time t′; i.e., $\underset{\sim}{\mathrm{r}}' = (1,0,0)$. Then F_{11}^{-1}, F_{12}^{-1}, and F_{13}^{-1} are the components of this embedded vector after the deformation; $\underset{\sim}{\mathrm{r}} = (\mathrm{F}_{11}^{-1},\mathrm{F}_{12}^{-1},\mathrm{F}_{13}^{-1})$. Here F_{ij}^{-1} is short for $(\underset{\approx}{\mathrm{F}}^{-1})_{ij}$. Since shearing does not stretch or rotate lines that lie in the plane of shear, $(\mathrm{r}_1,\mathrm{r}_2,\mathrm{r}_3) = (\mathrm{F}_{11}^{-1},\mathrm{F}_{12}^{-1},\mathrm{F}_{13}^{-1}) = (1,0,0)$. Likewise F_{21}^{-1}, F_{22}^{-1}, and F_{23}^{-1} are the coordinates of an embedded vector that at time t′ is a unit vector oriented in direction 2. This vector is rotated and stretched by the shearing so that $(\mathrm{F}_{21}^{-1},\mathrm{F}_{22}^{-1},\mathrm{F}_{23}^{-1}) = (\gamma,1,0)$. Finally, $(\mathrm{F}_{31}^{-1},\mathrm{F}_{32}^{-1},\mathrm{F}_{33}^{-1}) = (0,0,1)$.

1.4.1.2 Extensional Deformation As a second example, consider a rectilinear deformation that deforms a cube into a rectangular block, Figure 1-17. Rectilinear deformations are also called *extensions*. A material line oriented along any axis of the cube is stretched or compressed but not rotated. Thus $\underset{\approx}{\mathrm{F}}^{-1}$ contains only diagonal elements, and each diagonal element, F_{ii}^{-1}, is simply the stretch ratio, λ_i, the ratio of the block length in direction i after the deformation to that before. Furthermore, if the material is incompressible so that its volume is unchanged by the deformation, then $\lambda_1\lambda_2\lambda_3 = 1$. More generally, for any

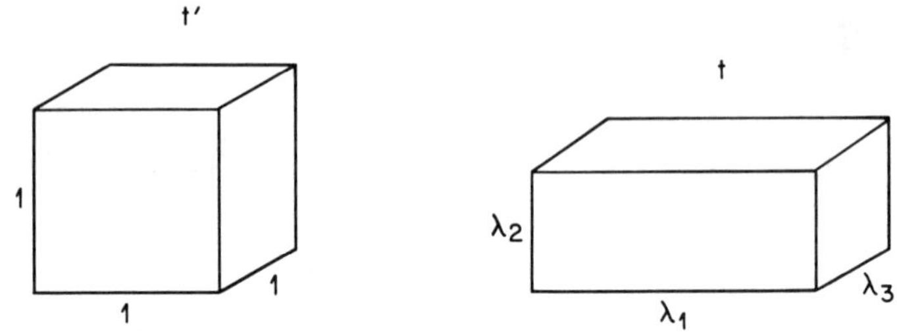

$$\underset{\approx}{F}^{-1} = \begin{bmatrix} \lambda_1 & 0 & 0 \\ 0 & \lambda_2 & 0 \\ 0 & 0 & \lambda_3 \end{bmatrix}$$

Fig. 1-17 The deformation gradient in an extensional deformation.

volume-preserving deformation,

$$\det \underset{\approx}{F}^{-1} = 1 \tag{1-8}$$

That is, the determinant of $\underset{\approx}{F}^{-1}$ is unity. For a simple elongational deformation, incompressibility implies that $\lambda_2 = \lambda_3 = \lambda_1^{-1/2}$, where λ_1 is the stretch ratio in the direction of elongation.

We can therefore use $\underset{\approx}{F}^{-1}(t,t')$ to describe the deformation that a material particle undergoes between some past time, t', and the present time, t. This deformation need not occur at a single instant between t' and t, but it could occur gradually and indeed might continue beyond time t. $\underset{\approx}{F}^{-1}(t,t')$ is the *cumulative deformation* that occurs between t' and t.

1.4.2 The Velocity Gradient

If we take the time derivative of $\underset{\approx}{F}^{-1}$, using the definition of $\underset{\approx}{F}^{-1}$ in Equation 1-6, and the chain rule from calculus, then

$$\frac{\partial}{\partial t}\underset{\approx}{F}^{-1} = \frac{\partial \dot{\underset{\sim}{r}}}{\partial \underset{\sim}{r}'} = \frac{\partial \underset{\sim}{r}}{\partial \underset{\sim}{r}'} \cdot \frac{\partial \dot{\underset{\sim}{r}}}{\partial \underset{\sim}{r}}$$

$$= \underset{\approx}{F}^{-1} \cdot \frac{\partial \underset{\sim}{v}}{\partial \underset{\sim}{r}} = \underset{\approx}{F}^{-1} \cdot \underset{\sim}{\nabla} \underset{\sim}{v} \tag{1-9}$$

Here $\underset{\sim}{\nabla}\underset{\sim}{v}$ is the *velocity gradient* tensor,

$$\underset{\sim\sim}{\nabla v} \equiv \begin{pmatrix} \dfrac{\partial v_1}{\partial r_1} & \dfrac{\partial v_2}{\partial r_1} & \dfrac{\partial v_3}{\partial r_1} \\ \dfrac{\partial v_1}{\partial r_2} & \dfrac{\partial v_2}{\partial r_2} & \dfrac{\partial v_3}{\partial r_2} \\ \dfrac{\partial v_1}{\partial r_3} & \dfrac{\partial v_2}{\partial r_3} & \dfrac{\partial v_3}{\partial r_3} \end{pmatrix} \tag{1-10}$$

The velocity gradient describes the steepness of velocity variation as one moves from point to point in the flow at a given instant in time. Thus if the velocity is $\underset{\sim}{v}_0$ at some point in the flow and the velocity gradient there is $\underset{\sim\sim}{\nabla v}$, then at a nearby point, $\underset{\sim}{r}$, the velocity is $\underset{\sim}{v}_0 + \underset{\sim}{r} \cdot \underset{\sim\sim}{\nabla v}$.

Equation 1-9 can be illustrated by the steady uniform elongational flow depicted in Figure 1-12. The "11" component of $\underset{\approx}{F}^{-1}$ is given by L/L_0. From Equation 1-4, $\dot{F}_{11}^{-1} = \dot{L}/L_0 = \dot{\epsilon}\exp(\dot{\epsilon}t) = \dot{\epsilon}L/L_0$. Thus $\dot{F}_{11}^{-1} = F_{11}^{-1}\dot{\epsilon}$. $\dot{\epsilon}$ is the "11" component of $\underset{\sim\sim}{\nabla v}$. Thus $\dot{F}_{11}^{-1} = F_{11}^{-1}\dot{\epsilon}$ is the "11" component of Equation 1-9. Equation 1-9 establishes the connection between the velocity gradient tensor and the deformation gradient tensor.

1.4.2.1 Shear In a simple shear flow, as depicted in Figure 1-16, $\underset{\sim\sim}{\nabla v}$ is

$$\underset{\sim\sim}{\nabla v} = \begin{pmatrix} 0 & 0 & 0 \\ \dot{\gamma} & 0 & 0 \\ 0 & 0 & 0 \end{pmatrix} \tag{1-11}$$

1.4.2.2 Extension In an extensional flow, as depicted in Figure 1-17,

$$\underset{\sim\sim}{\nabla v} = \begin{pmatrix} \dot{\epsilon}_1 & 0 & 0 \\ 0 & \dot{\epsilon}_2 & 0 \\ 0 & 0 & \dot{\epsilon}_3 \end{pmatrix} \tag{1-12}$$

where $\dot{\epsilon}_i$ is the velocity gradient, $\partial v_i/\partial r_i$, in direction i.

The condition of incompressibility, which implies that $\det\underset{\approx}{F}^{-1} = 1$, also implies that

$$\mathrm{tr}\underset{\sim\sim}{\nabla v} = 0 \tag{1-13}$$

That is, the trace of the velocity gradient tensor is zero.

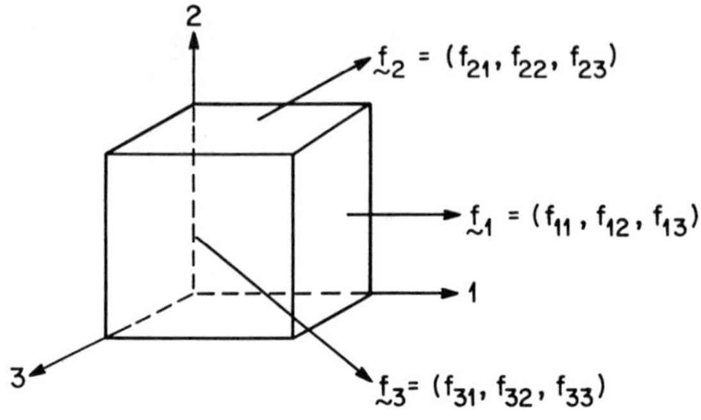

$$\text{THE STRESS TENSOR, } \underset{\approx}{\sigma} = \begin{bmatrix} f_{11} & f_{12} & f_{13} \\ f_{21} & f_{22} & f_{23} \\ f_{31} & f_{32} & f_{33} \end{bmatrix}$$

Fig. 1-18 The stress tensor.

1.4.3 The State-of-Stress Tensor

In our earlier discussion of a shearing deformation, we referred to the shear stress and the normal stresses. These are *components of the stress tensor.* In general, the ith row of the stress tensor, $\underset{\approx}{\sigma}$, is the force per unit area that material exterior to a unit cube exerts on a surface perpendicular to the ith coordinate axis; see Figure 1-18. If $\underset{\sim}{f}$ is the force per unit area acting on a surface perpendicular to an arbitrary unit vector, $\underset{\sim}{n}$, then

$$\underset{\sim}{f} = \underset{\sim}{n} \cdot \underset{\approx}{\sigma} \tag{1-14}$$

For the shearing deformation depicted in Figure 1-16, the shear stress is σ_{12}, and the primary normal stress difference is $\sigma_{11} - \sigma_{22}$. In all but the most unusual cases,[15] the principle of conservation of angular momentum for *isotropic materials* —that is, materials that have no preferred direction other than that imposed by the deformation— implies that $\underset{\approx}{\sigma}$ is symmetric, i.e., $\sigma_{ij} = \sigma_{ji}$. Except for some anisotropic fluids discussed in Chapter 10, all constitutive equations considered here involve a symmetric $\underset{\approx}{\sigma}$.

The state-of-stress, or total stress, $\underset{\approx}{T}$, is the stress tensor plus a pressure contribution:

$$\underset{\approx}{T} = \underset{\approx}{\sigma} - p\underset{\approx}{\delta} \tag{1-15}$$

where $\underset{\approx}{\delta}$ is the unit tensor:

$$\underset{\approx}{\delta} = \begin{pmatrix} 1 & 0 & 0 \\ 0 & 1 & 0 \\ 0 & 0 & 1 \end{pmatrix} \tag{1-16}$$

Tensors that are proportional to $\underset{\approx}{\delta}$ are called *isotropic*.

For an incompressible material, only gradients of p affect fluid motion. Thus a constant isotropic tensor of arbitrary magnitude can be added to $\underset{\approx}{T}$ without consequence. This implies that *$\underset{\approx}{\sigma}$ is determined only up to an additive isotropic term,* and the stress-free state is synonymous with the state of isotropic stress. It is for this reason that only the normal stress *difference*, $N_1 \equiv \sigma_{11} - \sigma_{22}$, and not σ_{11} and σ_{22} separately, can be measured in simple shear.

1.5 RELATING DEFORMATION AND STRESS

1.5.1 Viscoelastic Simple Fluids

We expect the stress tensor, $\underset{\approx}{\sigma}$, to be somehow related to the deformation gradient, $\underset{\approx}{F}^{-1}$. In 1961, Coleman and Noll[16] presented a mathematical framework for relating $\underset{\approx}{\sigma}$ to $\underset{\approx}{F}^{-1}(t,t')$ under general assumptions. These requirements define a class of materials they call "simple fluids." The stress in a small element of viscoelastic "simple fluid" is determined by the history of deformation, that is, by $\underset{\approx}{F}^{-1}(t,t')$, *for that fluid element*, and not by the deformation of neighboring elements. This is the principle of *local action*. Furthermore, the influence of deformation on the stress decreases as time progresses after the deformation. This is the principle of *fading memory*. The precise mathematical definition given to fading memory by Coleman and Noll carries the implications that *for sufficiently rapid deformations the material behaves as an elastic solid, while for sufficiently slow ones it behaves as a Newtonian liquid.*

1.5.2 The Newtonian Limit

A Newtonian liquid satisfies the constitutive equation

$$\underset{\approx}{\sigma} = 2\eta \underset{\approx}{D} \tag{1-17}$$

where $\underset{\approx}{D}$ is the rate-of-deformation tensor (i.e., the symmetric part of the velocity gradient):

$$2\underset{\approx}{D} = \underset{\sim}{\nabla} \underset{\sim}{v} + (\underset{\sim}{\nabla} \underset{\sim}{v})^T \tag{1-18}$$

The superscript T denotes matrix transpose. The antisymmetric part of the velocity gradient is called the vorticity tensor:

$$2\underset{\approx}{\omega} = \underset{\sim}{\nabla} \underset{\sim}{v} - (\underset{\sim}{\nabla} \underset{\sim}{v})^T \tag{1-19}$$

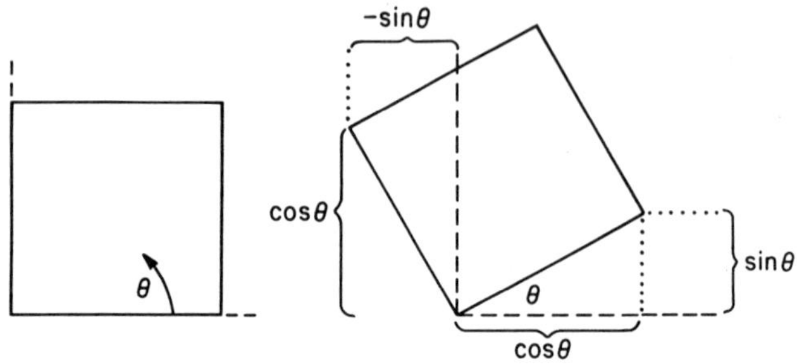

$$\underset{\approx}{F}^{-1} \equiv \begin{bmatrix} \cos\theta & \sin\theta \\ -\sin\theta & \cos\theta \end{bmatrix} = \underset{\approx}{Q}$$

Fig. 1-19 The deformation gradient for a rigid-body rotation.

The vorticity tensor is related to the angular velocity of the fluid element. For flows with no rotation, such as the extensional flow depicted in Figure 1-17, $\underset{\sim}{\nabla}\,\underset{\sim}{v}$ is symmetric and the vorticity tensor is zero. Viscoelastic simple fluids obey Equation 1-17 for sufficiently slow flows.

1.5.3 The Elastic Limit

For sufficiently rapid deformations, a viscoelastic simple liquid behaves elastically. An elastic body has a *natural state* or shape; it remembers this state perfectly and must return to it if the stress is to vanish. The stress tensor for an elastic material depends on the deformation, $\underset{\approx}{F}^{-1}$, relative to this natural state but not on the rate at which this deformation accumulates. Thus

$$\underset{\approx}{\sigma} = \underset{\approx}{\sigma}(\underset{\approx}{F}^{-1}) \tag{1-20}$$

with $\underset{\approx}{F}^{-1} = \partial\,\underset{\sim}{r}\,/\partial\,\underset{\sim}{r}'$ and $\underset{\sim}{r}'$ a vector embedded in the natural state.

1.5.4 Frame Invariance

Equation 1-20 is a general *constitutive equation* for elastic materials. Not all constitutive equations of this form are reasonable, however. To see this, let us consider a "foolish" constitutive law, $\underset{\approx}{\sigma} = G\underset{\approx}{F}^{-1}$, where G is a *modulus* or *rigidity* with dimensions of force per unit area. Suppose a block of stress-free material obeying this law is rotated through an angle, θ; see Figure 1-19. For this motion, the deformation gradient is

$$\underset{\approx}{F}^{-1} = \begin{pmatrix} \cos\theta & \sin\theta & 0 \\ -\sin\theta & \cos\theta & 0 \\ 0 & 0 & 1 \end{pmatrix} \equiv \underset{\approx}{Q} \tag{1-21}$$

$\underset{\approx}{F}^{-1}$ is just the rotation matrix, $\underset{\approx}{Q}$. Hence

$$\underset{\approx}{\sigma} = G \begin{pmatrix} \cos\theta & \sin\theta & 0 \\ -\sin\theta & \cos\theta & 0 \\ 0 & 0 & 1 \end{pmatrix} \tag{1-22}$$

The components of $\underset{\approx}{\sigma}$ therefore depend on the angle θ, and we find that a solid-body rotation induces a stress in the material! Put another way, we see that a rotation of the frame of reference through an angle θ induces a stress in the material. Hence the material is not *frame-invariant.*

This conundrum arises because $\underset{\approx}{F}^{-1}$ contains information about both *stretching* and *rotation* of material lines. Frame invariance requires that information about *rotation* of material lines *be excluded* from the constitutive equation. This can be done by making $\underset{\approx}{\sigma}$ depend not on $\underset{\approx}{F}^{-1}$ but on the Finger tensor, $\underset{\approx}{C}^{-1}$:

$$\underset{\approx}{C}^{-1} \equiv (\underset{\approx}{F}^{-1})^{T} \cdot \underset{\approx}{F}^{-1} \tag{1-23}$$

The Finger tensor naturally arises when one considers the *length* of a deformed material line. From Equation 1-7, the square of the length of an embedded vector after a deformation is

$$r^2 = \underset{\sim}{r}^{T} \cdot \underset{\sim}{r} = \underset{\sim}{r}'^{T} \cdot (\underset{\approx}{F}^{-1})^{T} \cdot \underset{\approx}{F}^{-1} \cdot \underset{\sim}{r}' = \underset{\sim}{r}'^{T} \cdot \underset{\approx}{C}^{-1} \cdot \underset{\sim}{r}' \tag{1-24}$$

$\underset{\approx}{C}^{-1}$, a symmetric tensor, contains information about the orientation of the three principle axes of stretch and the magnitudes of the three principle stretch ratios. But any such tensor $\underset{\approx}{C}^{-1}$ obtained by a deformation involving rotation can also be obtained by a rotation-free deformation; thus $\underset{\approx}{C}^{-1}$ contains no information about the rotation of material lines that occurs during the deformation. *The Finger tensor* $\underset{\approx}{C}^{-1}(t',t)$ *describes the change in shape of a small material element* between times t' and t.

For the simple constitutive equation

$$\underset{\approx}{\sigma} = G\underset{\approx}{C}^{-1} \tag{1-25}$$

a rotation through an angle θ gives

$$\underset{\approx}{\sigma} = G(\underset{\approx}{F}^{-1})^{T} \cdot \underset{\approx}{F}^{-1} = G\underset{\approx}{Q}^{T} \cdot \underset{\approx}{Q} = G\underset{\approx}{\delta} \tag{1-26}$$

The material therefore remains stress free (i.e., in a state of isotropic stress) no matter how it is rotated. Equation 1-25 preserves the stress-free state under rotations of frame. More generally, equations of the form, $\underset{\approx}{\sigma} = \underset{\approx}{\sigma}(\underset{\approx}{C}^{-1})$ satisfy the *principle of frame invariance*, regardless of the state of stress of the material. We shall have more to say about frame invariance later. For the time being we simply observe that the stress tensor of an elastic body must depend on $\underset{\approx}{C}^{-1}$, not on

$\underset{\approx}{F}^{-1}$

1.5.5 Examples of the Finger Tensor

From Equation 1-23, $\underset{\approx}{C}^{-1}$ can be computed if $\underset{\approx}{F}^{-1}$ is known. For shear, $\underset{\approx}{F}^{-1}$ is given in Figure 1-16, from whence

$$\underset{\approx}{C}^{-1} = \begin{pmatrix} 1+\gamma^2 & \gamma & 0 \\ \gamma & 1 & 0 \\ 0 & 0 & 1 \end{pmatrix} \tag{1-27}$$

For an extensional deformation for which $\underset{\approx}{F}^{-1}$ is given in Figure 1-17,

$$\underset{\approx}{C}^{-1} = \begin{pmatrix} \lambda_1^2 & 0 & 0 \\ 0 & \lambda_2^2 & 0 \\ 0 & 0 & \lambda_3^2 \end{pmatrix} \tag{1-28}$$

1.5.6 Relationship between the Finger Tensor and the Velocity Gradient

From the definition of $\underset{\approx}{C}^{-1}$ given in Equation 1-23, and the relationship of Equation 1-9, we find that

$$\dot{\underset{\approx}{C}}^{-1} \equiv \frac{\partial}{\partial t}\underset{\approx}{C}^{-1} = \underset{\sim}{\nabla}\underset{\sim}{v}^{T} \cdot \underset{\approx}{C}^{-1} + \underset{\approx}{C}^{-1} \cdot \underset{\sim}{\nabla}\underset{\sim}{v} \tag{1-29}$$

As we shall see, this identity is useful for relating integral and differential constitutive equations.

1.6 A SIMPLE VISCOELASTIC CONSTITUTIVE EQUATION

1.6.1 Integral Version

We have proposed Equation 1-25 as a frame-invariant constitutive equation for an elastic response to a deformation. We shall later show that this equation arises from a molecular model. Equation 1-25 is also a possible constitutive relationship for viscoelastic simple fluids in the limit of fast deformations, since viscoelastic simple fluids behave elastically in this limit. The shear stress, σ_{12}, in a rapid shearing deformation is $G\gamma$, according to this equation. For slow deformations, the contribution to the stress of strain increments occurring in the remote past must be weighted less heavily than those occurring in the recent past. Suppose that we weight each strain increment by $\exp((t'-t)/\lambda)$, where t is the present time, t′ the past time, and λ the relaxation time. The shear strain increment occurring during an interval of time dt′ at a past time, t′, is $\dot{\gamma}(t')dt'$. Thus the contribution, $d\sigma_{12}$, to the present shear stress resulting from this increment of strain is

$$d\sigma_{12} = Ge^{(t'-t)/\lambda}\,\dot{\gamma}(t')\,dt' \tag{1-30}$$

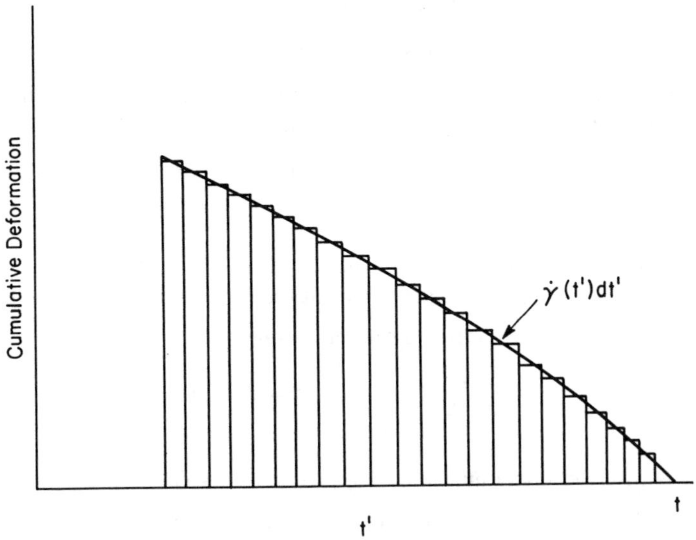

Fig. 1-20 Division of a strain history into strain increments.

The present shear stress is just the sum or integral of all these contributions from the remote past, which we take to be $t' = -\infty$, to the present time, t; see Figure 1-20:

$$\sigma_{12} = G \int_{-\infty}^{t} e^{(t'-t)/\lambda} \dot{\gamma}(t') \, dt' \tag{1-31}$$

Integrating this expression by parts gives

$$\sigma_{12} = G \int_{-\infty}^{t} \frac{1}{\lambda} e^{(t'-t)/\lambda} \gamma(t,t') \, dt' \tag{1-32}$$

where $\gamma(t,t')$ is the shear strain that accumulated between times, t' and t,

$$\gamma(t,t') = \int_{t'}^{t} \dot{\gamma}(t'') \, dt''$$

According to Equation 1-27, $\gamma(t,t')$ is the "12" component of the Finger tensor in shear. The complete stress tensor is

$$\underset{\approx}{\sigma} = G \int_{-\infty}^{t} \frac{1}{\lambda} e^{(t'-t)/\lambda} \underset{\approx}{C}^{-1}(t,t') \, dt' \tag{1-33}$$

Equation 1-33 is known as the "Lodge equation."

If all the strain is imposed just before the present time, t, then $\underset{\approx}{C}^{-1}(t,t')$, which is the strain that has accumulated between times t′ and t, is independent of t′. $\underset{\approx}{C}^{-1}(t,t')$ is therefore $\underset{\approx}{C}^{-1}$, a constant, and Equation 1-33 integrates to

$$\underset{\approx}{\sigma} = G\underset{\approx}{C}^{-1} \tag{1-34}$$

Thus the elastic limit is recovered. If, on the other hand, the deformation is very slow and $\underset{\approx}{C}^{-1}$ remains only a small perturbation away from $\underset{\approx}{\delta}$, then replacing $\underset{\approx}{C}^{-1}$ by $\underset{\approx}{\delta}$ on the right side of Equation 1-29, and integrating, we find that

$$\underset{\approx}{C}^{-1}(t,t') \sim \underset{\approx}{\delta} + 2(t-t')\underset{\approx}{D}(t) \tag{1-35}$$

When this is inserted into Equation 1-33, integration yields

$$\underset{\approx}{\sigma} = G\underset{\approx}{\delta} + 2G\lambda\underset{\approx}{D} \tag{1-36}$$

Thus, to within an isotropic term, the stress tensor is $2\eta\underset{\approx}{D}$, where the viscosity η, equals $G\lambda$. Hence the Newtonian limit is recovered for sufficiently slow flows, as required in simple fluid theory.

1.6.2 Differential Version

The simple constitutive equation, Equation 1-33, can be converted to an equivalent differential formulation, which for some purposes is more useful than the integral version. The present time, t, appears either explicitly or implicitly in the integral on the right side of Equation 1-33 in three places: in the upper limit of integration, in the exponential, and in the t-dependence of the Finger tensor, $\underset{\approx}{C}^{-1}(t,t')$. Thus when Equation 1-33 is differentiated with respect to t, the result is

$$\dot{\underset{\approx}{\sigma}} = \left\{\begin{matrix} G\dfrac{1}{\lambda}\underset{\approx}{C}^{-1}(t,t)\,e^{(t-t)/\lambda} \\[2ex] -G\displaystyle\int_{-\infty}^{t}\frac{1}{\lambda^2}e^{(t'-t)/\lambda}\,\underset{\approx}{C}^{-1}(t,t') \\[2ex] +G\displaystyle\int_{-\infty}^{t}\frac{1}{\lambda}e^{(t'-t)/\lambda}\,\dot{\underset{\approx}{C}}^{-1}(t,t') \end{matrix}\right\} = \left\{\begin{matrix} \dfrac{G}{\lambda}\underset{\approx}{\delta} \\[2ex] -\dfrac{1}{\lambda}\underset{\approx}{\sigma} \\[2ex] +\left(\underset{\sim}{\nabla}\underset{\sim}{v}^{T}\cdot\underset{\approx}{\sigma}+\underset{\approx}{\sigma}\cdot\underset{\sim}{\nabla}\underset{\sim}{v}\right) \end{matrix}\right\} \tag{1-37}$$

where we have used the identity in Equation 1-29 to replace $\dot{\underset{\approx}{C}}^{-1}$ with terms involving $\underset{\sim}{\nabla}\underset{\sim}{v}$. Thus the differential equation

$$\dot{\underset{\approx}{\sigma}} - \nabla \underset{\sim}{v}^{T} \cdot \underset{\approx}{\sigma} - \underset{\approx}{\sigma} \cdot \nabla \underset{\sim}{v} + \frac{1}{\lambda} \underset{\approx}{\sigma} = \frac{G}{\lambda} \underset{\approx}{\delta} \tag{1-38a}$$

is equivalent to the integral equation, Equation 1-33. This is written in more compact form as

$$\lambda \overset{\nabla}{\underset{\approx}{\sigma}} + \underset{\approx}{\sigma} = G \underset{\approx}{\delta} \tag{1-38b}$$

where ∇ denotes the ***upper-convected time derivative.*** Its operation on an arbitrary tensor, $\underset{\approx}{S}$, is defined as

$$\overset{\nabla}{\underset{\approx}{S}} \equiv \dot{\underset{\approx}{S}} - \nabla \underset{\sim}{v}^{T} \cdot \underset{\approx}{S} - \underset{\approx}{S} \cdot \nabla \underset{\sim}{v} \tag{1-39}$$

Equation 1-38 is called the ***upper-convected Maxwell equation.*** Another form of it can be obtained by noting that

$$\overset{\nabla}{\underset{\approx}{\delta}} = \dot{\underset{\approx}{\delta}} - \nabla \underset{\sim}{v}^{T} \cdot \underset{\approx}{\delta} - \underset{\approx}{\delta} \cdot \nabla \underset{\sim}{v} = -2\underset{\approx}{D} \tag{1-40}$$

Then, defining a new stress tensor

$$\underset{\approx}{\tau} = \underset{\approx}{\sigma} - G \underset{\approx}{\delta} \tag{1-41}$$

and substituting it for $\underset{\approx}{\sigma}$ in Equation 1-38 gives

$$\lambda \overset{\nabla}{\underset{\approx}{\tau}} + \underset{\approx}{\tau} = 2\eta \underset{\approx}{D} \tag{1-42}$$

where $\eta = G\lambda$. This form of the upper-convected Maxwell equation, which contains $\underset{\approx}{D}$ instead of $\underset{\approx}{\delta}$, is more often seen than Equation 1-38. The two are equivalent, however, since for an incompressible material, the stress is determined by the kinematics only up to an isotropic constant.

1.6.3 Predictions

1.6.3.1 Shear The rheological predictions of this simple equation can easily be seen from the integral version, Equation 1-33. The shear viscosity is reached after steady shearing has been imposed for a time much longer than the relaxation time. To compute η at steady state from the constitutive equation, we assume that steady shearing has continued from a time, t_0, in the remote past up to the present time, t. If $t - t_0 >> \lambda$, the computed value is insensitive to the exact value of t_0; that is, the material doesn't remember exactly when the shearing started. For convenience we take $t_0 = -\infty$. The shear strain, $\gamma(t,t')$, that has accumulated between times t′ and t is $\dot{\gamma}(t - t')$. To obtain the shear viscosity, we insert the appropriate component of the Finger tensor, $\underset{\approx}{C}^{-1}(t,t')$, from Equation 1-27 into the integral:

$$\eta \equiv \frac{\sigma_{12}}{\dot{\gamma}} = \frac{G}{\dot{\gamma}} \int_{-\infty}^{t} \frac{1}{\lambda} e^{(t'-t)/\lambda}\, \dot{\gamma}(t-t')\, dt' = G\lambda \tag{1-43}$$

Thus η is constant; there is no shear thinning.

The first normal stress difference, N_1, is obtained by inserting $C_{11}^{-1} - C_{22}^{-1} = \gamma^2 = \dot{\gamma}^2(t-t')^2$ into the constitutive equation. The result is

$$N_1 = 2G\lambda^2\dot{\gamma}^2 = 2\eta\lambda\dot{\gamma}^2 \tag{1-44}$$

Thus $\Psi_1 = N_1/\dot{\gamma}^2$, the first normal stress coefficient, is a constant, $2\eta\lambda$. Note that the ratio $\psi_1/2\eta$ is the relaxation time.

Another normal stress difference, $N_2 = \sigma_{22} - \sigma_{33}$, can be computed in steady-state shear. This is called the ***second normal stress difference***. For the Lodge equation, however, it is zero since $C_{22}^{-1} - C_{33}^{-1} = 0$.

1.6.3.2 Elongation In elongation, the viscosity, $\bar{\eta} = (\sigma_{11} - \sigma_{22})/\dot{\epsilon}$, is again obtained by inserting the appropriate components of the Finger tensor into Equation 1-33. Elongation is an extensional deformation with $\lambda_2 = \lambda_3 = \lambda_1^{-1/2}$, where "1" is the stretch direction. From Equation 1-28, $C_{11}^{-1} - C_{22}^{-1} = \lambda_1^2 - \lambda_1^{-1}$. For steady elongation, $\lambda_1(t,t') = \exp[\dot{\epsilon}(t-t')]$, where $\dot{\epsilon}$ is the velocity gradient in the flow direction; see Section 1.3.6. Thus

$$\sigma_{11} - \sigma_{22} = G \int_{-\infty}^{t} \frac{1}{\lambda} e^{(t'-t)/\lambda} \left[e^{2\dot{\epsilon}(t-t')} - e^{-\dot{\epsilon}(t-t')}\right] dt'$$

$$= G \int_{-\infty}^{t} \frac{1}{\lambda} \left[e^{(t'-t)(-2\dot{\epsilon}+1/\lambda)} - e^{(t'-t)(\dot{\epsilon}+1/\lambda)}\right] dt'$$

$$= \frac{G}{1-2\lambda\dot{\epsilon}} - \frac{G}{1+\lambda\dot{\epsilon}} \tag{1-45}$$

The elongational viscosity, $\bar{\eta} = (\sigma_{11} - \sigma_{22})/\dot{\epsilon}$, is

$$\bar{\eta} = \frac{3\lambda G}{(1-2\lambda\dot{\epsilon})(1+\lambda\dot{\epsilon})} = \lambda G \left(\frac{2}{1-2\lambda\dot{\epsilon}} + \frac{1}{1+\lambda\dot{\epsilon}}\right) \tag{1-46}$$

Notice that the elongational viscosity approaches infinity as $\dot{\epsilon} \rightarrow 1/2\lambda$.

Next, we calculate the elongational stresses from the upper-convected Maxwell equation to illustrate that the Maxwell equation is equivalent to Lodge integral equation. The velocity gradient, $\underset{\sim}{\nabla}\underset{\sim}{v}$, is given by Equation 1-12 with $\dot{\epsilon}_2 = \dot{\epsilon}_3 = -\dot{\epsilon}/2$, where $\dot{\epsilon} = \dot{\epsilon}_1$. At steady state, $\dot{\underset{\approx}{\sigma}} = 0$. From Equation 1-38a,

$$-\dot{\epsilon}\sigma_{11} - \sigma_{11}\dot{\epsilon} + \frac{1}{\lambda}\sigma_{11} = \frac{G}{\lambda}, \qquad \text{and}$$

$$+\frac{1}{2}\dot{\epsilon}\,\sigma_{22} + \sigma_{22}\,\frac{1}{2}\dot{\epsilon} + \frac{1}{\lambda}\sigma_{22} = \frac{G}{\lambda} \tag{1-47}$$

These equations are easily solved for σ_{11} and σ_{22}:

$$\sigma_{11} = \frac{G}{1 - 2\lambda\dot{\epsilon}}; \qquad \sigma_{22} = \frac{G}{1 + \lambda\dot{\epsilon}} \tag{1-48}$$

The stress difference, $\sigma_{11} - \sigma_{22}$, is therefore the same as that computed with the integral equation; see Equation 1-45.

Thus the simple Lodge equation or upper-convected Maxwell equation predicts important properties of viscoelastic fluids, such as a primary normal stress difference in shear and strain hardening in elongation. It also predicts the existence of stress relaxation after cessation of flow and elastic recoil. It is simplistic, however, in that it predicts that both the shear viscosity and the first normal stress coefficient are independent of shear rate, and that the steady-state elongational viscosity is infinite at a finite elongation rate.

The equation can predict qualitatively the phenomena of rod-climbing, extrudate swell, and spinning flows such as the tubeless siphon.

1.7 SUMMARY

Rheological constitutive equations are laws of flow. For viscoelastic materials, they relate stress to strain history and, if accurate, make possible the prediction of peculiar viscoelastic flow phenomena. These phenomena include the climbing of fluid up a rotating shaft, the large degree of swelling of extrudate as it emerges from a die, spinnability, and vortex formation in contraction flows. In simple flows, viscoelasticity manifests itself in shear thinning, normal stress differences, time-dependent viscosity, stress relaxation, elastic recovery, and sensitivity to deformation type.

An idealization of a viscoelastic fluid is the simple fluid, which behaves as an elastic solid in rapid deformations and as a Newtonian liquid in slow deformations. The concept that the constitutive equation should be the same in any frame of reference leads to a description of strain that is independent of the rotation of material lines. Perhaps the simplest example of a viscoelastic simple fluid is an elastic solid modified so that the influence of a strain on the stress is an exponentially decaying function of the time since the strain occurred. Integral and differential versions of a constitutive equation for this simple viscoelastic liquid can be developed. Although this simple equation predicts many of the basic phenomena, it does so only qualitatively.

References

1. C.A. Truesdell in the Proceedings of the 8th International Congress on Rheology, Naples, Sept. 1–5, 1980.

2. K. Weissenberg, Nature, *159*, 310 (1947).
3. R.B. Bird, R.C. Armstrong, and O. Hassager, *Dynamics of Polymeric Liquids*, Vol. 1, John Wiley & Sons, New York (1977).
4. R.I. Tanner, *Engineering Rheology*, Oxford, New York (1985).
5. G. Fano, Arch. Fisiol., *5*, 365 (1908).
6. S.T.J. Peng and R.F. Landel, J. Appl. Phys., *47*, 4255 (1976).
7. A.B. Metzner, E.A. Uebler, and C.F. Chan Man Fong, A.I.Ch.E. Journal, *15*, 750 (1969).
8. H. Nguyen and D.V. Boger, J. Non-Newt. Fluid Mech., *5*, 353 (1979).
9. H.M. Laun, Rheol. Acta, *17*, 1 (1978).
10. M.H. Wagner and H.M. Laun, Rheol. Acta, *17*, 138 (1978).
11. H.M. Laun, Rheol. Acta, *21*, 464 (1982).
12. J. Meissner, Rheol. Acta, *10*, 230 (1971).
13. J. Meissner, J. Applied Polym. Sci., *16*, 2877 (1972).
14. M.H. Wagner, T. Raible, and J. Meissner, Rheol. Acta, *18*, 427 (1979).
15. J.S. Dahler and L.E. Scriven, Nature, *192*, 36 (1961).
16. B.D. Coleman and W. Noll, Rev. Mod. Phys., *33*, 239 (1961).

CHAPTER 2

Classical Molecular Models

" *'Tis a gift to be simple*"

Shaker Hymn

2.1 INTRODUCTION

In the previous chapter, integral and differential versions of a simple constitutive equation were developed. The equation was presented as an example of a viscoelastic simple liquid — a fluid that is rubbery when deformed rapidly and Newtonian when deformed slowly. This chapter shows how this same equation arises out of the two simplest *molecular theories*, one for melts and the other for dilute solutions. We present these two models in the same chapter to illustrate the differences and similarities between the molecular dynamics of melts and of dilute polymer solutions.

In any polymeric system, dilute or concentrated, the stress depends on the conformations of the polymer molecules. As the orientation and degree of stretch of a molecule increases, so does its contribution to the stress. The total stress depends on contributions from all molecules in the fluid element, each of which has its own conformation. Molecular theories for stress must therefore account for the *distribution of conformations* that results from a deformation. The starting point for developing such a theory is to determine the distribution of conformations occurring at equilibrium (i.e., in the absence of deformation).

2.2 THE EQUILIBRIUM STATE

2.2.1 Configuration Distribution Function

Here we restrict the discussion to long flexible polymer molecules, which have an enormous number of conformations, owing to the large number of bonds in the backbone and their rotational freedom. Figure 2-1 shows a two-dimensional projection of a typical conformation of a polymethylene chain with 1000 bonds.[1] A random polymer conformation of this kind is often called a polymer "coil." Kuhn[2] noticed that details of the bond angles and rotational potentials that control the arrangement of nearest-neighbor atoms in the chain do not significantly affect molecular shape at distances of many atoms. Thus a typical molecular conformation on a gross, coarse-grained scale is the same as a typical three-dimensional random walk with a large number of steps (if we neglect interactions of the polymer chain with itself and with solvent). The random walk can also be

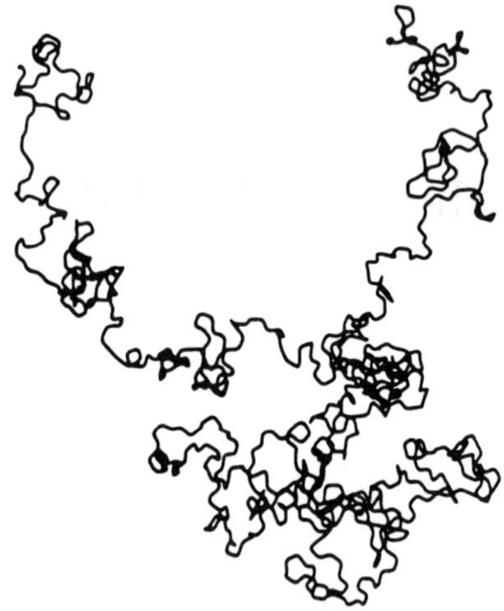

Fig. 2-1 Random walk formed by a model of a polymethylene chain with 1000 links (from Treloar,[1] reprinted with permission from *The Physics of Rubber Elasticity*, Third ed., Oxford University Press, 1975).

envisioned as a configuration of a *freely-jointed chain*, that is a chain made up of rigid links connected to each other at completely flexible joints. If the freely-jointed chain contains N_K links each of length b the average straight-line distance from one end of the walk to the other is $b\sqrt{N_K}$.

The freely-jointed chain can be used to describe the statistical properties of conformations of macromolecules if the parameters b and N_K are suitably chosen. One such choice is

$$b^2 N_K = \langle R^2 \rangle_0 \tag{2-1}$$

$$bN_K = \ell P \tag{2-2}$$

where ℓ is the length of a monomer unit in the molecule, P is the number of monomers in the molecule, and $\langle R^2 \rangle_0$ is the unperturbed mean square end-to-end length of the molecule. With these definitions, the fully-extended length of the molecule, ℓP, is the same as that of the freely-jointed chain, bN_K. Likewise the mean square end-to-end lengths of the molecule, $\langle R^2 \rangle_0$ and of the freely-jointed chain, b^2N_K, are identical.

The freely-jointed chain is then a simple model with the same gross statistical properties as the real molecule. With the above definitions, b, the effective step size, or "Kuhn length," depends on chain flexibility; stiff chains with less rotational freedom require a larger succession of monomers to make up an effective random step and so have larger values of b. Thus b depends on the chemical composition of the polymer. "N_K", the number of Kuhn steps, depends on the molecular weight M of the polymer molecule. For typical flexible polymers, b is 5-10 times ℓ, the monomer length. The ratio, b/ℓ, is called the "characteristic ratio," C_∞, of the polymer. A tabulation of characteristic ratios can be found in Flory[3].

If one end of the freely jointed chain with N_K links lies at the origin, the probability, $\psi_0 dR^3$, that the other end lies at a position between $\underset{\sim}{R}$ and $\underset{\sim}{R} + d\underset{\sim}{R}$, is approximately a Gaussian function:

$$\psi_0(\underset{\sim}{R}) = \left(\frac{\beta}{\sqrt{\pi}}\right)^3 \exp(-\beta^2 R^2) \tag{2-3}$$

where $\beta^2 \equiv 3/(2N_K b^2)$, and $R^2 = \underset{\sim}{R} \cdot \underset{\sim}{R} = R_1^2 + R_2^2 + R_3^2$. $\underset{\sim}{R}$ is called the end-to-end vector of the polymer chain. This formula, which implies that there is a finite probability for arbitrarily large $R = |\underset{\sim}{R}|$, is really only valid when R is much less than bN_K, the maximum extension of the chain.

From the distribution function, ψ_0, average geometric properties of the chain can be calculated by taking appropriate moments of the distribution. For example, $<R^2>_0$, the average square of the end-to-end distance of the chain, is

$$<R^2>_0 \equiv \int R^2 \psi_0(\underset{\sim}{R})\, dR^3 = \frac{3}{2}\beta^{-2} = N_K b^2 \tag{2-4}$$

where the integral $\int \cdot\, dR^3$ in Equation 2-4 represents 3 nested integrals over each coordinate, dR_i, taken from $-\infty$ to $+\infty$. The notation, $<\cdot>_0$, where "$\cdot$" represents any quantity, is an average of "$\cdot$" over ψ_0,

$$<\cdot>_0 = \int \psi_0 \cdot dR^3$$

2.2.2 Polymer Chains as Hookean Springs

The configuration distribution function, $\psi_0(\underset{\sim}{R})$, allows us to explain the elasticity of flexible polymer molecules. Suppose the ends of the polymer chain are held fixed so that the end-to-end vector of the chain is $\underset{\sim}{R}$. The number of internal configurations, Ω, of the chain that satisfy this constraint is $\Omega = c\psi_0(\underset{\sim}{R})$, with c a constant. From the celebrated relationship of statistical thermodynamics, the entropy, S, is

$$S = k\log(\Omega) = k\log(c\psi_0) = k[\log c + 3\log(\beta/\sqrt{\pi}) - \beta^2 R^2] \tag{2-5}$$

where we have used Equation 2-3 for ψ_0. The free energy, W, is $-TS$, where T is the absolute temperature.

Suppose we now pull on the chain ends to increase the end-to-end distance. The force one needs to exert is

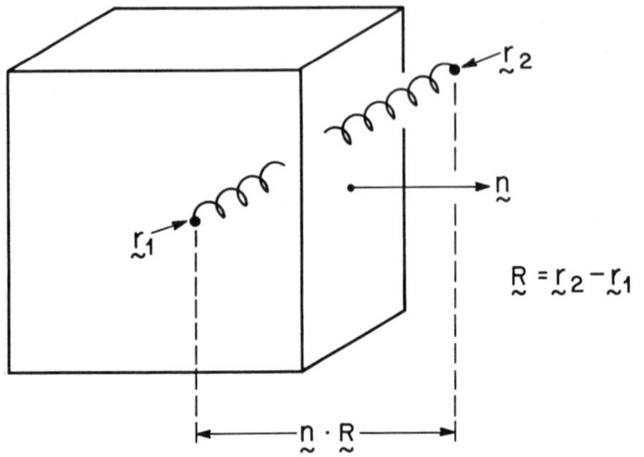

Fig. 2-2 Contribution of a Gaussian or springlike macromolecular chain to the stress on the face of a unit cube.

$$\underset{\sim}{F} = \frac{\partial W}{\partial \underset{\sim}{R}} = -T\frac{\partial S}{\partial \underset{\sim}{R}} = 2kT\beta^2\underset{\sim}{R} \tag{2-6}$$

An increase in the separation distance, R, reduces the number of allowed configurations; the molecule thus resists this action with a force proportional to $\underset{\sim}{R}$. *A Gaussian chain therefore acts like a Hookean spring.* Note that this force is proportional to the temperature, T, and, since $\beta^{-2} \propto N_K$, it is also inversely proportional to the molecular weight M of the polymer.

2.3 THE STRESS TENSOR

2.3.1 Derivation from Spring Force

We develop a simple formula for the stress tensor resulting from elastic forces in stretched polymer chains, following the approach of Bird et al.[4] Suppose there are ν such chains per unit volume of material. Guided by our finding that the elastic force can be represented by a spring force, we replace each polymer chain by a spring having the same end-to-end vector, $\underset{\sim}{R}$, and the same elastic parameter, β, as each original chain. Consider now a unit cube with one of its surfaces oriented perpendicular to a unit vector, $\underset{\sim}{n}$, Figure 2-2. The number of springs per unit cube with end-to-end vector between $\underset{\sim}{R}$ and $\underset{\sim}{R} + d\underset{\sim}{R}$ is $\nu\psi(\underset{\sim}{R})dR^3$. Here ψ need not equal ψ_0, the distribution function at equilibrium. ψ, for example could be a nonequilibrium distribution produced by a deformation. The projected length of one such spring in the direction perpendicular to the unit surface, and therefore the probability that its end-to-end vector intersects that surface, is $\underset{\sim}{n} \cdot \underset{\sim}{R}$, assuming $\underset{\sim}{n} \cdot \underset{\sim}{R} << 1$. Thus $(\underset{\sim}{n} \cdot \underset{\sim}{R})\,(\nu\psi(\underset{\sim}{R})dR^3)$ is the average number of springs that intersect the unit surface. The force contributed by these springs acting on the unit surface from outside the cube is $(\underset{\sim}{n} \cdot \underset{\sim}{R})\,(\nu\psi(\underset{\sim}{R})dR^3)\,\underset{\sim}{F}$, where for

Gaussian chains $\underset{\sim}{F}$ is given by Equation 2-6. Adding up the contributions from each $\underset{\sim}{R}$ gives the total force, $\underset{\sim}{f}$, acting on the surface,

$$\underset{\sim}{f} = \nu \underset{\sim}{n} \cdot \int \underset{\sim}{R}\underset{\sim}{F}\psi(\underset{\sim}{R})\, dR^3 = \nu \underset{\sim}{n} \cdot \langle \underset{\sim}{R}\underset{\sim}{F} \rangle \tag{2-7}$$

Here $\langle \cdot \rangle$ is an average over the nonequilibrium distribution, ψ,

$$\langle \cdot \rangle \equiv \int \psi \cdot dR^3$$

The force on the unit surface is $\underset{\sim}{n} \cdot \underset{\approx}{\sigma}$, where $\underset{\approx}{\sigma}$ is the stress tensor. Equation (2-7) therefore implies that

$$\underset{\approx}{\sigma} = \nu \langle \underset{\sim}{R}\underset{\sim}{F} \rangle \tag{2-8}$$

For Gaussian chains, using Equation 2-6, we obtain

$$\underset{\approx}{\sigma} = 2kT\nu\beta^2 \langle \underset{\sim}{R}\underset{\sim}{R} \rangle \tag{2-9}$$

Thus the viscoelastic stress tensor arises from *the distribution of polymer configurations*; specifically, it is the second moment of the configuration distribution function. Mechanical stresses can therefore be interpreted in terms of biases in molecular orientations. The positive first normal stress difference, $\sigma_{11} - \sigma_{22}$ in simple shearing, for example, implies that more molecules are oriented in the direction of flow, the "1" direction, than in the "2" direction, which is perpendicular to the shearing plane. The second normal stress difference, $\sigma_{22} - \sigma_{33}$, is usually seen to be negative in experiments; this implies that the orientation perpendicular to the shearing plane is *depleted* of polymer configurations relative to the orientation in the neutral direction "3."

2.3.2 Derivation from Virtual Work

A second approach[5] leads again to Equation 2-9, for the polymer contribution to the elastic stress tensor and elucidates the conditions under which the formula is valid. We return to the formula for the entropy of a single chain,

$$S = -k\beta^2R^2 + \text{constant} \tag{2-10}$$

The sum of these contributions over all chains in a unit volume is then

$$S = -\nu k \int \psi R^2\, dR^3 + \text{constant} \tag{2-11}$$

There is, in general, another, combinatoric, contribution to the entropy arising from a count of the number of ways a set of N chains can be chosen to satisfy the distribution function, ψ. Suppose we subdivide configuration space into a large number of small elements, each with volume dR^3. Let $\underset{\sim}{R}_p$ be centered in the pth configuration volume element. There are $N\psi(\underset{\sim}{R}_p)dR^3$ chains in the pth element. Let $\psi_p \equiv \psi(\underset{\sim}{R}_p)$. According to the combinatorial formula, there are

$$\Omega = \frac{N!}{\prod_p (N\psi_p dR^3)!} \tag{2-12}$$

ways of distributing N chains among the configuration-volume elements such that the pth element, has $N\psi_p dR^3$ chains. The combinatorial entropy per unit (physical) volume resulting from this distribution is

$$S_c = k \log \Omega / V \doteq -k\frac{N}{V}\sum_p \psi_p \, dR^3 \log \psi_p + \text{constant} \tag{2-13}$$

where V is the volume of fluid containing the N chains. Here Stirling's approximation, $\log(M!) \doteq M\log M$, has been used. Making the configuration volume elements, dR^3, arbitrarily small, the sum can be replaced by an integral,

$$S_c \doteq -k\nu \int \psi \log \psi \, dR^3 + \text{constant} \tag{2-14}$$

where $\nu = N/V$ is the number of chains per unit volume. In the present calculations this term contributes nothing to the stress and so can be dropped. Later, however (in Section 4.4.4) we shall need to refer to this contribution.

The free energy is therefore

$$W = -TS = \nu k T \beta^2 \int \psi R^2 \, dR^3 - \frac{3}{2}\nu k T \tag{2-15}$$

where the constant, $-(3/2)\nu kT$, has been chosen so that $W = 0$ at equilibrium. Thus Equation 2-15 is the free energy relative to the equilibrium state.

Suppose the material containing these chains is deformed, and the endpoints of the chains are convected with the macroscopic deformation. This is called an *affine motion* of the endpoints. The macroscopic deformation transforms an embedded vector from $\underset{\sim}{r}'$ to $\underset{\sim}{r}$ according to

$$\underset{\sim}{r} = \underset{\sim}{r}' \cdot \underset{\approx}{F}^{-1} \tag{2-16}$$

where $\underset{\approx}{F}^{-1}$ is the deformation gradient as discussed in Chapter 1. Our assumption of affine motion means that

$$\underset{\sim}{R} = \underset{\sim}{R}' \cdot \underset{\approx}{F}^{-1} \tag{2-17}$$

where $\underset{\sim}{R}'$ and $\underset{\sim}{R}$ are the end-to-end vectors before and after the deformation, respectively. End-to-end vectors are stretched and rotated by the deformation in the same way as are embedded macroscopic vectors. Now because $\partial \underset{\approx}{F}^{-1}/\partial t = \underset{\approx}{F}^{-1} \cdot \underset{\sim}{\nabla} \underset{\sim}{v}$, according to Equation 1-9, it follows that

$$\dot{\underset{\sim}{R}} = \underset{\sim}{R}' \cdot \frac{\partial}{\partial t} \underset{\approx}{F}^{-1} = \underset{\sim}{R}' \cdot \underset{\approx}{F}^{-1} \cdot \underset{\sim\sim}{\nabla v} = \underset{\sim}{R} \cdot \underset{\sim\sim}{\nabla v} \tag{2-18}$$

Therefore

$$\frac{\partial}{\partial t} R^2 = \underset{\sim}{R} \cdot \dot{\underset{\sim}{R}} + \dot{\underset{\sim}{R}} \cdot \underset{\sim}{R} = 2 \underset{\sim}{R} \cdot \underset{\sim\sim}{\nabla v} \cdot \underset{\sim}{R} = 2 \underset{\approx}{D} : \underset{\sim\sim}{RR} \tag{2-19}$$

Averaging over all chains gives

$$\frac{\partial}{\partial t} \langle R^2 \rangle = 2 \underset{\approx}{D} : \langle \underset{\sim\sim}{RR} \rangle \tag{2-20}$$

Hence, from Equations 2-15 and 2-20,

$$\dot{W} \equiv \nu k T \beta^2 \frac{\partial}{\partial t} \langle R^2 \rangle = \underset{\approx}{D} : 2\nu k T \beta^2 \langle \underset{\sim\sim}{RR} \rangle \tag{2-21}$$

But the power, or the rate of doing work, required to deform a macroscopic sample is

$$\dot{W} = \underset{\sim\sim}{\nabla v}^{T} : \underset{\approx}{\sigma} = \underset{\approx}{D} : \underset{\approx}{\sigma} \tag{2-22}$$

Comparing Equations 2-21 and 2-22, we see that the polymer contribution to the stress tensor is evidently

$$\underset{\approx}{\sigma} = 2\nu k T \beta^2 \langle \underset{\sim\sim}{RR} \rangle \tag{2-23}$$

The same formula for $\underset{\approx}{\sigma}$ was arrived at by the spring argument. The entropy argument, however, serves to emphasize the conditions under which Equation 2-23 is valid namely, that the chain ends move affinely and that the rest of the chain samples its entire equilibrium distribution, subject to the constraints on the chain ends. Thus Equation 2-23 is valid when the effect of the deformation is concentrated entirely at the chain ends. This assumption is most likely to be valid when the chain ends are anchored in some way to the bulk material, as for example in *cross-linked rubbers.*

2.4 RUBBER ELASTICITY THEORY

We are now prepared to consider the classical theory for the elasticity of rubbers made up of cross-linked polymer chains, schematized in Figure 2-3. This theory was developed by Wall,[6] Flory and Rehner,[7] James and Guth,[8] and Treloar.[9,1] Suppose that cross-link points are evenly spaced along each polymer chain a distance N_K Kuhn steps apart. The portion of polymer between two neighboring cross-links is called a *strand.* For the simplistic theory considered here, nothing need be said about the functionality— that is, the coordination number of the cross-link points. We suppose that the cross-linking reactions that produced the rubber occurred in such a way that in the absence of deformation the probability, $\psi_0(\underset{\sim}{R}')dR'^3$, that a strand has an end-to-end vector between $\underset{\sim}{R}'$

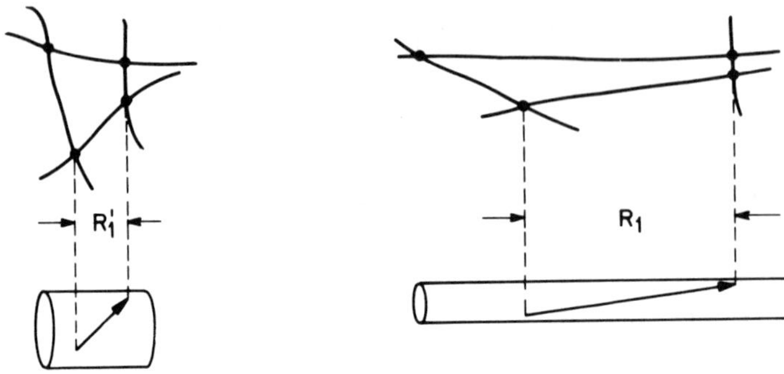

Fig. 2-3 Cross-linked rubber network before deformation (left) and after affine motion of network strands (right).

and $\underset{\sim}{R}' + d\underset{\sim}{R}'$ is the same as would be the case without the cross-links; that is,

$$\psi_0(\underset{\sim}{R}') = (\frac{\beta}{\sqrt{\pi}})^3 \exp(-\beta^2 R'^2) \qquad (2\text{-}24)$$

where $\beta^2 = 3/(2N_K b^2)$.

We now consider an extensional deformation of the rubber network, with deformation axes oriented along the coordinate axes, and stretch ratios λ_1, λ_2, and λ_3 in directions "1," "2," and "3," respectively. The deformation depicted in Figure 2-3, for example, is an elongation for which $\lambda_2 = \lambda_3 = \lambda_1^{-1/2}$. Suppose that the cross-link points are convected with the macroscopic deformation; that is, the cross-links move *affinely*. Because the deformation is extensional, each component, R_i, of $\underset{\sim}{R}$ is given by

$$R_i = R_i'\lambda_i \qquad (2\text{-}25)$$

Each strand vector, $\underset{\sim}{R}'$, is transformed by the deformation according to Equation 2-25, $R_i' = R_i/\lambda_i$. Before the deformation there were $N\psi_0(\underset{\sim}{R}')dR'^3$ polymer strands with end-to-end vectors between $\underset{\sim}{R}'$ and $\underset{\sim}{R}' + d\underset{\sim}{R}'$, where N is the total number of strands. After the deformation, there are the same number of strand vectors between $\underset{\sim}{R} = (R_i'\lambda_i,\ i = 1,2,3)$ and $\underset{\sim}{R} + d\underset{\sim}{R}$ $= (R_i'\lambda_i + dR_i'\lambda_i,\ i = 1,2,3)$. Thus the distribution function, $\psi(\underset{\sim}{R})$, after the deformation is related to that before the deformation by

$$\psi(\underset{\sim}{R})dR^3 = \psi_0(\underset{\sim}{R}')dR'^3 = \psi_0(R_i/\lambda_i,\ i = 1,2,3)\, dR'^3 \qquad (2\text{-}26)$$

But

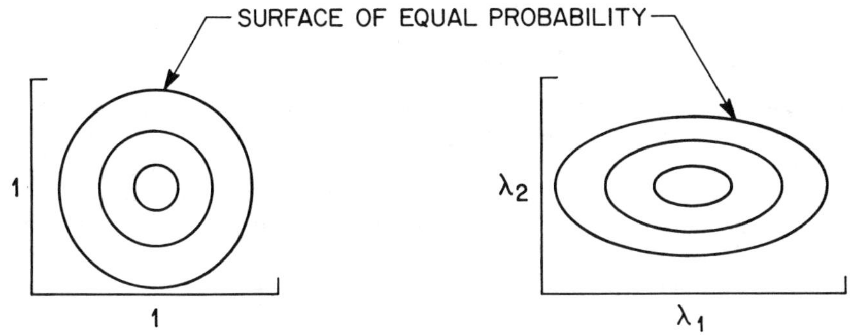

Fig. 2-4 Surfaces of equal probability, ψ, before (left) and after (right) deformation.

$$dR^3 = dR'^3\,\lambda_1\lambda_2\lambda_3 = dR'^3 \tag{2-27}$$

because of volume conservation. Thus

$$\psi(\underset{\sim}{R}) = \psi_0\left(\frac{R_i}{\lambda_i},\ i=1,2,3\right) = \left(\frac{\beta}{\sqrt{\pi}}\right)^3 \exp\left[-\beta^2\left(\frac{R_1^2}{\lambda_1^2}+\frac{R_2^2}{\lambda_2^2}+\frac{R_3^2}{\lambda_3^2}\right)\right] \tag{2-28}$$

Each surface of equal probability is therefore deformed from a sphere to an ellipsoid with principle axes λ_1, λ_2, and λ_3 times the diameter of the undeformed sphere, Figure 2-4.

The components of the stress tensor are

$$\begin{aligned}\sigma_{ii} &= 2\nu kT\beta^2 \langle R_iR_i\rangle = 2\nu kT\beta^2 \int R_i^2 \left(\frac{\beta}{\sqrt{\pi}}\right)^3 \exp\left[-\beta^2\sum_j \frac{R_j^2}{\lambda_j^2}\right] dR^3 \\ &= \nu kT\lambda_i^2\end{aligned} \tag{2-29}$$

where ν is the number of strands per unit volume. Thus

$$\underset{\approx}{\sigma} = \nu kT \begin{pmatrix}\lambda_1^2 & 0 & 0\\ 0 & \lambda_2^2 & 0\\ 0 & 0 & \lambda_3^2\end{pmatrix} = G\,\underset{\approx}{C}^{-1} \tag{2-30}$$

where $G = \nu kT$ is the modulus. Hence for extensional deformations, we obtain from rubber elasticity theory the relationship between stress and the Finger tensor proposed in Equation 1-25. The relationship between $\underset{\approx}{\sigma}$ and $\underset{\approx}{C}^{-1}$ is not affected by rotations of frame; hence Equation 2-30 also holds for nonextensional deformations. Thus classical rubber elasticity theory predicts $\underset{\approx}{\sigma} = G\underset{\approx}{C}^{-1}$ for *any* volume-preserving deformation.

2.5 THE TEMPORARY NETWORK MODEL

2.5.1 Derivation of Constitutive Equation

In 1946, Green and Tobolsky[10] proposed a theory appropriate for molten polymeric liquids that was inspired by rubber elasticity theory. This model is based on the notion that polymer chains that are not chemically cross-linked nevertheless interact with each other to form a *transient network*. In this "liquid" network, a strand is not permanently bound to the same cross-links, but occasionally breaks free of its constraints and is immediately recaptured by new constraints, so that the concentration of junction points, ν, is always constant. In the language of Green and Tobolsky, strands "break" and "re-form" at equal rates.

These temporary cross-links or constraints are not bonds that occur at specific sites along the polymer chain, but they are believed to be caused by *entanglements*; that is, topological constraints arising because one chain cannot pass through another.

As a result of the constraints, the strand end-to-end vector jumps impulsively from one value to another. If the material remains free of deformation, the strand will simply sample the equilibrium distribution, ψ_0, of end-to-end vectors. Suppose that the probability, per unit time, that the strand changes from one set of constraints to another is $1/\lambda$. Then the probability, $P_{t',t}$, that a strand retains the same end-to-end vector from time t′ to time t obeys the differential equation

$$\frac{d}{dt} P_{t',t} = -\frac{1}{\lambda} P_{t',t} \tag{2-31}$$

with $P_{t',t'} = 1$. Hence

$$P_{t',t} = e^{(t'-t)/\lambda} \tag{2-32}$$

Now suppose that material is being deformed. As long as a strand does not escape its constraints, it is stretched and rotated affinely. But when it does escape, it takes on a relaxed configuration, that is, *a configuration typical of equilibrium*, even though deformation is occurring. Thus, Green and Tobolsky assumed that a strand can do only two things: deform affinely or re-form into a configuration typical of equilibrium.

The probability that a strand re-forms in an interval of time between t′ and t′ + dt′ is dt'/λ. The probability that it survives without breaking from time t′ to time t is $P_{t',t}$. The contribution to the stress from those strands that meet both of these stipulations is then from Equation 2-30 equal to

$$d\underset{\approx}{\sigma} = \frac{dt'}{\lambda} P_{t',t}\, G\, \underset{\approx}{C}^{-1}(t',t) = G \frac{1}{\lambda} e^{(t'-t)/\lambda}\, \underset{\approx}{C}^{-1}(t',t)\, dt' \tag{2-33}$$

where $G = \nu kT$, as before. The Finger tensor, $\underset{\approx}{C}^{-1}(t',t)$, enters because the strands are deformed affinely like strands in a cross-linked rubber between times t′ and t. The total stress produced by strands re-formed at all past time, t′, is

$$\underset{\approx}{\sigma} = G \int_{-\infty}^{t} \frac{1}{\lambda} e^{(t'-t)/\lambda} \underset{\approx}{C}^{-1}(t',t)\, dt' \tag{2-34}$$

This is the simple constitutive equation put forward at the end of Chapter 1.

2.5.2 Assumptions of the Green-Tobolsky Model

The key assumptions of this temporary network of Green and Tobolsky are:

1. Strands are Gaussian.
2. Strands deform affinely until they break.
3. Strands break with a constant probability per unit time, independent of the network deformation.
4. Strands re-form as fast as they break.
5. Strands re-form in configurations typical of equilibrium.

We have also tacitly assumed that there is only one relaxation process, with a single relaxation time, λ. Green and Tobolsky clearly recognized that their approach could be extended to multiple relaxation processes. The extension was carried out independently by Lodge[11] and by Yamamoto.[12] Lodge gives the clearest derivation of the model with multiple relaxation processes, and the constitutive *equation* resulting from the molecular *model* of Green and Tobolsky has thereby come to be named after him:

$$\underset{\approx}{\sigma} = \int_{-\infty}^{t} m(t-t')\, \underset{\approx}{C}^{-1}(t,t')\, dt' \qquad \text{(Lodge Equation)} \tag{2-35}$$

The function $m(t-t')$ can accommodate any number of *independent* relaxation processes:

$$m(t-t') = \sum_i \frac{G_i}{\lambda_i} e^{-(t-t')/\lambda_i} \tag{2-36}$$

The inclusion of multiple relaxation times, λ_i, is necessary because stress relaxation in polymeric systems usually occurs over several decades of time; see for example Fig. 1-8. With one relaxation time, that is, a single exponential decay, relaxation takes place in only about a decade of time.

2.5.3 Successes and Limitations of the Green-Tobolsky Model

The Lodge equation predicts many of the qualitative phenomena of viscoelastic flow. It predicts a positive primary normal stress difference in shear, gradual stress relaxation after cessation of flow, and elastic recovery of strain. It predicts strain hardening in steady elongation, with a vengeance: in the Lodge equation, the time-dependent elongational viscosity rises without limit whenever the elongation rate, $\dot{\epsilon}$, exceeds $1/2\lambda_1$, where λ_1 is the *longest* relaxation time. A steep rise in elongational viscosity reasonably describes some melts such as Melt I; see Figure 2-5,[13] but in real melts $\bar{\eta}$ must eventually level off or reach a maximum at long times. A hint of this is seen in Figure 2-5 at strain rates of 1.0 sec^{-1} and 0.2

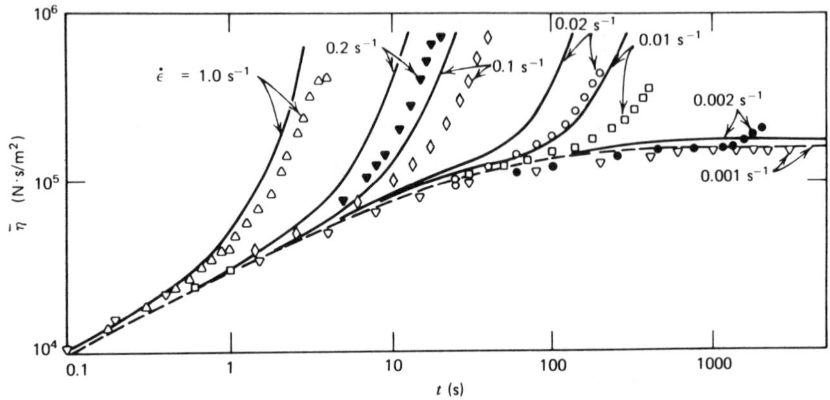

Fig. 2-5 Predictions of the temporary network model (lines) compared to experimental data (symbols) for elongational stress growth in Melt I low-density polyethylene;[13] figure from Bird et al.,[4] reprinted from *Dynamics of Polymeric Liquids,* Vol. 2 ©1977 by permission of John Wiley & Sons, Inc.).

$\sec^{-1}$.

In quantitative detail, the Lodge equation is often inaccurate. We have seen that, contrary to experiment, it predicts unlimited growth in elongational viscosity. In steady shearing, it predicts that both the shear viscosity, η, and the first normal stress coefficient, Ψ_1, are independent of shear rate; in experiments both decrease with increasing $\dot{\gamma}$; see Fig. 1-6. It also overpredicts elastic recovery. The Lodge equation thus overemphasizes the elasticity of viscoelastic fluids — it overpredicts elastic stresses and elastic recoverable strains.

2.6 THE ELASTIC DUMBBELL MODEL

The model of Green and Tobolsky was developed to describe polymer melts or solutions concentrated enough that the polymer coils overlap and entangle with each other. Rouse,[14] in 1953 developed a theory for polymer solutions that are sufficiently dilute that each molecule is isolated from the others. Rouse's model of the polymer coil can be thought of as a series of "beads" equally spaced along the polymer backbone and connected one to the next by springs; see Figure 2-6a. Each spring models the elastic forces generated by a portion of the macromolecule called a submolecule. Thus it is assumed that in any deformation the internal motions of each submolecule are sufficiently rapid that they are in equilibrium, subject, however, to the constraints on the endpoints of each submolecule. Arguments presented in Section 2.3 imply that under these conditions the submolecule generates entropic forces equivalent to those of a spring.

Although the spacing between the beads is somewhat arbitrary, if the beads are spaced far enough apart that the submolecules have enough degrees of freedom to develop random walk statistics, and yet are spaced close enough that the

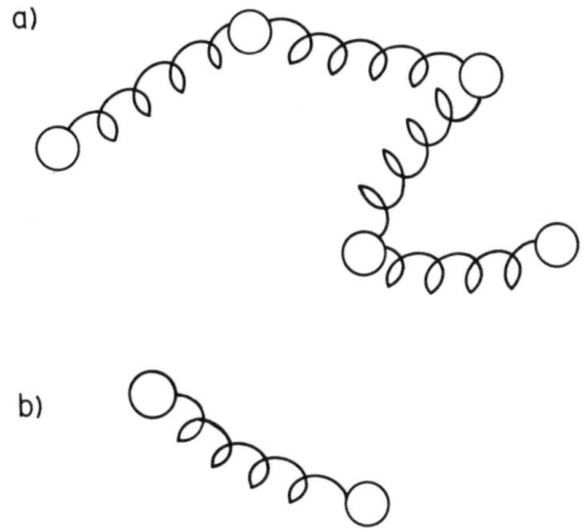

Fig. 2-6 (a) The Rouse beads and springs model for a polymer molecule; (b) the elastic dumbbell.

motions interior to a submolecule are fast enough to be in equilibrium during a deformation, then the exact choice of spacing doesn't affect the viscoelastic properties.

The viscous drag force that in reality is distributed continuously along the chain is, in the Rouse model, concentrated at the beads. As with the springs, the representation of the viscous force by a force concentrated at discrete points along the chain is reasonable for the calculation of properties controlled by stretching of the polymer chain that occurs over much longer distances than the interbead distance. This is true when deformations are considered that are slow enough that short-range conformational changes, which are controlled by fast relaxation processes, are not disturbed. Long-range changes in conformation are controlled by highly cooperative, and therefore slow, relaxation processes; these are well represented by the bead-spring model.

The slowest and in many ways most important relaxation process is the relaxation of the entire molecule. This process affects the end-to-end vector, $\underset{\sim}{R}$, of the entire molecule, and its dynamics are captured by a simple model with just one spring having a bead at each end, Figure 2-6b. This is the elastic dumbbell model, first proposed by W. Kuhn[15].

2.6.1 The Langevin Equation

Consider an elastic dumbbell immersed in solvent, where the bead diameter is small enough that the inertia of both the beads and solvent dragged by the beads is negligible. There are four forces: inertia, drag, the elastic spring force, and a random Brownian force. The balance of these forces gives

$$m\dot{\underset{\sim}{u}} + \varsigma(\underset{\sim}{u} - \underset{\sim}{R} \cdot \underset{\sim}{\nabla} \underset{\sim}{v}) + 4kT\beta^2 \underset{\sim}{R} = \underset{\sim}{A}$$

(inertia) (drag) (spring) (Brownian) (2-37)

Here the second term is the force on the spring resulting from the drag the solvent exerts on the beads. The drag force is proportional to $\underset{\sim}{u}$, the rate at which the the beads are moving away from each other, *minus* $\underset{\sim}{R} \cdot \underset{\sim}{\nabla} \underset{\sim}{v}$, the *ambient* solvent velocity near one bead relative to that near the other. The *ambient* solvent velocity near a bead is just the velocity the solvent would have in the absence of that bead. Equation 2-37 implies that the drag force is zero only if the dumbbell is deformed affinely; that is,

$$\underset{\sim}{u} = \underset{\sim}{R} \cdot \underset{\sim}{\nabla} \underset{\sim}{v} \tag{2-38}$$

If the dumbbell is deformed affinely, that is, in direct proportion to the macroscopic solvent velocity, then the dumbbell does not disturb the solvent velocity field and there is no drag. Any discrepancy between the ambient solvent velocity and that of the beads causes drag. The term $\varsigma(\underset{\sim}{u} - \underset{\sim}{R} \cdot \underset{\sim}{\nabla} \underset{\sim}{v})$ assumes that the drag force is directly proportional to the discrepancy between the macroscopic solvent velocity and the bead velocities. The friction coefficient ς is the proportionality constant. For spherical beads, $\varsigma = 6\pi\eta_s a$ according to Stokes' law, where a is the bead radius and η_s is the solvent viscosity.

Neglected in Equation 2-37 is *hydrodynamic interaction*, the effect the motion of one bead has on the ambient solvent velocity near the other bead. The effect of hydrodynamic interaction is taken up in Chapter 8. We have also assumed that the velocity gradient, $\underset{\sim}{\nabla} \underset{\sim}{v}$, is uniform or nearly so over the length of the dumbbell.

The third term of Equation 2-37 is the spring force, $4kT\beta^2 \underset{\sim}{R}$. The inertial term, $m\dot{\underset{\sim}{u}}$, is the bead mass times the acceleration of one bead relative to the other. $\underset{\sim}{A}$ is a stochastic Brownian force. It represents the bombardment of the bead by the surrounding solvent molecules.

Equation 2-37 is a *Langevin equation.* A Langevin equation combines forces from the macroscopic world (the drag forces) with forces from the microscopic world (the Brownian forces). Because of the random Brownian term, a Langevin equation is not solvable in the conventional deterministic sense. We can, however, consider an *ensemble* of dumbbells and seek the probability, $\psi(\underset{\sim}{R},t)$, that a dumbbell has an end-to-end vector, $\underset{\sim}{R}$ at a given time, t.

This distribution function is obtained more easily by noting that bead *velocities* fluctuate much more rapidly than bead *positions*. This means that a bead thoroughly samples its distribution of velocities in a time short compared to the time required for it to drift very far in position. Indeed, the time required for the bead to equilibrate its velocity distribution can be estimated from the ratio of the coefficients of $\dot{\underset{\sim}{u}}$ and $\underset{\sim}{u}$; this velocity equilibration time is $t_v = m/6\pi\eta_s a \sim 10^{-12}$ seconds or less. The time required for the dumbbell to sample its distribution of end-to-end vectors $\underset{\sim}{R}$ is ten orders of magnitude slower. Thus the Langevin equation can be solved first on a short time scale, the time scale of velocity equilibration, and then on a long time scale, the time scale of configuration equilibration. On the short time scale the dominant forces are inertia, drag, and the Brownian force; the spring force and the flow term are negligible. Thus

$$m\dot{\underset{\sim}{u}} + \varsigma \underset{\sim}{u} = \underset{\sim}{A} \qquad \text{(short time scale)} \tag{2-39}$$

Over an interval of time comparable to the velocity equilibration time t_v each bead establishes a steady-state velocity distribution $\Xi(\underset{\sim}{u})$. If the beads are to be in thermal equilibrium with their surroundings, this velocity distribution must be that of particles equilibrated at temperature, T. Thus they must have a *Maxwellian distribution of velocities,*

$$\Xi(\underset{\sim}{u}) = \frac{m}{2\pi k T^{3/2}} \exp\left(-\frac{mu^2}{2kT}\right)$$

It can be shown[16] that if Equation 2-39 is to yield this Maxwellian distribution then the Brownian force averaged over a time long compared to t_v must equal $-2kT\,\partial \log\psi/\partial \underset{\sim}{R}$. Thus if Equation 2-37 is averaged over a time long compared to t_v, but short compared to the time required for significant change in $\underset{\sim}{R}$, the result is

$$\varsigma(\underset{\sim}{J}/\psi - \underset{\sim}{R}\cdot\underset{\sim}{\nabla}\underset{\sim}{v}) + 4kT\beta^2 \underset{\sim}{R} + 2kT\frac{\partial \log\psi}{\partial \underset{\sim}{R}} = \underset{\sim}{0} \tag{2-40}$$

The inertial term, when averaged over a time scale long compared to t_v, is negligible and does not appear in Equation 2-40. $\dot{\underset{\sim}{R}} \equiv \underset{\sim}{J}/\psi$ is the average of $\underset{\sim}{u}$ over the above time scale. $\underset{\sim}{J}$ is the ***probability flux.*** It is the probability per unit area per unit time that the end-to-end vector of a dumbbell crosses a small surface element in $\underset{\sim}{R}$ - space. $\underset{\sim}{J}$ is analogous to a mass flow-rate vector, $\rho\underset{\sim}{v}$, in real space. The mass density ρ is analogous to the probability ψ, and the velocity $\underset{\sim}{v}$ is analogous to $\dot{\underset{\sim}{R}}$. Just as $\rho\underset{\sim}{v}$ and ρ are related to each other by a mass balance equation in real space,

$$\frac{\partial}{\partial t}\rho + \frac{\partial}{\partial \underset{\sim}{r}}\cdot(\rho\underset{\sim}{v}) = 0 \tag{2-41}$$

so $\underset{\sim}{J} = \psi\dot{\underset{\sim}{R}}$ and ψ are related to each other by a probability balance equation in $\underset{\sim}{R}$-space,

$$\frac{\partial}{\partial t}\psi + \frac{\partial}{\partial \underset{\sim}{R}} \cdot \underset{\sim}{J} = 0 \tag{2-42}$$

Rewriting Equation 2-40, we have

$$\underset{\sim}{J} \equiv \underset{\sim}{R} \cdot \nabla \underset{\sim}{v} \psi - \frac{4kT\beta^2}{\varsigma} \underset{\sim}{R} \psi - \frac{2kT}{\varsigma} \frac{\partial \psi}{\partial \underset{\sim}{R}} \tag{2-43}$$

The first term on the right side is the probability flux arising from convective flow. The second is migration due to the spring force. The third term is the diffusive flux. The diffusive flux is analogous to Fick's law of diffusion of concentration, c, in real space, where

$$\underset{\sim}{j} = -D_c \frac{\partial c}{\partial \underset{\sim}{r}} \tag{2-44}$$

Here $\underset{\sim}{j}$ is the diffusive mass flux in real space.

2.6.2 The Smoluchowski Equation

We substitute $\underset{\sim}{J}$ from Equation 2-43 into the probability continuity equation, Equation 2-42. The result is the *Smoluchowski equation*

$$\frac{\partial \psi}{\partial t} + \frac{\partial}{\partial \underset{\sim}{R}} \cdot \left[\underset{\sim}{R} \cdot \nabla \underset{\sim}{v} \psi - \frac{4kT\beta^2}{\varsigma} \underset{\sim}{R} \psi - \frac{2kT}{\varsigma} \frac{\partial \psi}{\partial \underset{\sim}{R}} \right] = 0 \tag{2-45}$$

The Smoluchowski equation is an equation of change for the distribution function ψ.

2.6.3 The Constitutive Equation

If the Smoluchowski equation is solved for ψ, the average, $<\underset{\sim\sim}{RR}> \equiv \int \psi \underset{\sim\sim}{RR}\, dR^3$, can be evaluated, thereby giving the stress tensor. Fortunately ψ does not have to be obtained explicitly to calculate $<\underset{\sim\sim}{RR}>$. Multiplying Equation 2-45 by $\underset{\sim\sim}{RR}$ and integrating over $\underset{\sim}{R}$-space gives the following terms:

$$\int \frac{\partial \psi}{\partial t} \underset{\sim\sim}{RR}\, dR^3 = \frac{\partial}{\partial t} \int \psi \underset{\sim\sim}{RR}\, dR^3 = \frac{\partial}{\partial t} <\underset{\sim\sim}{RR}>$$

$$\int \underset{\sim\sim}{RR} \frac{\partial}{\partial \underset{\sim}{R}} \cdot (\underset{\sim}{R} \cdot \nabla \underset{\sim}{v} \psi)\, dR^3 = -\nabla \underset{\sim\sim}{v}^T \cdot <\underset{\sim\sim}{RR}> - <\underset{\sim\sim}{RR}> \cdot \nabla \underset{\sim\sim}{v}$$

$$-\frac{4kT\beta^2}{\varsigma}\int \underset{\sim}{R}\underset{\sim}{R}\frac{\partial}{\partial \underset{\sim}{R}}\cdot(\underset{\sim}{R}\psi)\,dR^3 = \frac{8kT\beta^2}{\varsigma}\langle\underset{\sim}{R}\underset{\sim}{R}\rangle$$

$$-\frac{2kT}{\varsigma}\int \underset{\sim}{R}\underset{\sim}{R}\frac{\partial}{\partial \underset{\sim}{R}}\cdot\frac{\partial\psi}{\partial \underset{\sim}{R}}\,dR^3 = -\frac{4kT}{\varsigma}\underset{\approx}{\delta} \tag{2-46}$$

The right sides of the last three equations were obtained by integrating the left sides by parts. (The boundary terms vanish because ψ goes to zero at large $|\underset{\sim}{R}|$.)

Since $\underset{\approx}{\sigma} = 2\nu kT\beta^2\langle\underset{\sim}{R}\underset{\sim}{R}\rangle$, where ν is the number of dumbbells per unit volume, we have, after collecting terms and multiplying by $2\nu kT\beta^2$,

$$\dot{\underset{\approx}{\sigma}} - \underset{\sim}{\nabla}\underset{\sim}{v}^{T}\cdot\underset{\approx}{\sigma} - \underset{\approx}{\sigma}\cdot\underset{\sim}{\nabla}\underset{\sim}{v} + \frac{1}{\lambda}(\underset{\approx}{\sigma} - G\underset{\approx}{\delta}) = \underset{\approx}{0} \tag{2-47}$$

with $G = \nu kT$ and $\lambda = \varsigma/8kT\beta^2$. This is just the upper-convected Maxwell equation derived from the Lodge equation in Chapter 1. Thus the simplest model for melts, the temporary network model of Green and Tobolsky, and the simplest model for dilute solutions, the dumbbell model, give the same constitutive equation for the polymer contribution to the stress! That the same constitutive equation can be obtained from different molecular models is a recurrent theme in this field. It is a heartening signal in the sense that one can hope that a valuable constitutive equation might emerge even if the molecular model is dubious. It is a warning signal, however, against taking the accuracy of the constitutive equation as proof of the validity of the assumptions of the molecular model.

In one sense, it is not too surprising that different constitutive models give the same equation for stress. The stress, after all, is the second moment of a distribution function ψ. The details of ψ are averaged out in the stress tensor. As is discussed in Chapter 6, the temporary network model and the dilute dumbbell model do *not* predict the same distribution functions, ψ, even though the second moments of the distributions, and therefore the stresses, are identical.

Left out of the above constitutive equation was the direct contribution of the solvent to the stress tensor. Even in the absence of any polymer (i.e., $\nu = 0$), there is stress in a flowing solvent. Assuming that the solvent is Newtonian, this contribution is

$$\underset{\approx}{\sigma}_s = 2\eta_s\underset{\approx}{D} \tag{2-48}$$

The total stress tensor is then

$$\underset{\approx}{\sigma} = \underset{\approx}{\sigma}_p + \underset{\approx}{\sigma}_s \tag{2-49}$$

where $\underset{\approx}{\sigma}_p$ is the polymer contribution to the stress tensor that satisfies Equation 2-47. Equation 2-49 is equivalent to the early constitutive equation called the "Oldroyd-B" equation.

2.7 THE ROUSE MODEL

2.7.1 The Langevin Equation

We now return to the Rouse multiple-spring-bead model of Figure 2-6a. In this case, the Langevin equation for the end-to-end vector of a given spring is coupled to the equations for the neighboring springs, because beads attached to the given spring feel forces from the neighboring springs. The average Langevin equation, the counterpart to Equation 2-40 for the dumbbell model, is

$$\varsigma(\dot{\underset{\sim}{R}}_i - \underset{\sim}{R}_i \cdot \underset{\sim}{\nabla} \underset{\sim}{v}) + kT\sum_{i=j}^{N} A_{ij} \left(2\beta^2 \underset{\sim}{R}_j + \frac{\partial \log \psi}{\partial \underset{\sim}{R}_j}\right) = \underset{\sim}{0} \tag{2-50}$$

where

$$A_{ij} = \begin{cases} 2 & \text{if } i=j \\ -1 & \text{if } i=j+1 \text{ or } i=j-1 \\ 0 & \text{otherwise} \end{cases} \tag{2-51}$$

and $\underset{\sim}{R}_j$ is the end-to-end vector of the jth spring in the chain. N is the number of springs. Equation 2-50 can be thought of as a matrix equation with off-diagonal terms.

2.7.2 Normal Mode Transformation

Rouse discovered a transformation that diagonalized this matrix equation into uncoupled equations, that is, into N *normal modes*:

$$\varsigma(\dot{\underset{\sim}{R}}'_i - \underset{\sim}{R}'_i \cdot \underset{\sim}{\nabla} \underset{\sim}{v}) + 4kT \sin^2\left(\frac{i\pi}{2(N+1)}\right) \left[2\beta^2 \underset{\sim}{R}'_i + \frac{\partial \log \psi}{\partial \underset{\sim}{R}'_i}\right] = \underset{\sim}{0} \tag{2-52}$$

where $\underset{\sim}{R}'_i$ is a transformed end-to-end vector. The first mode, i=1, corresponds to cooperative relaxation of the entire chain. The other normal modes, i=2,3,..., correspond to relaxations of successively smaller portions of the chain. An analogy can be drawn to a vibrating string, which has a fundamental mode corresponding to vibration of the whole string, and a series of higher harmonics that correspond to vibrations of successively smaller fractions of the string.

The probability flux is the sum over the contributions from all modes:

$$\underset{\sim}{J}/\psi = \sum_{i=1}^{N} \dot{\underset{\sim}{R}}'_i \tag{2-53}$$

which can be substituted into the probability continuity equation, Equation 2-42. Because each mode is independent, the probability distribution has a factorable solution:

$$\psi(\underset{\sim}{R}'_1, \underset{\sim}{R}'_2, \ldots, \underset{\sim}{R}'_N) = \prod_{i=1}^{N} \psi_i(\underset{\sim}{R}'_i) \tag{2-54}$$

Each ψ_i satisfies its own Smoluchowski equation:

$$\frac{\partial \psi_i}{\partial t} + \frac{\partial}{\partial \underset{\sim}{R}'_i} \cdot \left\{ \underset{\sim}{R}'_i \cdot \nabla \underset{\sim}{v} \psi_i - \frac{4kT}{\varsigma} \sin^2\left(\frac{i\pi}{2(N+1)}\right) \left[2\beta^2 \underset{\sim}{R}'_i + \frac{\partial \log \psi_i}{\partial \underset{\sim}{R}'_i} \right] \right\} \tag{2-55}$$

2.7.3 The Stress Tensor and Constitutive Equation

The polymeric contribution to the stress tensor is a sum of contributions from all the springs:

$$\underset{\approx}{\sigma}_p = 2\nu kT\beta^2 \sum_{i=1}^{N} <\underset{\sim}{R}_i \underset{\sim}{R}_i> \tag{2-56}$$

It is a property of the normal-mode decomposition that

$$\sum_{i=1}^{N} <\underset{\sim}{R}_i \underset{\sim}{R}_i> = \sum_{i=1}^{N} <\underset{\sim}{R}'_i \underset{\sim}{R}'_i> \tag{2-57}$$

Thus the stress can be broken down into a sum of contributions,

$$\underset{\approx}{\sigma}_p = \sum_{i=1}^{N} \underset{\approx}{\sigma}_i \tag{2-58}$$

each contribution coming from a distinct normal mode,

$$\underset{\approx}{\sigma}_i = 2\nu kT\beta^2 <\underset{\sim}{R}'_i \underset{\sim}{R}'_i> \tag{2-59}$$

Because the distribution function, ψ_i, associated with mode i satisfies a Smoluchowski equation similar to the Smoluchowski equation for dumbbells, $\underset{\approx}{\sigma}_i$ obeys an upper-convected Maxwell equation (see Section 1.6.2),

$$\overset{\nabla}{\underset{\approx}{\sigma}_i} + \frac{1}{\lambda_i}(\underset{\approx}{\sigma}_i - G\underset{\approx}{\delta}) = \underset{\approx}{0} \tag{2-60}$$

where

$$\lambda_i = \frac{\varsigma}{16kT\beta^2 \sin^2(i\pi/2(N+1))}, \qquad G = \nu kT \tag{2-61}$$

Equations 2-58, 2-60, and 2-61 therefore constitute *the constitutive equation of the Rouse model.*

2.7.4 Approximation for Slow Modes

In the Rouse constitutive equation, *there are N independent contributions to the stress tensor, each having its own relaxation time,* λ_i. For N large, there are many long relaxation times for which $i << N$, and therefore

$$\sin^2\left(\frac{i\pi}{2(N+1)}\right) \approx \frac{i^2\pi^2}{4(N+1)^2} \tag{2-62}$$

The longer relaxation times are therefore approximately

$$\lambda_i \approx \frac{\varsigma(N+1)^2}{4\pi^2\,\beta^2\,i^2\,kT}, \qquad G_i = G = \nu kT \tag{2-63}$$

This equation for λ_i shows an *apparent* dependence on N, the number of submolecules into which the chain is subdivided. N is not a purely molecular property, but depends on how many beads and springs one chooses to describe the molecule. Fortunately, the apparent dependence of λ_i on N disappears. This is so because for a given polymer molecule, as one increases N, the number of submolecules, the fully extended length of each submolecule decreases as 1/N. Then β^{-2}, which is proportional to the mean square end-to-end distance of the submolecule, is proportional to 1/N. Also ς, the bead friction coefficient, is inversely proportional to N, since as one increases the number of beads used to describe a given polymer molecule, the amount of friction that must be assigned to each bead decreases. Equation 2-63 can therefore be rearranged to a form that shows no dependence on N:

$$\lambda_i \approx \frac{\varsigma_0\,P^2\ell^2}{6\pi^2 i^2 kT} = \frac{6\,(\eta-\eta_s)}{\pi^2 i^2 \nu kT} \tag{2-64}$$

Here P is the number of monomers in the whole chain, ℓ is the monomer length and ς_0 is the friction coefficient per monomer. $\eta - \eta_s$ is the polymer contribution to the shear viscosity. The second equality in 2-64 is derived from the relationship $\eta - \eta_s = \nu kT \sum \lambda_i$ and the identity $\sum \lambda_i = \; = \lambda_1 \sum 1/i^2 = \pi^2 \lambda_1/6$.

This equation for λ_i is only valid if i/N is small. For large i/N, that is, for modes that describe the relaxation of only a few submolecules, the approximation in Equation 2-64 does not hold, and λ_i for each of these modes depends on N, the arbitrary number of submolecules. Thus the Rouse model is only valid for the longer relaxation modes.

Polymer motions not described by the Rouse model include bond rotations and coordinated movements of several backbone atoms. These motions are sometimes called "glassy modes" since they account for much of the relaxation that occurs in bulk polymers that have been vitrified, i.e., cooled to the point of solidification into a glass. The temperature at which a fluid vitrifies is called the *glass transition temperature.*

2.7.5 Assumptions of the Rouse Model

The assumptions underlying the Rouse model are:

1. Excluded volume and polymer-solvent interactions are neglected; the coil has a Gaussian equilibrium distribution of configurations.

2. The deformations are not so severe that they distort the chain conformation beyond the limit of Gaussian behavior. The distortions can thus be modeled by the displacements of a discrete set of frictional beads connected by submolecules

that act as springs.

3. The frictional drag on a bead is proportional to the velocity of the bead minus the ambient solvent velocity, that is, the velocity the solvent would have in the absence of that bead.

4. The influence of the drag that other beads have on the ambient solvent velocity around a given bead is neglected; i.e., hydrodynamic interaction is not accounted for.

5. Brownian forces keep each bead in thermal equilibrium with the solvent; each bead maintains the Maxwell distribution of velocities corresponding to the solvent temperature.

Finally, we have tacitly assumed that

6. The topological restrictions that prohibit one part of the chain from passing through another are neglected; the Rouse molecule is therefore a "phantom chain."

For modest deformations, the weakest of these assumptions are the neglect of solvent effects and the neglect of hydrodynamic interaction. Corrections for these are discussed in Chapter 8. The effects of the phantom chain assumption are uncertain.[17]

2.8 LINEAR VISCOELASTICITY

2.8.1 Distribution of Relaxation Times

2.8.1.1 The Rouse spectrum The Rouse model is more complete than the model of Green and Tobolsky, because it yields values for the relaxation times, λ_i, and moduli, $G_i = G = \nu kT$, for all i. These values constitute what is called the *distribution* or *spectrum of relaxation times.* The Rouse relaxation spectrum is a distinctive feature of the model that can be tested experimentally. The simplest and most accurate type of experiment that tests this spectrum is the *small-amplitude oscillatory deformation.* This can be performed in shear, for example in the cone and plate geometry of Fig. 1-5, by an oscillatory rotation of the cone about its axis of symmetry. In this way the shear becomes a sinusoidal function of time:

$$\gamma = \gamma_0 \sin \omega t \tag{2-65}$$

here γ_0, the maximum shear strain, is called the strain *amplitude*; ω is the *frequency.*

The response of the shear stress for the upper-convected Maxwell equation to such a strain history, and hence the response of a single mode, say mode i, of the Rouse model, is

$$\sigma = \gamma_0 \left[G_i \frac{\omega^2\lambda_i^2}{\omega^2\lambda_i^2 + 1} \sin \omega t + G_i \frac{\omega\lambda_i}{\omega^2\lambda_i^2 + 1} \cos \omega t\right] \tag{2-66}$$

The shear stress is therefore proportional to the strain amplitude, γ_0, no matter how large the value of γ_0. This result is a peculiar, and unrealistic, feature of the upper-convected Maxwell equation. Nevertheless for all constitutive equations that obey the postulates of viscoelastic simple fluid theory, and perhaps more to the point, for virtually all real polymer melts and solutions, the shear stress *is* proportional to γ_0, *if* γ_0 *is small.* The region of small γ_0 where σ is proportional to γ_0 is called the ***regime of linear viscoelasticity***, since σ in this regime is linear in γ_0. Note in Equation 2-66 that the linear viscoelastic shear stress has two terms.* In general,

$$\sigma = \gamma_0 \left[G'(\omega) \sin \omega t + G''(\omega) \cos \omega t\right] \tag{2-67}$$

where G′ and G″ (the prime here does *not* denote differentiation) are called the *storage modulus* and the *loss modulus*, respectively. These names derive from the fact that the term involving G′ is *in phase with the strain*; that involving G″ is *in phase with the strain rate*. For a perfectly elastic material that *stores* all mechanical energy imparted to it, $G'' = 0$. For a Newtonian liquid, which instantly *loses* all mechanical energy, $G' = 0$.

For the Rouse model, Equation 2-60, we find

$$G'(\omega) = \sum_i G_i \frac{\omega^2 \lambda_i^2}{\omega^2 \lambda_i^2 + 1} \tag{2-68}$$

$$G''(\omega) = \sum_i G_i \frac{\omega \lambda_i}{\omega^2 \lambda_i^2 + 1} \tag{2-69}$$

G′ and G″ are thus determined by the values of G_i and λ_i, that is by the relaxation spectrum.

For the Rouse model, G_i and λ_i are given by Equation 2-63. For materials that do not obey the Rouse model, Equations 2-68 and 2-69 can be *fit* to the measured $G'(\omega)$ and $G''(\omega)$ by appropriately choosing a set of relaxation times, $\{\lambda_i\}$, and moduli, $\{G_i\}$. These sets of relaxation times and moduli constitute a *discreet representation* of the relaxation spectrum. The measurement of G′ and G″ as functions of frequency is thus a means of determining the distribution of relaxation times. Note that the contribution of mode i to G″ is maximum when $\omega = 1/\lambda_i$. Thus G″ is sensitive to relaxation processes with relaxation times around $1/\omega$. By varying ω over a wide range of frequencies, the strengths, G_i, of relaxation mechanisms occurring over a correspondingly wide range of times are probed.

Figure 2-7 shows $G'(\omega)$ and $G''(\omega)$ for the Rouse model. The model predicts that at low frequencies,

$$G' \propto \omega^2 \ ; \qquad G'' \propto \omega \tag{2-70}$$

---

* Outside of the linear viscoelastic regime, higher powers of γ_0 appear, as well as ***higher harmonics***, that is higher powers of $\sin \omega t$ and $\cos \omega t$.

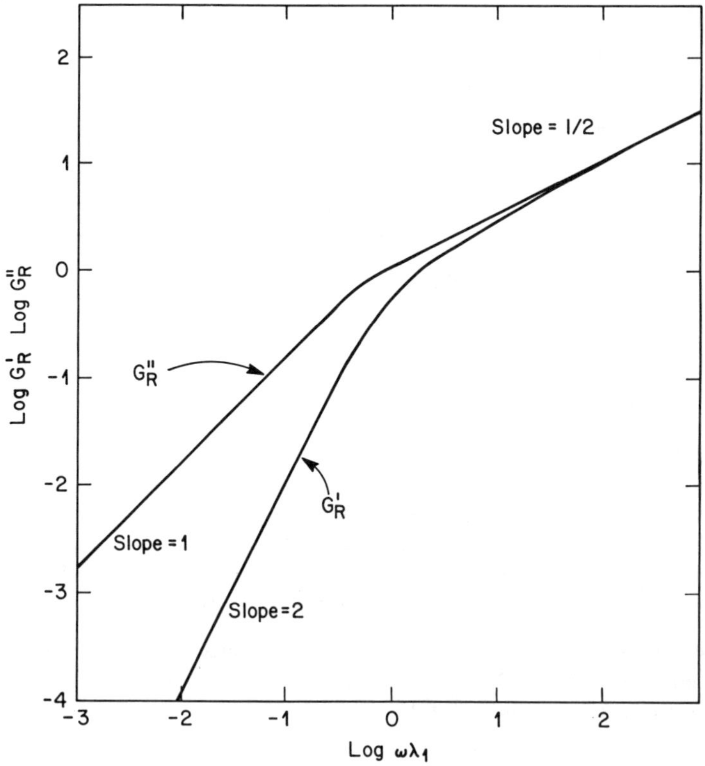

Fig. 2-7 The Rouse predictions for the dimensionless storage and loss moduli G'_R and as functions of dimensionless frequency, $\omega\lambda_1$, where λ_1 is the longest relaxation time. The moduli are made dimensionless by dividing G' and G'' by the modulus $G = \nu kT$ (from Ferry[18] reprinted from *Viscoelastic Properties of Polymers*, 3rd Ed., ©1980 by permission of John Wiley & Sons, Inc.).

This result is not unique to the Rouse model, but is true any time there is a longest relaxation time that dominates at small ω. The region of frequency in which this limiting behavior occurs is called the *terminal zone.*[18]

At high frequencies, the Rouse model gives the distinctive prediction

$$G' = G'' \propto \omega^{1/2} \tag{2-71}$$

This prediction is a consequence of the spacing of relaxation times in the Rouse model, namely, $\lambda_i \propto 1/i^2$; see Equation 2-63. In general, if $\lambda_i \propto 1/i^p$, and $G_i = G$ is independent of i, then

$$G' \propto G'' \propto \omega^{1/p} \tag{2-72}$$

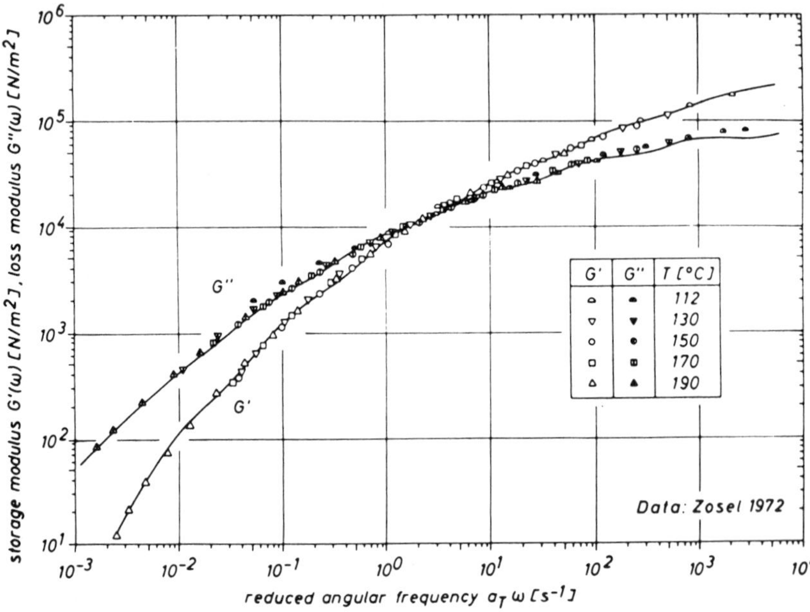

Fig. 2-8 Storage and loss moduli for Melt I obtained by time-temperature shifting to the reference temperature of 150 °C. The lines are fits to the data (from H.M. Laun,[20] reproduced with permission from Rheol. Acta)

2.8.1.2 The effects of polydispersity The Rouse theory presented in Section 2.7 applies to a *monodisperse* solution of polymer molecules, that is, a solution in which all the molecules have the same molecular weight. The molecular weight, M, of a polymer molecule is the degree of polymerization, P, times the monomer molecular weight M_0. In practice polymeric samples are always polydisperse; commercial materials often extremely so. Since in the Rouse model each molecule acts independently of the others, the effect of polydispersity can be accounted for simply by adding together contributions to the stress from each molecular-weight component, with weighting given to each component in proportion to its concentration in the mixture. According to Equation 2-64, each relaxation time in the Rouse model for a monodisperse solution is proportional to M^2. Thus the transition from terminal to high-frequency behavior, Figure 2-7, in the Rouse model occurs at a frequency, ω, that depends on molecular weight. In a polydisperse solution, the position of the transition is therefore different for each component, and the sum of all contributions thus ***broadens the transition*** over a range of frequency the breadth of which depends on the degree of polydispersity.

A measure of the degree of polydispersity is given by the ratio of weight-averaged to number-averaged molecular weights, M_w/M_n, where

$$M_n = \sum_{j=1}^{p} n_j M_j \; ; \qquad M_w = \sum_{j=1}^{p} c_j M_j \tag{2-73}$$

Here n_j is the number fraction, and c_j is the weight fraction of polymer molecules having molecular weight M_j. For a perfectly monodisperse solution, $M_w/M_n = 1$. The ratio increases with increasing polydispersity.

2.8.2 Time-Temperature Superposition

Note in Equation 2-64 that each λ_i and G_i depend on absolute temperature T. For λ_i, the most important temperature dependence is not the explicit 1/T dependence evident in Equation 2-63, but the indirect dependence through the dependence of the friction coefficient ζ_0 on T. Nevertheless, the dependence is *the same for each* λ_i! This means that on a log-log plot, a change in temperature shifts G′ and G″ equally along the frequency axis. The moduli are also shifted through the dependence of G_i on T and density ν (Equation 2-63). The shift along the modulus axis is often negligible, since increasing T decreases ν by volume expansion and these two effects at least partially cancel each other. Thus the curves $G'(\omega)$ and $G''(\omega)$ measured at different temperatures *can be superposed* by shifting along the frequency axis.

This procedure, called *time-temperature superposition*, is not restricted to fluids obeying the Rouse model, but holds for melts and concentrated solutions generally. Deviations are rarely found except near the glass transition temperature.[18] For dilute solutions, solvent quality — that is the strength of the solvent's interactions with the polymer — can be a function of temperature. Hence *time-temperature superposition* cannot always be carried out for dilute solutions. Fluids for which G′ and G″ at different temperatures do superpose are said to be *thermorheologically simple*.

Thermorheological simplicity can be put to use in the measurement of G′ and G″ as functions of ω. There are upper and lower limits on the values of ω obtainable in mechanical experiments. Frequencies higher than the maximum can be accessed, however, by lowering the temperature, thereby shifting the high-frequency response to a frequency low enough to be achieved in the laboratory. Likewise, the low-frequency response can be shifted into a measurable range by raising the temperature. Figure 2-8 shows $G'(\omega)$ and $G''(\omega)$ for Melt I at $T = 150\,^\circ C$, which Laun[20] obtained by shifting results from temperatures ranging from $112\,^\circ C$ to $190\,^\circ C$. These curves are called *master curves*. The abscissa, $a_T\omega$, is called the *reduced frequency*; it is the frequency shifted by an amount a_T. Since relaxation is a thermally activated process, a_T is roughly an exponential function of inverse temperature, except near the glass transition temperature, where it has an even stronger dependence on T. Note that seven decades of frequency are plotted; only about three or four of these are obtained at any one temperature.

From the master curves, the distribution of relaxation times is obtained by choosing a set of λ_i's and adjusting the corresponding G_i's until a fit to $G'(\omega)$ and $G''(\omega)$ is obtained. Figure 2-9, for example, shows how the contributions from each mode add together, according to Equation 2-68, to give $G'(\omega)$ for Melt I. The curve obtained in Figure 2-9 closely fits the G′ data; see Figure 2-8. The values of λ_i and G_i that give this fit are listed in Table 2-1[20]

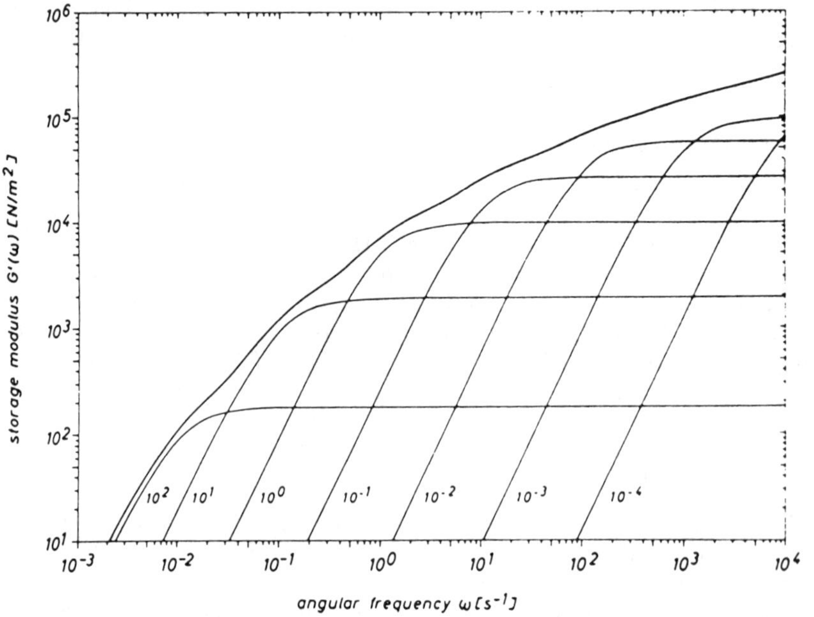

Fig. 2-9 Fitting of eight relaxation modes to the storage modulus of Melt I (from Laun,[20] reproduced with permission from Rheol. Acta)

Table 2-1 Relaxation Spectrum for Melt I

i	λ_i [sec.]	G_i [N/m^2]
1	10^3	1.00×10^0
2	10^2	1.80×10^2
3	10^1	1.89×10^3
4	10^0	9.80×10^3
5	10^{-1}	2.67×10^4
6	10^{-2}	5.86×10^4
7	10^{-3}	9.48×10^4
8	10^{-4}	1.29×10^5

Notice that the lines in Figures 2-8 and 2-9 show some slight oscillation. This oscillation is an artifact of the spacing chosen for the relaxation times. The oscillation is eliminated if relaxation times are spaced more closely than the factor-of-10 spacing chosen for this fit.

Why are melts and solutions generally, including those that don't follow the Rouse model, thermorheologically simple? In the Rouse model, rheological simplicity holds because all relaxation processes, no matter how slow, arise finally from the motion of simple beads. The slower processes simply require more ***cooperativity***, or coordination of the movement of many beads. So the rates of the slower

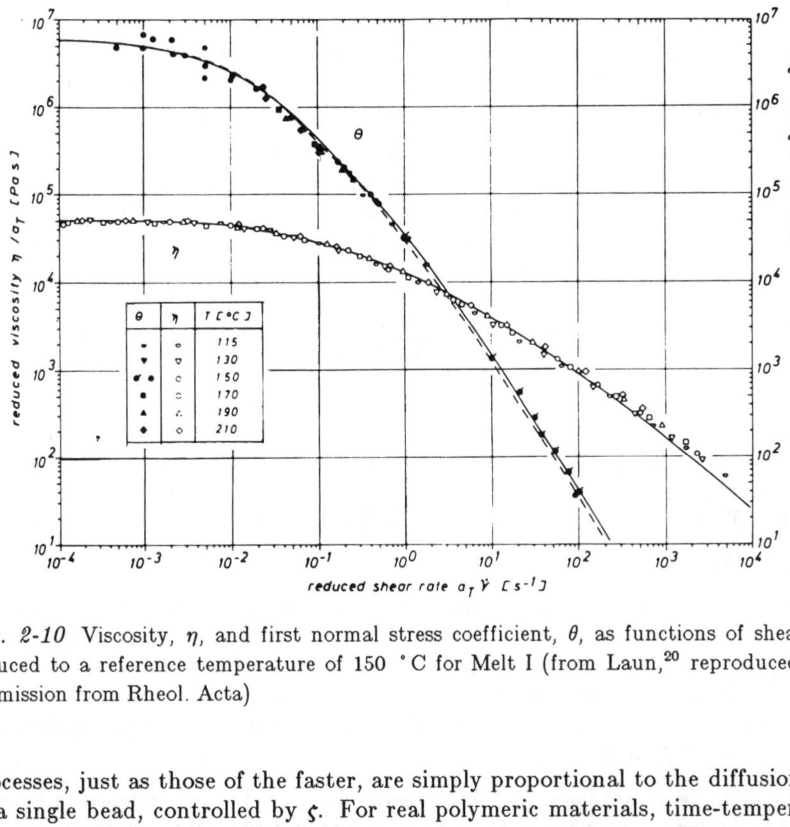

Fig. 2-10 Viscosity, η, and first normal stress coefficient, θ, as functions of shear rate reduced to a reference temperature of 150 °C for Melt I (from Laun,[20] reproduced with permission from Rheol. Acta)

processes, just as those of the faster, are simply proportional to the diffusion rate of a single bead, controlled by ς. For real polymeric materials, time-temperature shifting works for modes from the slowest to the fastest glassy modes, encompassing as many as 15 decades or more of time.[18] This is explained by noting that the slow processes are merely cooperative combinations of the faster glassy processes. Thus no matter how cooperative they may be, and how much slower than the glassy processes they are, their rates are proportional to the rates of the fast glassy modes and change with temperature to the same extent as do the rates of the glassy modes.

2.8.3 Nonlinear Superposition

Not only do the linear viscoelastic properties G′ and G″ show time-temperature superposition, but so do nonlinear rheological functions. All the data shown in Chapter 1 for shear viscosity, primary normal stress differences, stress growth and relaxation, and elastic recoil have been measured at other temperatures and been have found to superpose onto the data shown when the shift factors of linear viscoelasticity are applied to them.[20] Shear rates and extension rates shift in the same direction, and by the same magnitude, as frequency; time shifts in the opposite direction. Since stress has its origin in entropy, it is expected to shift in proportion to T, but this shifting is small compared to the exponential dependence of a_T on temperature and is usually neglected. Thus viscosity, which is stress over strain rate, must be shifted in proportion to $1/a_T$. Figure 2-10 shows the curves for reduced shear viscosity and first normal stress coefficient as functions of reduced shear rate for Melt I at 150 °C, that Laun[20] obtained by shifting data obtained at temperatures of 115 °C to 210 °C. It illustrates that time-temperature superposition works for nonlinear as well as linear viscoelastic properties.

Thermorheological simplicity obviously reduces enormously the task of characterizing polymeric liquids. Apart from determining the shift factors, rheological studies on materials that are thermorheologically simple need only be carried out at a single temperature. Once a constitutive equation is found that fits the material at the experimental temperature, it can be used to describe the material at any temperature by making each relaxation time proportional to the shift factor. Nonisothermal polymer processing problems can then be tackled by solving the constitutive equation together with the usual mass and momentum balance equations, and the energy balance equation.[21]

The relaxation spectrum can also be affected by changes in pressure, p, if these pressure changes are very large — hundreds of atmospheres.[18,21] As with temperature, a simple shifting procedure can be used to account for pressure effects.[18] Thus in the most general case, each relaxation time is shifted by $a_{T,p}$.

2.9 SUMMARY

The earliest and most elementary constitutive equations have been presented, namely, the temporary network model of Green and Tobolsky for melts and the bead-spring model of Rouse for dilute solutions. In both models, the stress tensor is proportional to the second moment of the configuration distribution function. Both models assume an affine motion of strands or portions of macromolecules in response to a deformation.

The temporary network model for melts accounts for entanglements between macromolecules in an ad hoc fashion by supposing that entanglements act like temporary cross-links. The entanglements continually break and re-form so that the structure, though rubbery because of the cross-links, is fluid because any given cross-link is transient.

The Rouse bead-spring model accounts for the elasticity or resistance to stretching of an isolated coil, its Brownian motion, and the viscous drag exerted on the coil by solvent. The polymer molecule is represented as a chain of beads and springs; the springs embody the elastic forces, and the beads are subject to Brownian forces and viscous drag exerted by the solvent.

Both the temporary network model and the Rouse model yield the constitutive equation of Lodge or, equivalently, the upper-convected Maxwell equation.

The Rouse model, however, predicts a specific distribution of relaxation times that determines the viscoelastic response in small-amplitude oscillatory deformations. It also predicts that when the temperature is changed, the entire distribution of relaxation times shifts uniformly without changing the relative magnitude of any one relaxation with respect to another. This result, which allows time-temperature shifting, is usually valid for polymeric fluids.

References

1. L.R.G. Treloar, *The Physics of Rubber Elasticity*, 3rd Ed., Clarendon Press, Oxford (1975).
2. W. Kuhn, Kolloid Z., *76*, 258 (1936); *87*, 3 (1939).
3. P.J. Flory, *Principles of Polymer Chemistry*, Cornell University Press, Ithaca, N.Y. (1953).
4. R.B. Bird, O. Hassager, R.C. Armstrong, and C.F. Curtiss, *Dynamics of Polymeric Liquids*, Vol. 2, John Wiley & Sons, New York (1977).
5. R.G. Larson, J. Non-Newt. Fluid Mech., *13*, 279 (1983).
6. F.T. Wall, J. Chem. Phys., *10*, 485 (1942).
7. P.F. Flory and J. Rehner, J. Chem. Phys., *11*, 512 (1943).
8. H.M. James and E. Guth, J. Chem. Phys., *11*, 455 (1943).
9. L.R.G. Treloar, Trans. Faraday Soc., *37*, 36; 241 (1943).
10. M.S. Green and A.V. Tobolsky, J. Chem. Phys., *14*, 80 (1946).
11. A.S. Lodge, Trans. Faraday Soc., *52*, 120 (1956); Rheol. Acta, *7*, 379 (1968).
12. M. Yamamoto, J. Phys. Soc. Japan, *11*, 413 (1956); *12*, 1148 (1957); *13*, 1200 (1958).
13. Data from J. Meissner, Rheol. Acta, *10*, 230 (1971); theory from H. Chang and A.S. Lodge, Rheol. Acta, *11*, 127 (1972).
14. P.E. Rouse Jr., J. Chem. Phys., *21*, 1272 (1953).
15. W. Kuhn, Kolloid-Z., *68*, 2 (1934).
16. S. Chandrasekhar, Rev. Mod. Phys., *15*, 1 (1943).
17. P.G. de Gennes and F. Brochard, Macromolecules, *10*, 1157 (1977).
18. J.D. Ferry, *Viscoelastic Properties of Polymers*, 3rd ed., John Wiley & Sons, New York (1980).
19. M.H. Wagner and H.M. Laun, Rheol Acta, *17*, 138 (1978).
20. H.M. Laun, Rheol. Acta, *17*, 1 (1978).
21. R.I. Tanner, *Engineering Rheology*, Oxford, New York (1985).

CHAPTER 3

Continuum Theories

It is the office of theoretical investigation to give the form in which the results of experiment may be expressed.

J.W. Gibbs, 1881*

3.1 INTRODUCTION

In the preceding chapter, the simplest molecular models for viscoelastic fluids were presented, namely Green and Tobolsky's temporary network model for melts and Rouse's beads and springs model for dilute solutions. Both of these models lead to the upper-convected Maxwell equation and all the limitations inherent to this equation, such as its inability to predict shear thinning.

To improve these *molecular* models, one must make more precise assumptions about entanglement couplings for melts or of solvent interactions for dilute solutions. Thus one must *define* the material more precisely. In so doing, however, the generality of the constitutive equation is decreased. Improvements can be made without further material definition if one is willing to allow parameters in the constitutive equation that are not immediately known from molecular properties but are determined by fits to rheological data. Thus, for example, the Rouse spectrum of relaxation times, which is accurate for only a few materials, as discussed in the next chapter, can be replaced by a distribution that is measured in small-amplitude oscillatory shear. With the measured spectrum, one can predict the stresses in certain other flows, namely flows for which the *strain or the strain rate* is small. This is the regime of *linear viscoelasticity*.

For large strains, or large strain rates, linear viscoelasticity fails, and rheological behavior is not completely predictable from the relaxation spectrum. There are *principles of continuum mechanics*, however, that limit the possible kinds of nonlinear behavior. Chief among these is the principle of *frame invariance*, which excludes constitutive equations that don't obey certain tensorial transformation

---

* Quoted by M. Rukeyser in "Willard Gibbs," p. 267, Doubleday & Co., Inc., Garden City, N.Y., 1942.

rules in rotations of frame. With this principle, and some simplifying mathematical approximations and assumptions, one can find useful continuum mechanical constitutive equations without making any molecular assumptions. Before moving into the more difficult nonlinear regime, however, we first present some of the important fruits of simple linear viscoelasticity.

3.2 THE CONSTITUTIVE EQUATION OF LINEAR VISCOELASTICITY

A linear viscoelastic response is seen not only in oscillatory shear, but in *any deformation history when the total strain is small.* Consider, for example, the Lodge equation, Equation 1-33. If the total strain is small, the Finger tensor is

$$\underset{\approx}{C}^{-1}(t',t) \doteq \underset{\approx}{\delta} + 2\underset{\approx}{\gamma}(t,t') \tag{3-1}$$

where the *linear strain tensor,* $\underset{\approx}{\gamma}$, is defined as the strain rate tensor integrated over an interval of time:

$$\underset{\approx}{\gamma}(t,t') \equiv \int_{t'}^{t} \underset{\approx}{D}(t'')\, dt'' \tag{3-2}$$

Equation 3-1 is valid as long as $\underset{\approx}{\gamma}$ is small. Therefore to within an isotropic constant the Lodge equation in this limit is

$$\underset{\approx}{\sigma} = 2\int_{-\infty}^{t} m(t-t')\underset{\approx}{\gamma}(t,t')\, dt' \tag{3-3}$$

Differentiating by parts gives

$$\underset{\approx}{\sigma} = 2\int_{-\infty}^{t} G(t-t')\underset{\approx}{D}(t')\, dt' \tag{3-4}$$

where

$$G(s) = \int_{s}^{\infty} m(s')\, ds' \tag{3-5}$$

is called the *relaxation modulus.* From it, G' and G'' can easily be obtained:

$$G' = \omega \int_{0}^{\infty} G(t) \sin \omega t\, dt \tag{3-6}$$

$$G'' = \omega \int_{0}^{\infty} G(t) \cos \omega t\, dt \tag{3-7}$$

G' and G'' are respectively the storage and loss moduli in small amplitude oscillatory deformations; see Section 2.8.1.1.

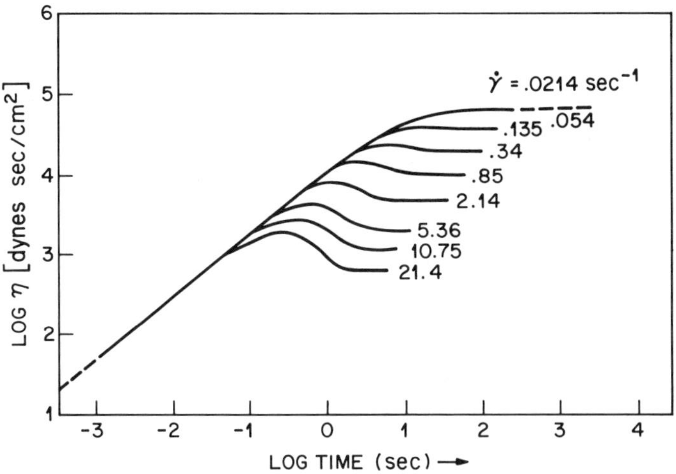

Fig. 3-1a Growth of shear viscosity as a function of time after start-up of steady shearing, at various shear rates $\dot{\gamma}$. The dashed line is the prediction of linear viscoelasticity (from Menezes and Graessley[1], reproduced with permission from Rheologica Acta).

The linear viscoelastic constitutive equation, Equation 3-4, has been found valid for polymer melts and solutions in general for *any strain rate*, as long as the total strain remains small. Although we have derived it from the Lodge equation, Equation 3-4 holds for most constitutive equations and indeed, for all equations that satisfy the postulates of simple fluid theory. A constitutive equation is considered suspect if it does not satisfy the linear viscoelastic constitutive equation at all strain rates when the strain is small.

3.2.1 Shear

3.2.1.1 Deviations from Linear Viscoelasticity If a sample of polymeric liquid at equilibrium is subjected to steady straining, deviations from linear viscoelasticity can occur when the strain exceeds a limiting value. The higher the strain rate, the smaller the time, t, at which this critical strain is exceeded. Figure 3-1a illustrates this phenomenon in shear. Note that at small t, the shear viscosity, η, is the same for all shear rates, $\dot{\gamma}$, and equals the linear viscoelastic prediction. As $\dot{\gamma}$ increases, the deviations from linear viscoelasticity occur at progressively smaller t, but the strain, γ_d, at which the deviations occur remains the same, $\gamma_d \sim 1$. The first normal stress coefficient follows the same pattern, Figure 3-1b. At small enough shear rates, linear viscoelasticity holds through the time required to attain a steady-state viscosity. This linear viscoelastic steady-state viscosity is η_0, the low-shear-rate plateau observed in Figure 2-10. Its value can therefore be predicted from linear viscoelasticity. For start-up of steady shear, $\dot{\gamma}(t,t')$ is zero before $t' = 0$ and constant thereafter. Equation 3-4 then gives

$$\eta = \frac{\sigma}{\dot{\gamma}} = \int_0^t G(s)\, ds \tag{3-8}$$

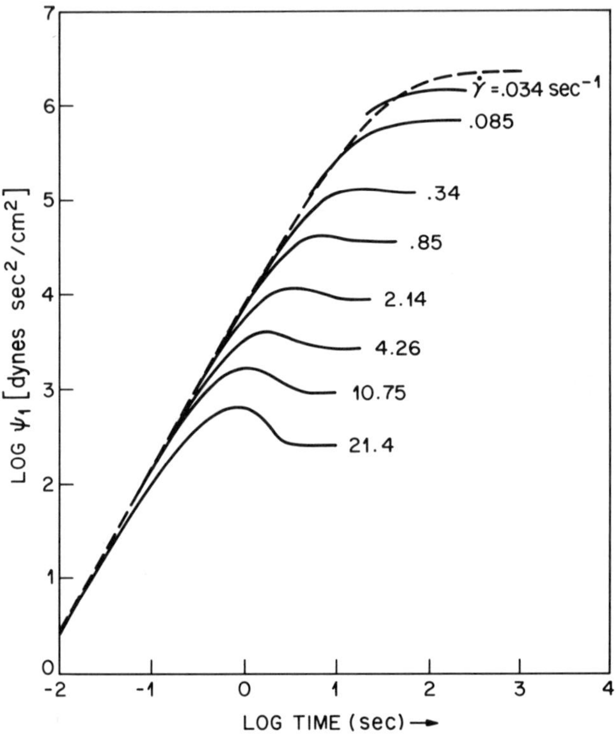

Fig. 3-1b Growth of first normal stress coefficient as a function of time after start-up of steady shearing at various shear rates $\dot{\gamma}$ for a nearly monodisperse polystyrene solution. The dashed line is the prediction of linear viscoelasticity. (from Menezes and Graessley[1], reproduced with permission from Rheologica Acta)

which is the linear viscoelastic response shown in Figure 3-1a. The zero-shear viscosity is the asymptote of this response as $t \rightarrow \infty$:

$$\eta_0 = \int_0^\infty G(s)\, ds \tag{3-9}$$

Thus *the viscosity in the Newtonian limit is the integral of the linear relaxation modulus.*

Maxwell[2] in 1867 appears to have been the first to interpret the viscosity of fluids in terms of relaxation processes that are rapid compared to the deformation rate. He proposed the linear equation

$$\sigma + \lambda \frac{\partial \sigma}{\partial t} = G\lambda\dot{\gamma} \tag{3-10}$$

which has an elastic limit $\sigma = G\gamma$ for rapid deformations and a viscous limit $\sigma = G\lambda\dot{\gamma}$ for slow ones.

Boltzmann[3] in 1876 proposed the more general equation of linear viscoelasticity, Equation 3-4, in which the influence of strains occurring at difference times in the past are added linearly to give the stress at the present time. This idea is now called the *Boltzmann superposition principle.* If $G(t-t')$ is a single exponential, then Equation 3-4 is equivalent to Maxwell's differential equation.

3.2.1.2 Normal Stresses Figure 3-1b shows that the time-dependent primary normal stress coefficient, Ψ_1, can also be predicted from the linear viscoelastic shear modulus, if the strain is small. For then,

$$N_1 = \Psi_{1,0}(t)\,\dot{\gamma}^2 = 2\dot{\gamma}^2 \int_0^t sG(s)\,ds \tag{3-11}$$

The subscript, 0, denotes the low-shear-rate limit. Equation 3-11 is not strictly a consequence of the theory of linear viscoelasticity, since N_1 is quadratic in $\dot{\gamma}$. It appears instead to be a result of *frame invariance*[4] that the lowest-order term in the expansion for N_1 (i.e., the quadratic term) is related to the shear modulus, $G(s)$, by Equation 3-11.

The second normal stress coefficient in the limit of low shear rates, $\Psi_2(t)$, can*not* be predicted from $G(s)$ alone, but is sensitive to the choice of constitutive assumptions. $\Psi_{2,0}(t)$ is not necessarily even related to $G(s)$ by an integral over $sG(s)$, as $\Psi_{1,0}$ is.

3.2.2 Other Deformations

Within the strictly linear regime, the Boltzmann superposition principle, embodied in Equation 3-4, can be used to predict stresses in deformations other than shear. Consider, for example, steady elongation, for which

$$\underset{\approx}{D} = \begin{pmatrix} \dot{\epsilon} & 0 & 0 \\ 0 & -\tfrac{1}{2}\dot{\epsilon} & 0 \\ 0 & 0 & -\tfrac{1}{2}\dot{\epsilon} \end{pmatrix} \tag{3-12}$$

and $\dot{\epsilon} = \partial v_1/\partial x_1$ is the velocity gradient in the axial direction. The measurable quantity is $\Delta\sigma = \sigma_{11} - \sigma_{22}$, the normal stress difference between the axial direction and the radial direction. In the linear limit $\Delta\sigma$ is given by

$$\overline{\eta}\dot{\epsilon} \equiv \Delta\sigma = 3\dot{\epsilon}\int_0^t G(t-t')\,dt' = 3\eta\dot{\epsilon} \tag{3-13}$$

Thus, in the linear limit, $\overline{\eta}(t) = 3\eta(t)$; the linear elongational viscosity is three times the linear shear viscosity. This result, which holds at any strain rate for Newtonian fluids, is known as *Trouton's Rule.*[5] Note in Figure 1-13 that the experimental $\overline{\eta}(t)$ and $\eta(t)$ *are* parallel to each other in the limit of small strain rates or small strains; they are spaced apart by a factor of 3, the Trouton ratio.

3.3 FRAME INVARIANCE

The constitutive equation of linear viscoelasticity cannot hold for large strains, even in principle. This is because it violates the principal of *frame invariance.* As discussed in Chapter 1, for elastic solids frame invariance, also called

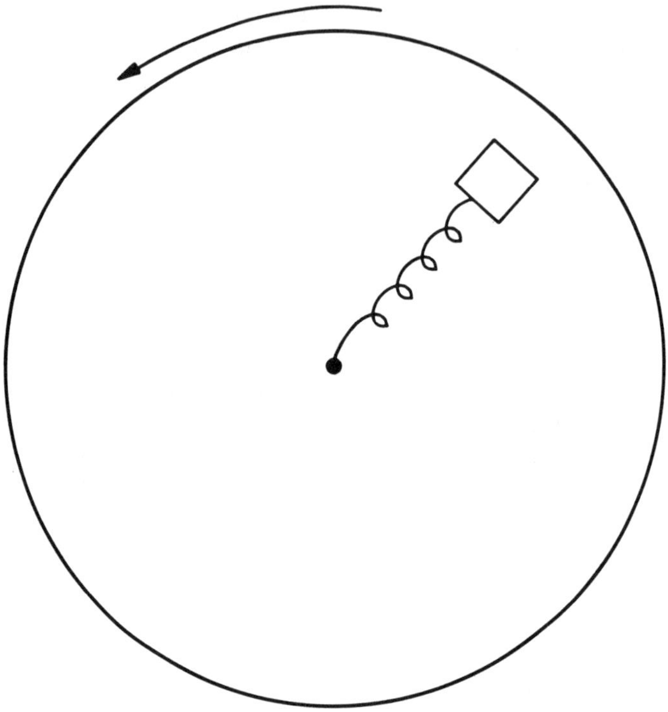

Fig. 3-2 A weight attached by a spring to the axis of a rotating turntable. The surface of the turntable is frictionless.

material objectivity, requires that the *relationship* between the stress tensor and the deformation gradient tensor be independent of any rotation, $\underset{\approx}{Q}$, of the body or of the observer. For viscoelastic liquids, *the constitutive equation must be independent of any* **time-dependent** *rotation* $\underset{\approx}{Q}(t)$. This is a much more powerful statement, since seemingly important characteristics of a flow can be changed by superimposing a time-dependent rotation on the flow. For example, vorticity can be added to a vorticity-free extensional flow by superimposing a time-dependent rotation. Extensional components appear in otherwise purely shearing flows if the frame is made to rotate.

Frame invariance does *not* imply that the stress tensor is independent of rotations of frame; it implies only that the *relationship* between the stress tensor and the deformation history is independent. Truesdell and Noll[6] illustrate the principle by considering a spring lying on a rotating turntable with one end attached to the turntable axis and the other to a weight, Figure 3-2. Both the extension of the spring and the force that the weight exerts depend on the rate at which the turntable rotates, but *the relationship between the extension and the force* is a property only of the spring and *is independent of the rotation rate.*

The principle of frame invariance is not a fundamental law as is the principle of momentum balance. It holds for fluids in which the stress at the level of the microstructure is determined by simple viscous and elastic springlike forces. It doesn't hold if electromagnetic or inertial forces are significant for the individual elements of the microstructure. In the turntable example, if the rotation rate is suddenly increased the inertia of the weight might cause a delay before the spring extension reaches a steady state. This would be a violation of frame invariance. In a suspension of spheres if inertia in the flow becomes so large that it influences the movement of individual spheres, then frame invariance doesn't hold.[7,8] The frame invariance of the constitutive equation is *not* violated when inertia or other forces are significant only at the continuum level. When, for example, inertia is significant at the continuum but not the microstructural level, it must be introduced into the continuum momentum balance, but does not affect the frame invariance of the constitutive equation. For macromolecular fluids, no violation of frame invariance has ever been claimed.

We can define a rotation of frame by its effect on an arbitrary vector, $\underset{\sim}{\mathrm{x}}$. Letting $\hat{\underset{\sim}{\mathrm{x}}}$ be the vector in the rotated frame, we have

$$\hat{\underset{\sim}{\mathrm{x}}} = \underset{\sim}{\mathrm{x}} \cdot \underset{\approx}{\mathrm{Q}} \tag{3-14}$$

where $\underset{\approx}{\mathrm{Q}}$ is a unitary matrix, i.e.,

$$\underset{\approx}{\mathrm{Q}}^{\mathrm{T}} = \underset{\approx}{\mathrm{Q}}^{-1} \tag{3-15}$$

$\underset{\approx}{\mathrm{Q}}$ contains the angles through which the axes of the frame are rotated. See Figure 1-19 for example. ($\underset{\approx}{\mathrm{Q}}$ can also be an *improper rotation*, describing a change in the direction of one or more of the axes.)

Quantities in the rotated frame, denoted by "^," are related to those in the unrotated frame by

$$\hat{\underset{\approx}{\mathrm{F}}}^{-1}(\mathrm{t},\mathrm{t}') = \underset{\approx}{\mathrm{Q}}^{\mathrm{T}}(\mathrm{t}') \cdot \underset{\approx}{\mathrm{F}}^{-1}(\mathrm{t},\mathrm{t}') \cdot \underset{\approx}{\mathrm{Q}}(\mathrm{t})$$

$$\hat{\underset{\approx}{\mathrm{C}}}^{-1}(\mathrm{t},\mathrm{t}') = \underset{\approx}{\mathrm{Q}}^{\mathrm{T}}(\mathrm{t}') \cdot \underset{\approx}{\mathrm{C}}^{-1}(\mathrm{t},\mathrm{t}') \cdot \underset{\approx}{\mathrm{Q}}(\mathrm{t})$$

$$\hat{\underset{\sim}{\nabla} \underset{\sim}{\mathrm{v}}} = \underset{\approx}{\mathrm{Q}}^{\mathrm{T}} \cdot \underset{\sim}{\nabla} \underset{\sim}{\mathrm{v}} \cdot \underset{\approx}{\mathrm{Q}} + \underset{\approx}{\mathrm{Q}}^{\mathrm{T}} \cdot \dot{\underset{\approx}{\mathrm{Q}}}$$

$$\hat{\underset{\approx}{\mathrm{D}}} = \underset{\approx}{\mathrm{Q}}^{\mathrm{T}} \cdot \underset{\approx}{\mathrm{D}} \cdot \underset{\approx}{\mathrm{Q}}$$

$$\hat{\underset{\approx}{\sigma}} = \underset{\approx}{\mathrm{Q}}^{\mathrm{T}} \cdot \underset{\approx}{\sigma} \cdot \underset{\approx}{\mathrm{Q}} \tag{3-16}$$

According to the principle of frame invariance, any constitutive equation that relates stress to the history of the deformation gradient or the velocity gradient, say

$$\underset{\approx}{\sigma} = \underset{\approx}{\Sigma}\{\underset{\sim\sim}{\nabla \mathrm{v}}, \underset{\approx}{\mathrm{F}}^{-1}\} \tag{3-17a}$$

must give the same relationship between the rotated quantities,

$$\underset{\approx}{\hat{\sigma}} = \underset{\approx}{\Sigma}\{\underset{\sim\sim}{\hat{\nabla} \mathrm{v}}, \underset{\approx}{\hat{\mathrm{F}}}^{-1}\} \tag{3-17b}$$

3.4 OLDROYD'S CONSTITUTIVE EQUATIONS

3.4.1 Convected Time Derivatives

As mentioned earlier, the linear viscoelastic constitutive equation, Equation 3-3 or 3-4, is *not* frame invariant. Neither is the simple linear Maxwell equation, Equation 3-10. Oldroyd[9] proposed that frame invariance could be recovered by writing the constitutive equation in a frame of reference that is *convected* or *deformed* with the material elements. To do so, one must reinterpret the time derivative in the Maxwell equation as a time derivative, $\delta/\delta \mathrm{t}$, in a convected coordinate system:

$$\lambda \frac{\delta}{\delta \mathrm{t}} \underset{\approx}{\sigma} + \underset{\approx}{\sigma} = 2\eta \underset{\approx}{\mathrm{D}} \tag{3-18}$$

There are two simple convected coordinate systems. In one, the base coordinate vectors are parallel to material *lines*; they are stretched and rotated with the material lines. With these, the "contravariant" base vectors, we have

$$\frac{\delta}{\delta \mathrm{t}} \underset{\approx}{\sigma} = \underset{\approx}{\overset{\nabla}{\sigma}} \equiv \underset{\approx}{\dot{\sigma}} - \underset{\sim\sim}{\nabla \mathrm{v}}^{\mathrm{T}} \cdot \underset{\approx}{\sigma} - \underset{\approx}{\sigma} \cdot \underset{\sim\sim}{\nabla \mathrm{v}} \tag{3-19}$$

This is the *upper-convected derivative* discussed at the end of Chapter 1.

The second simple choice is a set of base vectors that are normal to material *planes*; in a deformation each base vector rotates to remain normal and stretches so that its length remains proportional to the area of the material plane to which it is normal.[10] With these "covariant" base vectors,

$$\frac{\delta}{\delta \mathrm{t}} \underset{\approx}{\sigma} = \underset{\approx}{\overset{\Delta}{\sigma}} \equiv \underset{\approx}{\dot{\sigma}} - \underset{\sim\sim}{\nabla \mathrm{v}} \cdot \underset{\approx}{\sigma} - \underset{\approx}{\sigma} \cdot \underset{\sim\sim}{\nabla \mathrm{v}}^{\mathrm{T}} \tag{3-20}$$

This is the *lower-convected time derivative*. The terms, "upper-convected" and "lower-convected," were originally given because superscripts were used for contravariant base vectors, subscripts for covariant base vectors.

In Equations 3-19 and 3-20 $\underset{\approx}{\dot{\sigma}}$ is the *substantial time derivative*, $\underset{\approx}{\dot{\sigma}} \equiv \mathrm{D}\underset{\approx}{\sigma}/\mathrm{Dt}$, that is the time derivative in a frame that translates with the material particle. $\underset{\approx}{\dot{\sigma}}$ is related to the time derivative, $\partial \underset{\approx}{\sigma}/\partial \mathrm{t}$, in a fixed frame by

$$\frac{D}{Dt}\underset{\approx}{\sigma} = \frac{\partial}{\partial t}\underset{\approx}{\sigma} + \underset{\sim}{v}\cdot\underset{\sim}{\nabla}\,\underset{\approx}{\sigma} \tag{3-21}$$

If the stress $\underset{\approx}{\sigma}$ is homogeneous, that is, spatially uniform, then the convective transport term, $\underset{\sim}{v}\cdot\underset{\sim}{\nabla}\,\underset{\approx}{\sigma}$, is zero.

3.4.2 Upper- and Lower-Convected Maxwell Equations

These two choices for convected time derivatives give two frame-invariant Maxwell equations:

$$\lambda\overset{\nabla}{\underset{\approx}{\sigma}} + \underset{\approx}{\sigma} = 2\eta\underset{\approx}{D} \qquad \text{(UCM)} \tag{3-22}$$

$$\lambda\overset{\Delta}{\underset{\approx}{\sigma}} + \underset{\approx}{\sigma} = 2\eta\underset{\approx}{D} \qquad \text{(LCM)} \tag{3-23}$$

The first of these is our now-familiar upper-convected Maxwell equation (UCM). The second is the lower-convected Maxwell equation (LCM).

As discussed in Section 1.5.6, the upper-convected derivative is related to the Finger tensor by the relation

$$\frac{D}{Dt}\underset{\approx}{C}^{-1} - \underset{\sim}{\nabla}\underset{\sim}{v}^{T}\cdot\underset{\approx}{C}^{-1} - \underset{\approx}{C}^{-1}\cdot\underset{\sim}{\nabla}\underset{\sim}{v} \equiv \overset{\nabla}{\underset{\approx}{C}^{-1}} = \underset{\approx}{0} \tag{3-24}$$

Similarly, the lower-convected derivative is related to a strain tensor

$$\overset{\Delta}{\underset{\approx}{C}} = \underset{\approx}{0} \tag{3-25}$$

where $\underset{\approx}{C}$ is the *Cauchy* tensor; it is the matrix inverse of $\underset{\approx}{C}^{-1}$. Because of Equation 3-24, there is an integral equivalent to the UCM equation (see Section 1.6.2):

$$\underset{\approx}{\sigma} = \int_{-\infty}^{t} \frac{\eta}{\lambda^2}\, e^{(t'-t)}\, \underset{\approx}{C}^{-1}(t,t')\, dt' \tag{3-26}$$

There is likewise an integral equivalent to the LCM equation that arises from Equation 3-25:

$$\underset{\approx}{\sigma} = -\int_{-\infty}^{t} \frac{\eta}{\lambda^2}\, e^{(t'-t)}\, \underset{\approx}{C}(t,t')\, dt' \tag{3-27}$$

On purely continuum-mechanical grounds we have no reason to prefer one of these Maxwell equations to the other; both satisfy frame invariance. In shearing, both show no shear thinning and quadratic primary normal stress differences. The secondary normal stress difference, $\sigma_{22} - \sigma_{33}$, of the LCM equation, however, is equal in magnitude to the primary normal stress difference, $\sigma_{11} - \sigma_{22}$, and of opposite sign. This can be seen from the Cauchy tensor in shear:

$$\underset{\approx}{C} = \begin{pmatrix} 1 & -\gamma & 0 \\ -\gamma & 1+\gamma^2 & 0 \\ 0 & 0 & 1 \end{pmatrix} \quad \text{(shear)} \tag{3-28}$$

When inserted into Equation 3-27, this gives a negative sign for $N_2 \equiv \sigma_{22} - \sigma_{33}$, which is consistent with all experimental data. The magnitude, $N_2/N_1 = -1$, is much too large, however, since experiments usually show $N_2/N_1 \approx -1/10$. Because of this, and because the LCM equation has no molecular basis, while the UCM equation does (see Chapter 2), the LCM equation is regarded as no more than a curiosity.

3.4.3 Oldroyd's Simple Equations

Oldroyd's original proposal was to render frame invariant a linear equation more general than Maxwell's, namely

$$\lambda \frac{\delta}{\delta t} \underset{\approx}{\sigma} + \underset{\approx}{\sigma} = 2\eta \left(\underset{\approx}{D} + \lambda_r \frac{\delta}{\delta t} \underset{\approx}{D}\right) \tag{3-29}$$

This equation, called the *Jeffreys equation,*[11] appends a term, $2\eta\lambda_r(\delta/\delta t)\underset{\approx}{D}$, to Maxwell's equation. This term can be regarded as arising from stresses in the solvent. When made frame invariant in the two ways proposed by Oldroyd, the results are

$$\lambda \overset{\nabla}{\underset{\approx}{\sigma}} + \underset{\approx}{\sigma} = 2\eta(\underset{\approx}{D} + \lambda_r \overset{\nabla}{\underset{\approx}{D}}) \quad \text{(Oldroyd-B equation)} \tag{3-30}$$

and

$$\lambda \overset{\Delta}{\underset{\approx}{\sigma}} + \underset{\approx}{\sigma} = 2\eta(\underset{\approx}{D} + \lambda_r \overset{\Delta}{\underset{\approx}{D}}) \quad \text{(Oldroyd-A equation)} \tag{3-31}$$

The Oldroyd-A equation is not used because it inherits the problems of the LCM equation. The Oldroyd-B equation, however, is equivalent to the equation

$$\underset{\approx}{\sigma} = \underset{\approx}{\sigma}_p + \underset{\approx}{\sigma}_s \tag{3-32}$$

where $\underset{\approx}{\sigma}_p$, the "polymer stress," satisfies the UCM equation

$$\lambda \overset{\nabla}{\underset{\approx}{\sigma}}_p + \underset{\approx}{\sigma}_p = 2\eta_p \underset{\approx}{D} \tag{3-33}$$

and $\underset{\approx}{\sigma}_s$, the solvent stress, satisfies

$$\underset{\approx}{\sigma}_s = 2\eta_s \underset{\approx}{D} \tag{3-34}$$

λ_r and η in Equation 3-30 are then related to η_p, η_s, and λ in Equations 3-33 and 3-34 by

$$\lambda_r = \frac{\eta_s \lambda}{\eta_s + \eta_p}, \qquad \eta = \eta_p + \eta_s \tag{3-35}$$

This equation is exactly the result obtained from the elastic dumbbell in a solvent; see Equations 2-47 to 2-49. λ_r is called the "retardation time."

3.4.4 Corotational Maxwell Equation

Equations 3-19 and 3-20 are not the only frame-invariant time derivatives. Oldroyd recognized that infinitely many could be constructed. Frame invariance requires that

$$\underset{\approx}{0} = \dot{\underset{\approx}{\sigma}} - \underset{\approx}{\omega}^{T} \cdot \underset{\approx}{\sigma} - \underset{\approx}{\sigma} \cdot \underset{\approx}{\omega} - \underset{\approx}{f}(\underset{\approx}{\sigma}, \underset{\approx}{D}) \tag{3-36}$$

where $\underset{\approx}{f}$ is a frame-invariant tensor function of $\underset{\approx}{\sigma}$ and $\underset{\approx}{D}$. Thus the vorticity dependence is fixed by frame invariance. In the simplest case, $\underset{\approx}{f} = \lambda^{-1}(\underset{\approx}{\sigma} - 2\eta_0 \underset{\approx}{D})$. Then

$$\lambda \overset{0}{\underset{\approx}{\sigma}} + \underset{\approx}{\sigma} = 2\eta_0 \underset{\approx}{D} \tag{3-37}$$

where we have defined

$$\overset{0}{\underset{\approx}{\sigma}} \equiv \dot{\underset{\approx}{\sigma}} - \underset{\approx}{\omega}^{T} \cdot \underset{\approx}{\sigma} - \underset{\approx}{\sigma} \cdot \underset{\approx}{\omega} \tag{3-38}$$

$\overset{0}{\underset{\approx}{\sigma}}$ is called the *corotational* or *Jaumann* time derivative of $\underset{\approx}{\sigma}$. It is the time derivative with respect to a frame that *rotates with the angular velocity of the fluid element.*[10] The angular velocity, which is defined as the curl of the velocity $\underset{\sim}{v}$, is the rate of rotation that a small inertialess sphere would experience if placed in the flow. The corresponding Maxwell equation, Equation 3-37, is called the "corotational Maxwell equation." Like the upper- and lower- convected Maxwell equations, the corotational Maxwell equation can be converted to integral form. The integral form, with an arbitrary spectrum of relaxation times, is the "Goddard-Miller" equation.[12]

3.4.4.1 Shear Thinning The corotational Maxwell equation (CRM) finally breaks the monotony of the constitutive equations we have considered thus far; it shows shear thinning and a nonconstant primary normal stress coefficient:

$$\eta(\dot{\gamma}) = \frac{\eta_0}{1 + \lambda^2 \dot{\gamma}^2} \tag{3-39}$$

$$\Psi_1(\dot{\gamma}) = \frac{2\lambda\eta_0}{1 + \lambda^2 \dot{\gamma}^2} \tag{3-40}$$

According to the CRM equation, the shear stress, $\eta\dot{\gamma}$, goes through a maximum at

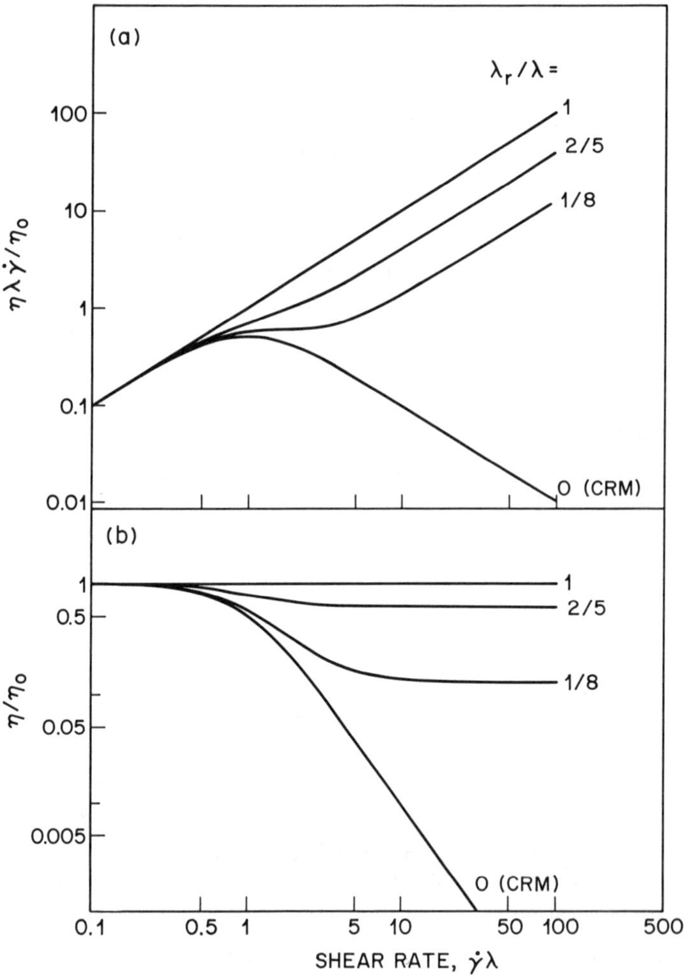

Fig. 3-3 The shear stress (a) and shear viscosity (b) versus shear rate $\dot{\gamma}$ for the corotational Jeffreys equation in steady-state shearing. When $\lambda_r/\lambda < 1/8$, the shear thinning is excessive.

$\lambda\dot{\gamma} = 1$; this is never seen experimentally. The excessive shear thinning can be corrected by adding a retardation term

$$\lambda \overset{0}{\underset{\approx}{\sigma}} + \underset{\approx}{\sigma} = 2\eta_0(\underset{\approx}{D} + \lambda_r \overset{0}{\underset{\approx}{D}}) \tag{3-41}$$

Here $\overset{0}{\underset{\approx}{D}}$ is the corotational time derivative acting on $\underset{\approx}{D}$:

$$\overset{0}{\underset{\approx}{D}} \equiv \dot{\underset{\approx}{D}} - \underset{\approx}{\omega}^{T} \cdot \underset{\approx}{D} - \underset{\approx}{D} \cdot \underset{\approx}{\omega} \tag{3-42}$$

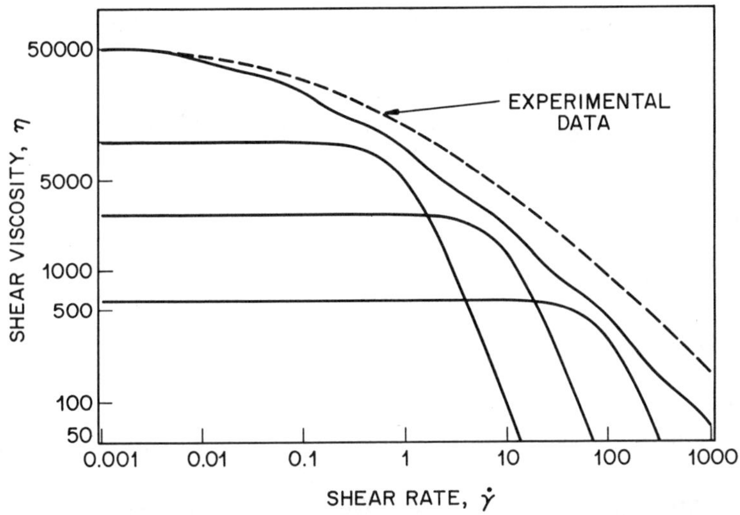

Fig. 3-4 The shear viscosity versus shear rate for the corotational Maxwell equation with a distribution of relaxation times. The spectrum of relaxation times is that of Melt I from Table 2-1. The contributions from three of the modes are also shown. The dashed line gives the experimental result.

Equation 3-41 is called the corotational Jeffreys equation. The retardation term can be regarded as arising from a solvent stress with viscosity $\eta_s = \eta_0 \lambda_r / \lambda$. If $\eta_s \geq \eta_0/8$, no maximum appears in the shear stress, $\dot{\gamma}\eta$, Figure 3-3a. However, the shear viscosity then approaches a constant, η_∞, as $\dot{\gamma} \rightarrow \infty$; see Figure 3-3b. If $\eta_s = \eta_0/8$, the minimal value needed to avoid excessive shear thinning,[10] then η_∞, the asymptotic viscosity, takes a value of $1/9$ times the zero-shear viscosity, η_0.

An asymptotically constant shear viscosity at high shear rates may describe dilute solutions (see Chapter 8); for melts, however, if η_∞ exists at all, it must be decades lower in magnitude than η_0. See for example Figure 2-10. Thus if we wish to describe melts or most solutions, a retardation time is not an adequate cure for the excessive shear thinning of the corotational Maxwell equation.

A much more appropriate answer is to include the spectrum of relaxation times characteristic of the material. This spectrum can easily be obtained from oscillatory shear measurements of G′ and G″, as discussed at the end of Chapter 2. Thus

$$\underset{\approx}{\sigma} = \sum_{i} \underset{\approx}{\sigma}_i \tag{3-43}$$

where each $\underset{\approx}{\sigma}_i$ satisfies Equation 3-41 with values of $\eta_0 = \eta_{0i}$ and $\lambda = \lambda_i$ from the relaxation spectrum. Figure (3-4) shows how inclusion of a relaxation spectrum, namely that for Melt I, broadens the shear-thinning response.

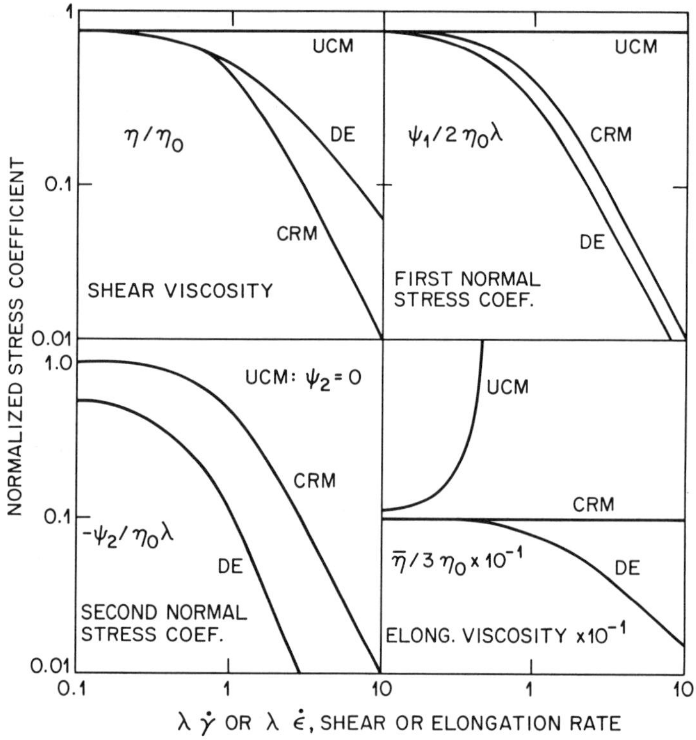

Fig. 3-5 Shear viscosity, primary normal stress coefficient, secondary normal stress coefficient, and elongational viscosity for the upper-convected Maxwell (UCM), the corotational Maxwell (CRM), and the Doi-Edwards (DE) constitutive equations.

A retardation term can be thought of as a fast relaxation mode. If the relaxation time, λ_i, of mode i is much smaller than $1/\dot{\gamma}$ for any experimentally realized $\dot{\gamma}$, then mode i gives a nearly Newtonian response, $\underset{\approx}{\sigma}_i \approx 2\eta_i \underset{\approx}{D}$. Thus the retardation term in the CRM equation, or in any other constitutive equation, can be thought of as arising from a fast relaxation mode.

3.4.4.2 Other Predictions The CRM equation not only predicts shear thinning but is also "softer" than the UCM equation in other deformations, including *extensional* deformations. In steady elongation, where the UCM equation gives a viscosity singularity at a strain rate $\dot{\epsilon} = 1/2\lambda$, the CRM equation predicts the Trouton value

$$\overline{\eta}(\dot{\epsilon}) = 3\eta \tag{3-44}$$

for all $\dot{\epsilon}$. This behavior, which is not affected by the relaxation spectrum, is too soft; it cannot predict the strain hardening typically seen, in Figure 1-13, for example.

The second normal stress coefficient is predicted by the CRM equation to be

$$\Psi_2 = \frac{-\lambda\eta_0}{1+\lambda^2\dot{\gamma}^2} \tag{3-45}$$

which is $-1/2$ times Ψ_1. This magnitude is too large to agree with most data. Figure 3-5 compares the predictions of the UCM equation and the CRM equation for $\eta(\dot{\gamma})$, $\Psi_1(\dot{\gamma})$, $\Psi_2(\dot{\gamma})$, and $\overline{\eta}(\dot{\epsilon})$. Also included are predictions of the Doi-Edwards (DE) equation. The DE equation is a fairly realistic molecular-based equation of the *Kaye-BKZ* type, to be discussed shortly. Note that for these steady-state quantities, the CRM equation is much closer to the DE predictions than is the UCM equation. Note also that the biggest divergences between the equations occurs in their description of elongational flow. Accurate stresses in elongational flow are particularly difficult to obtain.

3.4.5 Oldroyd Eight-Constant Equation

Oldroyd sought to produce a more general constitutive equation, subject to the constraints of frame invariance. He did this by letting $\underset{\approx}{f}(\underset{\approx}{\sigma},\underset{\approx}{D})$ in Equation 3-36 contain terms that are at most linear in $\underset{\approx}{\sigma}$ and at most quadratic in $\underset{\approx}{D}$ and $\underset{\approx}{\sigma}$ combined. The result is[13]

$$\lambda\overset{0}{\underset{\approx}{\sigma}} + \underset{\approx}{\sigma} + \mu_0 \mathrm{tr}(\underset{\approx}{\sigma})\underset{\approx}{D} - \mu_1(\underset{\approx}{\sigma}\cdot\underset{\approx}{D} + \underset{\approx}{D}\cdot\underset{\approx}{\sigma}) + \nu_1(\underset{\approx}{\sigma}:\underset{\approx}{D})\underset{\approx}{\delta}$$

$$= 2\eta_0\left[\underset{\approx}{D} + \lambda_r\overset{0}{\underset{\approx}{D}} - 2\mu_2\underset{\approx}{D}^2 + \nu_2\underset{\approx}{D}:\underset{\approx}{D}\,\underset{\approx}{\delta}\right] \tag{3-46}$$

The terms that appear in Equation 3-46 emerge as a result of a ***representation theorem*** for tensors. As a simpler example, suppose $\underset{\approx}{f}$ is a function of $\underset{\approx}{D}$ only, $\underset{\approx}{f}(\underset{\approx}{D})$. Suppose also that $\underset{\approx}{f}$ can be expanded in a power series

$$\underset{\approx}{f} = a_0\underset{\approx}{\delta} + a_1\underset{\approx}{D} + a_2\underset{\approx}{D}^2 + a_3\underset{\approx}{D}^3 + \cdots \tag{3-47}$$

There is a relationship known as the *Cayley-Hamilton theorem* between powers of any symmetric three-dimensional tensor $\underset{\sim}{A}$:

$$\underset{\sim}{A}^3 - I_A\underset{\sim}{A}^2 + II_A\underset{\sim}{A} - III_A\underset{\approx}{\delta} = \underset{\approx}{0} \tag{3-48}$$

where I_A, II_A, and III_A are the scalar ***invariants*** of $\underset{\sim}{A}$:

$$I_A = \mathrm{tr}\underset{\sim}{A}$$

$$II_A = \frac{1}{2}[(\mathrm{tr}\underset{\sim}{A})^2 - \mathrm{tr}\underset{\sim}{A}^2]$$

$$III_A = \det\underset{\sim}{A} \tag{3-49}$$

These are called ***invariants*** because their numerical values are unchanged by rotations of frame. (The components of tensors such as $\underset{\approx}{D}$ and $\underset{\approx}{C}^{-1}$ are changed by rotations of frame.) Because of the Cayley-Hamilton theorem, $\underset{\sim}{A}^3$ can be expressed in terms of $\underset{\sim}{A}^2$, $\underset{\sim}{A}$, $\underset{\approx}{\delta}$, and the invariants. If one multiplies Equation 3-48 by $\underset{\sim}{A}$, one finds that $\underset{\sim}{A}^4$ can be expressed in terms of $\underset{\sim}{A}^3$, $\underset{\sim}{A}^2$, and $\underset{\sim}{A}$. But $\underset{\sim}{A}^3$ can be eliminated by substitution in favor of $\underset{\sim}{A}^2$, $\underset{\sim}{A}$, and $\underset{\approx}{\delta}$. In this way, any integer power of $\underset{\sim}{A}$ can be reduced to a sum of terms $\underset{\sim}{A}^2$, $\underset{\sim}{A}$, and $\underset{\approx}{\delta}$, with coefficients that are functions of the invariants. This means that a power series, such as Equation 3-47, can be expressed as

$$\underset{\approx}{f} = \alpha_0\underset{\approx}{\delta} + \alpha_1\underset{\approx}{D} + \alpha_2\underset{\approx}{D}^2 \tag{3-50}$$

where each α_i depends on the invariants, I_D, II_D, and III_D. The invariant $I_D = \mathrm{tr}\underset{\approx}{D}$ is zero because of incompressibility. Equation 3-50 is the ***representation theorem*** for $\underset{\approx}{f}(\underset{\approx}{D})$. If, in the dependence of $\underset{\approx}{f}$ on $\underset{\approx}{D}$, only terms that are first and second order in $\underset{\approx}{D}$ are allowed, the result is

$$\underset{\approx}{f} = b_0\mathrm{tr}(\underset{\approx}{D}^2)\underset{\approx}{\delta} + b_1\underset{\approx}{D} + b_2\underset{\approx}{D}^2 \tag{3-51}$$

where each b_i is a constant.

Using the analogous representation theorem for $\underset{\approx}{f}(\underset{\approx}{D}, \underset{\approx}{\sigma})$, Oldroyd obtained all terms involving $\underset{\approx}{\sigma}$ and $\underset{\approx}{D}$ that appear in his eight-constant equation. The terms $\overset{0}{\underset{\approx}{\sigma}}$ and $\overset{0}{\underset{\approx}{D}}$ are the only frame-invariant time derivatives that are at most quadratic in $\underset{\approx}{\sigma}$ and $\underset{\approx}{D}$ taken together and are linear in $\underset{\approx}{\sigma}$.

The Oldroyd eight-constant equation contains the upper-convected, lower-convected, and corotational Maxwell equations as special cases. The requirement that the behavior in shear be reasonable places some restrictions on the ranges of values of the parameters, as discussed in Bird, Armstrong, and Hassager.[10]

The usefulness of the Oldroyd equation, with all eight constants, is somewhat limited, mainly because dealing with all the terms is tedious and difficult. Furthermore the behavior of the equation in elongation is unrealistic, generally showing singularities at finite $\dot{\epsilon}$, except for special values of the parameters. There are special cases of this equation that are of considerable interest, but these were first noticed because they were derived in other ways, and we will defer discussion of them until Chapter 5.

Experience with the Oldroyd eight-constant equation points out the limitation of purely continuum mechanical derivations of constitutive equations. As one systematically increases the order of the equation to improve its ability to describe viscoelastic fluid behavior, the number of constants grows so rapidly that the equation quickly becomes unwieldy. Already with only quadratic terms, the stress terms being at most linear, there are eight constants. It seems that frame invariance in itself is not restrictive enough to keep the number of terms manageable. Even with eight constants, the Oldroyd equation lacks a term that would cause the extensional viscosity to increase with extension rate, and yet be bounded at all extension rates. The simplest term that does so is one *quadratic in stress*. Consider, for example

$$\lambda \overset{\nabla}{\underset{\approx}{\sigma}} + \underset{\approx}{\sigma} + a\underset{\approx}{\sigma}^2 = 2\eta \underset{\approx}{D} \tag{3-52}$$

Equation 3-52 is the UCM equation with an extra quadratic stress. It was proposed by Giesekus[14] from *molecular* arguments. Though the arguments are simplistic, they produce an equation more useful in many ways than Oldroyd's, and having some basis in the molecular behavior of polymers. Oldroyd's continuum-mechanical approach would not have produced such an equation unless he had gone out to higher order, in which case there would have been many more than eight constants.

3.5 THE KAYE-BKZ CLASS OF EQUATIONS

Just as there are continuum-mechanical generalizations of the UCM differential constitutive equation, namely Oldroyd's equations, so there are continuum mechanical generalizations of the Lodge equation, which is the integral equivalent of the UCM equation. The most important of these is the *class* of equations developed independently by Kaye[15] (1962) and by Bernstein, Kearsley, and Zapas[16] (1963).

3.5.1 The Strain Energy Function

Like the Lodge equation, the development of the Kaye-BKZ class of equations was motivated by ideas of rubber elasticity theory. Rather than limiting themselves to a particular model for rubber, however, these authors considered instead the most general *strain energy*, really a free energy, W, which is a function of the elastic strain imposed on the rubber. In the classical theory of rubber elasticity,

$$W = \frac{1}{2} G \,\mathrm{tr}\underset{\approx}{C}^{-1} \tag{3-53}$$

Letting the dependence of W on the strain be general, we have

$$W = W(\underset{\approx}{C}^{-1}) \tag{3-54}$$

Since W, a scalar, must be independent of the frame of reference, it can only depend on the properties of $\underset{\approx}{C}^{-1}$ that are independent of rotation. Thus W depends on *the invariants* of $\underset{\approx}{C}^{-1}$:

$$I_1 = \mathrm{tr}\underset{\approx}{C}^{-1}$$

$$I_2 = \frac{1}{2}[(\mathrm{tr}\underset{\approx}{C}^{-1})^2 - \mathrm{tr}\underset{\approx}{C}^{-2}]$$

$$I_3 = \det\underset{\approx}{C}^{-1} \tag{3-55}$$

These are analogous to the invariants I_D, II_D, and III_D of the strain rate, discussed earlier. Because of incompressibility, $\det\underset{\approx}{C}^{-1} = 1$ and

$$I_2 = \mathrm{tr}\underset{\approx}{C} \tag{3-56}$$

Thus $W = W(I_1, I_2)$.

Consider now a unit cube of the rubber in an extensional deformation; see Figure 1-17. For the moment, we drop the incompressibility constraint and note that W is then a function of three invariants or, equivalently, a function of the three principle stretch ratios λ_1, λ_2, and λ_3. Consider an increment, $d\lambda_1$, of the deformation in direction "1", holding λ_2 and λ_3 fixed. The increment of work required to produce the deformation is

$$dW = f_1 d\lambda_1 \tag{3-57}$$

where f_1 is the normal force in direction 1. The stress is

$$\sigma_{11} = f_1 A^{-1} = \frac{dW}{d\lambda_1} A^{-1} \tag{3-58}$$

where A is the cross-sectional area, $A = \lambda_2\lambda_3$. If we simultaneously deform in all directions,

$$\sigma_{ii} = \frac{\partial W}{\partial \lambda_i} \lambda_j^{-1} \lambda_k^{-1} \tag{3-59}$$

where i,j, and k are all different. We now impose volume conservation. Thus $\lambda_j^{-1}\lambda_k^{-1} = \lambda_i$. Hence

$$\sigma_{ii} = \frac{\partial W}{\partial \lambda_i}\lambda_i = \frac{\partial W}{\partial \epsilon_i} \tag{3-60}$$

where

$$\epsilon_i = \log \lambda_i \tag{3-61}$$

and ϵ_i is called the "Hencky strain."

In tensor notation,

$$\underset{\approx}{\sigma} = \frac{\partial W}{\partial \underset{\approx}{\epsilon}} \qquad (3\text{-}62)$$

where for an extensional deformation,

$$\underset{\approx}{\epsilon} = \begin{pmatrix} \log\lambda_1 & 0 & 0 \\ 0 & \log\lambda_2 & 0 \\ 0 & 0 & \log\lambda_3 \end{pmatrix} \qquad (3\text{-}63)$$

By frame invariance, Equation 3-62 holds for nonextensional strains as well. The Hencky strain tensor in a nonextensional frame, $\hat{\underset{\approx}{\epsilon}}$, can be found by rotating to a frame in which the deformation is extensional, forming $\underset{\approx}{\epsilon}$ as a diagonal tensor, and then rotating back. In the non-extensional frame, the Hencky tensor is

$$\hat{\underset{\approx}{\epsilon}} = \underset{\approx}{Q}^{T} \cdot \underset{\approx}{\epsilon} \cdot \underset{\approx}{Q} \qquad (3\text{-}64)$$

where $\underset{\approx}{Q}$ is the rotation matrix.

Equation 3-62 can be expressed in terms of the strain tensors $\underset{\approx}{C}^{-1}$ and $\underset{\approx}{C}$. In a frame in which the deformation is extensional,

$$\frac{\partial W}{\partial \epsilon_i} = \frac{\partial W}{\partial I_1}\frac{\partial I_1}{\partial \epsilon_i} + \frac{\partial W}{\partial I_2}\frac{\partial I_2}{\partial \epsilon_i} \qquad (3\text{-}65)$$

But

$$I_1 = \sum_i \lambda_i^2 = \sum_i \exp(2\epsilon_i)$$

$$I_2 = \sum_i \lambda_i^{-2} = \sum_i \exp(-2\epsilon_i) \qquad (3\text{-}66)$$

Therefore

$$\frac{\partial I_1}{\partial \epsilon_i} = 2\exp(2\epsilon_i)$$

$$\frac{\partial I_2}{\partial \epsilon_i} = -2\exp(-2\epsilon_i) \qquad (3\text{-}67)$$

and

$$\frac{\partial W}{\partial \epsilon_i} = 2\frac{\partial W}{\partial I_1}\exp(2\epsilon_i) - 2\frac{\partial W}{\partial I_2}\exp(-2\epsilon_i)$$

$$= 2\frac{\partial W}{\partial I_1}\lambda_i^2 - 2\frac{\partial W}{\partial I_2}\lambda_i^{-2} \tag{3-68}$$

In tensor form,

$$\underset{\approx}{\sigma} = \frac{\partial W}{\partial \underset{\approx}{\epsilon}} = 2\frac{\partial W}{\partial I_1}\underset{\approx}{C}^{-1} - 2\frac{\partial W}{\partial I_2}\underset{\approx}{C} \quad \text{(elastic body)} \tag{3-69}$$

to within an isotropic term. This is the general equation for an incompressible elastic body. For the classical theory of rubber elasticity, wherein $W = \frac{1}{2}GI_1$, Equation 3-69 gives $\underset{\approx}{\sigma} = 2G\underset{\approx}{C}^{-1}$, as expected.

3.5.2 The History Integral

In the Kaye-BKZ equation for viscoelastic fluids, elastic energy and stress are allowed to relax. W therefore depends on the *history* of I_1 and I_2. Let us make an *empirical assumption* that W is a *history integral* over a function of I_1, I_2, and $t-t'$:

$$W = \int_{-\infty}^{t} u(I_1, I_2, t-t')\, dt' \tag{3-70}$$

The corresponding stress tensor is

$$\underset{\approx}{\sigma} = \int_{-\infty}^{t} \left[2\frac{\partial u}{\partial I_1}\underset{\approx}{C}^{-1}(t,t') - 2\frac{\partial u}{\partial I_2}\underset{\approx}{C}(t,t')\right] dt' \tag{3-71}$$

This is the Kaye-BKZ *class* of constitutive equations. It is a *class* of equations, because it admits a great variety of behavior, depending on the choice of the kernel, $u(I_1, I_2, t-t')$. Our definition of a constitutive equation, as opposed to a class of equations, is that it is defined by a finite number of nonlinear parameters. A nonlinear parameter is one that defines behavior outside the linear viscoelastic regime. Thus the Lodge equation has no nonlinear parameters. A specific constitutive equation of the Kaye-BKZ class can be obtained by choosing a specific form for $u(I_1, I_2, t-t')$, with a few nonlinear parameters.

3.5.3 Shear

In a shearing deformation, the shear stress of any Kaye-BKZ equation simplifies to

$$\sigma = \int_{-\infty}^{t} 2\left(\frac{\partial u}{\partial I_1} - \frac{\partial u}{\partial I_2}\right)\gamma(t,t')\, dt' \tag{3-72}$$

The quantity $\phi \equiv (\partial u/\partial I_1 - \partial u/\partial I_2)$, is a function of the shear strain, $\gamma(t,t')$, and of $t-t'$. Hence

$$\sigma = \int_{-\infty}^{t} \phi[\gamma(t-t'), t-t']\, \gamma(t,t')\, dt' \tag{3-73}$$

3.5.3.1 Step Strains The simplest way to obtain a specific constitutive equation of the Kaye-BKZ class is to fit the function $u(I_1, I_2, s)$ to *step-strain* data. In a step strain, say step shear, the material in equilibrium is suddenly at time t=0 subjected to a shear strain of some magnitude γ. (To be approximated as a step, the strain must be imposed in an interval of time short compared to the duration of time one waits after the strain imposition before measuring the stress.) The strain history in a step shear is

$$\gamma(t,t') = \begin{cases} 0, & t' > 0 \\ \gamma, & t' < 0 \end{cases} \tag{3-74}$$

Equation 3-73 therefore becomes

$$\sigma = \gamma \int_{0}^{t} \phi(\gamma, t-t')\, dt' = \gamma \int_{0}^{t} \phi(\gamma, s)\, ds \equiv \gamma \Phi(\gamma, t) \tag{3-75}$$

Thus the kernel function $\phi(\gamma,s)$ can be obtained from the experimental response to a step strain by differentiating the relaxing stress $\sigma(t)$ with respect to t:

$$\gamma\phi(\gamma,t) = \frac{\partial}{\partial t}\sigma(t) \tag{3-76}$$

Once $\phi(\gamma,t)$ is known, the shear stress in *any shear history* can be predicted if the Kaye-BKZ equation holds.

Actually, the differentiation in Equation 3-76 is rarely carried out. More commonly, the validity of the Kaye-BKZ class of equations is tested experimentally. This can be done by performing double-step strains with the first step, γ_1, separated from the second, γ_2, by an interval of time, t_1. The total strain imposed is $\gamma_1 + \gamma_2$, and γ_2 can have either the same sign or the opposite sign as γ_1. According to the Kaye-BKZ equation, the response at a time t after the imposition of the second strain is determined by the response to the single-step strain:

$$\sigma(t, t_1, \gamma_1, \gamma_2) = \sigma(t, \gamma_2) + \sigma(t+t_1, \gamma_1+\gamma_2) - \sigma(t+t_1, \gamma_2) \tag{3-77}$$

All the terms on the right side are of the form $\sigma(s, \gamma)$ where $\sigma(s, \gamma)$ is the shear stress at time s after a single- step strain γ.

The results of these tests are mixed. The Kaye-BKZ equation closely predicts the stress if the two strains *are in the same direction*, but is sometimes poor if the strains are in opposite direction to each other.[17,18] Possible corrections for this failing are discussed in Chapter 5.

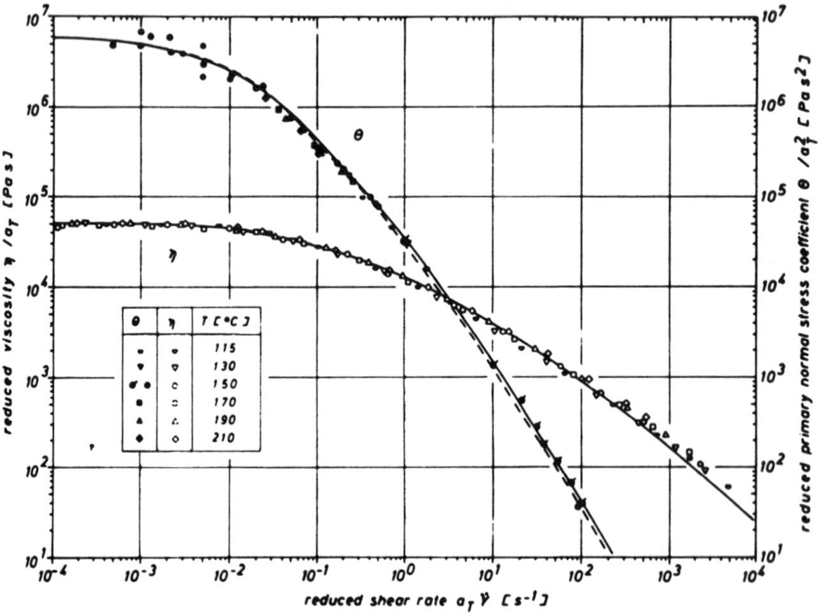

Fig. 3-6 Steady-state shear viscosity and primary normal stress coefficient for Melt I and from the Kaye-BKZ equation with the double exponential damping function (solid lines) and with the single exponential (dashed lines) (from Laun[19], reproduced with permission from Rheologica Acta).

3.5.3.2 Steady-State, Stress Growth, and Stress Relaxation The reversing double-step strain is a severe deformation history. Therefore the inaccuracy of the Kaye-BKZ equation in describing it should be kept in perspective. Many interesting polymer flows have no strain reversal in them; if one is interested in these flows, the reversing double-step strain is not a reasonable test of the constitutive equation. Less stringent experimental tests are the measurement of steady-state viscosity and first normal stress difference, stress growth following the onset of steady shearing, and stress relaxation following cessation of steady-state shearing. For these tests, the Kaye-BKZ equation generally does well, as Figure 3-6 to 3-8 show.[19] The experimental results were obtained with Melt I, a low-density polyethylene.

3.5.4 Time-Strain Separability

The Kaye-BKZ kernel function that allows the predictions in these figures was obtained from relaxation after single-step shear.* These step shear results are

* The predictions shown in Figures 3-6 to 3-8 were actually obtained with Wagner's equation, discussed at the end of the chapter. In predictions of shear and first normal stress differences, Wagner's equation is identical to the Kaye-BKZ equation.

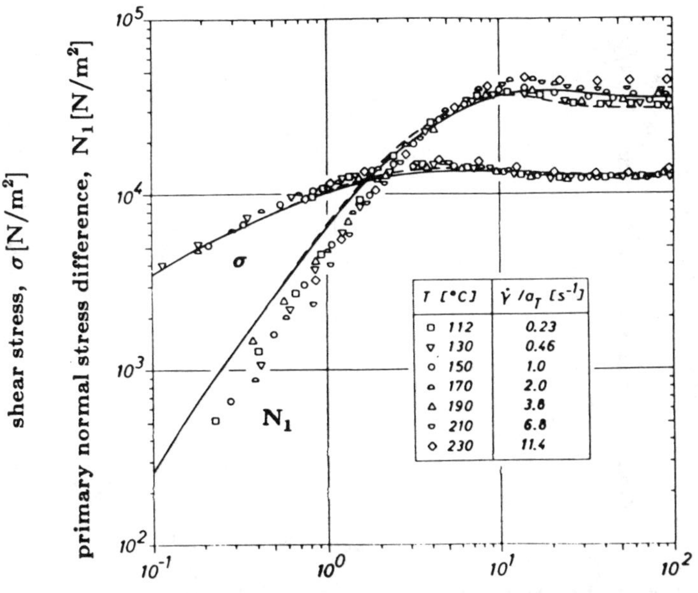

Fig. 3-7 Growth of shear viscosity and primary normal stress coefficient after start-up of steady shearing of Melt I compared to predictions of the Kaye-BKZ equation with the double exponential damping function (solid lines) and with the single exponential (dashed lines) (from Laun[19], reproduced with permission from Rheologica Acta).

shown in Figure 3-9, where $\sigma(\gamma, t)/\gamma$ is plotted against time, t, for various strains, γ. Note that on a log-log plot, the stress relaxation curves are all parallel. This means that the time dependence of $\sigma(\gamma, t)$ is factorable from the strain dependence; i.e.,

$$\sigma(\gamma, t) = G(t)\,h(\gamma)\,\gamma \tag{3-78}$$

G(t) is the linear relaxation modulus. Such factorability can only occur if the potential function $u(I_1, I_2, s)$ is factorable:

$$u(I_1, I_2, s) = m(s)\,U(I_1, I_2) \tag{3-79}$$

The Kaye-BKZ equation then takes the form

$$\underset{\approx}{\sigma} = \int_{-\infty}^{t} m(t-t')\left[2\frac{\partial U}{\partial I_1}\underset{\approx}{C}^{-1} - 2\frac{\partial U}{\partial I_2}\underset{\approx}{C}\right] dt' \tag{3-80}$$

This is called the *separable* or *factorable* Kaye-BKZ equation, because the time effects, described by $m(t-t')$, are separated from the strain effects, described by

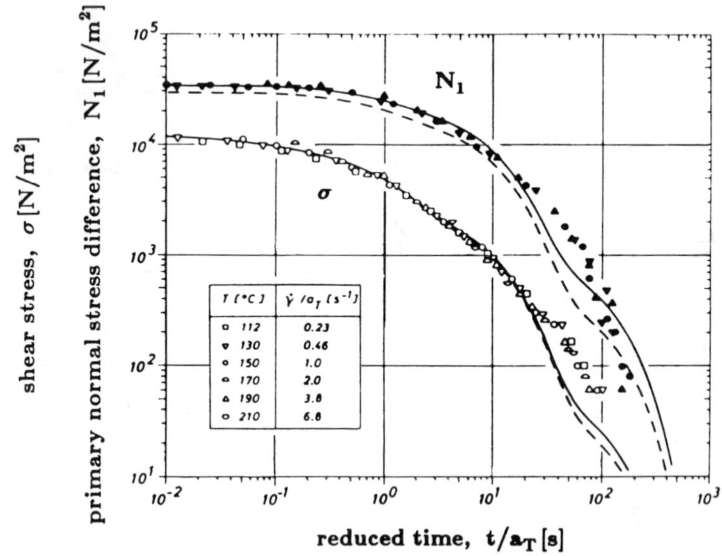

Fig. 3-8 Relaxation of shear stress and primary normal stress difference after cessation of steady-state shearing of Melt I compared to predictions of the Kaye-BKZ equation with the double exponential damping function (solid lines) and with the single exponential (dashed lines) (from Laun[19], reproduced with permission from Rheologica Acta).

the term in brackets. (Of course the bracketed term depends on time *implicitly*, in that $\underset{\approx}{C}^{-1}$, $\underset{\approx}{C}$, I_1, and I_2 are functions of t and t'.) In shear, the shear stress of the separable Kaye-BKZ equation is

$$\sigma = \int_{-\infty}^{t} m(t-t')\, h(\gamma)\, \gamma(t,t')\, dt' \tag{3-81}$$

The function $h(\gamma)$ is called the "damping function." It is unity when γ is small — that is, within the limits of linear viscoelasticity. As γ, the step in shear, increases, $h(\gamma)$ decreases, which can be inferred from Figure 3-9. $h(\gamma)$ can be thought of as the nonlinearity of the relaxation modulus. It can be obtained graphically by finding the amount of vertical shift required on a log-log plot to superpose the relaxation curve for each nonlinear strain γ onto the linear curve. Figure 3-10 plots $h(\gamma)$ for Melt I. As the strains becomes large, $h(\gamma)$ falls almost two orders of magnitude below unity!

Wagner[20] proposed a simple empirical fit to $h(\gamma)$:

$$h(\gamma) = \exp(-n\gamma) \tag{3-83}$$

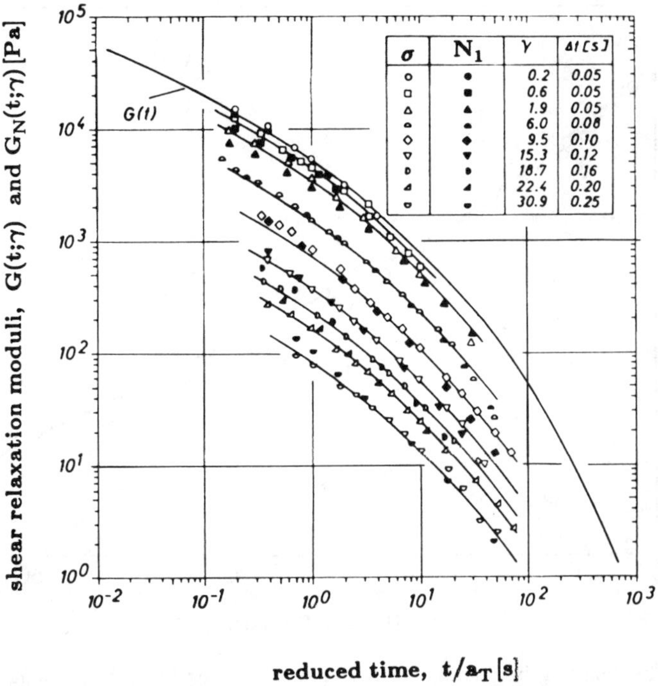

Fig. 3-9 Modulus of shear stress, σ/γ (open symbols), and of the primary normal stress difference, N_1/γ_1 (closed symbols) of Melt I, as functions of time after a step shear (from Laun[19], reproduced with permission from Rheologica Acta).

Osaki[21] proposed a double exponential:

$$h(\gamma) = f_1 \exp(-n_1\gamma) + f_2 \exp(-n_2\gamma) \tag{3-84}$$

Laun found that Equation 3-84 fits the Melt I data almost perfectly, as shown in Figure 3-10, when $f_1 = 0.57$, $f_2 = 0.43$, $n_1 = 0.31$, and $n_2 = 0.106$.

Laun used these two empirical damping functions in the calculations of shear viscosities and normal stress differences in Figures 3-6 — 3-8. The small differences in the predictions attest that the sensitivity to $h(\gamma)$ is not high for these material functions.

The memory function, $m(t-t')$, used in the predictions was a sum of exponentials with relaxation times and moduli given in Table 2-1. The wide spacing between the relaxation times accounts for the slight oscillation in predicted stresses seen in Figures 3-6 — 3-8.

3.5.5 Lodge-Meissner Relationship

For the relaxation of N_1, the first normal stress difference, after a step shear, the separable Kaye-BKZ equation gives

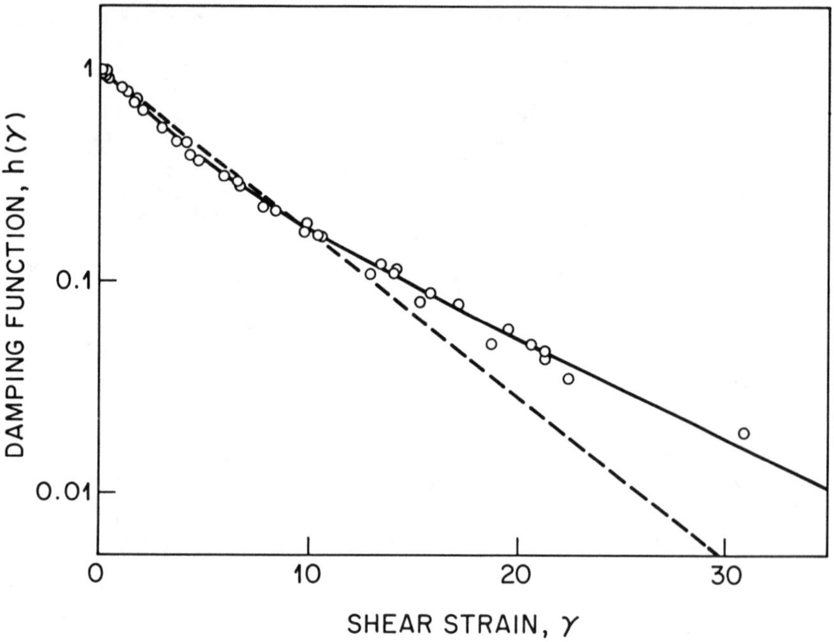

Fig. 3-10 The damping function for Melt I in step shearing fitted by the single exponential of Equation 3-83 and the double exponential of Equation 3-84 (from Laun[19], reproduced with permission from Rheologica Acta).

$$N_1(\gamma, t) \;=\; G(t)\, h(\gamma)\, \gamma^2 \tag{3-85}$$

which according to Equation 3-78 is γ times the shear stress. Thus

$$\frac{N_1}{\sigma} = \gamma \tag{3-86}$$

for relaxation after a step shear. This result is called the *Lodge-Meissner relationship,*[22] after the two authors who first proposed Equation 3-85 for viscoelastic liquids. Rivlin[23] had earlier derived this result for elastic solids. The Lodge-Meissner relationship for viscoelastic liquids can be derived by assuming that during a step strain, the material behaves as an elastic solid.

For an elastic solid, *the stress is a function of the total strain,* i.e., $\underset{\approx}{\sigma} = \underset{\approx}{f}[\underset{\approx}{C}^{-1}]$. Frame invariance implies that for any rotation, $\underset{\approx}{Q}$,

$$\underset{\approx}{\hat{\sigma}} \;\equiv\; \underset{\approx}{Q}^{T} \cdot \underset{\approx}{\sigma} \cdot \underset{\approx}{Q} \;=\; \underset{\approx}{f}[\underset{\approx}{Q}^{T} \cdot \underset{\approx}{C}^{-1} \cdot \underset{\approx}{Q}] \tag{3-87}$$

If $\underset{\approx}{Q}$ is chosen so that $\hat{\underset{\approx}{C}}^{-1} \equiv \underset{\approx}{Q}^T \cdot \underset{\approx}{C}^{-1} \cdot \underset{\approx}{Q}$ is diagonal, then for *isotropic materials** — that is materials having no orientation dependence except that induced by the strain — $\underset{\approx}{f}[\hat{\underset{\approx}{C}}^{-1}] = \hat{\underset{\approx}{\sigma}}$ must be diagonal. Thus $\underset{\approx}{\sigma}$ is diagonalized by the same rotation that diagonalizes $\underset{\approx}{C}^{-1}$. The *principal axes of stress and strain therefore remain parallel to each other*, that is *they rotate together in elastic deformations.* In a simple shear with velocity in direction 1 and velocity gradient in direction 2, the tensor $\underset{\approx}{Q}$ needed to diagonalize $\underset{\approx}{C}^{-1}$ is

$$\underset{\approx}{Q} = \begin{pmatrix} \cos\theta & -\sin\theta & 0 \\ \sin\theta & \cos\theta & 0 \\ 0 & 0 & 1 \end{pmatrix} \tag{3-88}$$

where θ is the *clockwise* angle through which axes 1 and 2 must be rotated. θ is related to the shear strain γ by

$$2\cot(2\theta) = \gamma \tag{3-89}$$

Since $\underset{\approx}{\sigma}$ is diagonalized by this same rotation,

$$\underset{\approx}{\sigma} = \underset{\approx}{Q} \cdot \hat{\underset{\approx}{\sigma}} \cdot \underset{\approx}{Q}^T = \underset{\approx}{Q} \cdot \begin{pmatrix} \hat{\sigma}_1 & 0 & 0 \\ 0 & \hat{\sigma}_1 & 0 \\ 0 & 0 & \hat{\sigma}_1 \end{pmatrix} \cdot \underset{\approx}{Q}^T \tag{3-90}$$

When the above tensor multiplications are carried out, the components of $\underset{\approx}{\sigma}$ are related to those of $\hat{\underset{\approx}{\sigma}}$ by

$$\begin{aligned} 2\sigma_{11} &= (\hat{\sigma}_1 + \hat{\sigma}_2) + (\hat{\sigma}_1 - \hat{\sigma}_2)\cos(2\theta) \\ 2\sigma_{22} &= (\hat{\sigma}_1 + \hat{\sigma}_2) - (\hat{\sigma}_1 - \hat{\sigma}_2)\cos(2\theta) \\ 2\sigma_{12} &= (\hat{\sigma}_1 - \hat{\sigma}_2)\sin(2\theta) \end{aligned} \tag{3-91}$$

Taking the ratio of $\sigma_{11} - \sigma_{22}$ to σ_{12} gives $2\cot(2\theta)$ which equals γ, according to Equation 3-89. This ratio remains constant even if $\hat{\sigma}_1$ and $\hat{\sigma}_2$ relax with time after the step strain. Thus the Lodge-Meissner relationship holds for viscoelastic liquids with an initially elastic response to step shears.

Note in Figure 3-9 that σ/γ and N_1/γ^2 following a step shear are coincident for Melt I. Thus the Lodge-Meissner relationship holds for this material. It appears to hold generally for melts and concentrated solutions, except when the

---

* Isotropy is assumed throughout Chapters 1-9 of this book.

polymer is of ultrahigh molecular weight,[24] thereby giving evidence of the elastic character of these materials during single-step deformations.

3.5.6 Other Types of Deformation

To predict the response of the Kaye-BKZ equation to any shear deformation, one must first measure the relaxation of stress for the material of interest after a series of step shears. This allows one to obtain the Kaye-BKZ kernel function, $h(\gamma, t)$. Then, predictions can be made for *any shear deformation*.

The rub comes when one tries to determine the response to an arbitrary type of deformation. The kernel function, u, depends on *two* invariants, I_1 and I_2. In shear, only a one-dimensional line in two-dimensional invariant space is sampled, namely the line along which $I_1 = I_2 = 3 + \gamma^2$. Thus the function $h(\gamma, t)$ obtained in shearing deformations does not allow one to determine the dependence of u on I_1 *and* I_2, unless $I_1 = I_2$.

3.5.6.1 Planar Deformations I_1 and I_2 are equal to each other only in *planar deformations*. Besides shear the simplest planar deformation is *planar extension*, also called *pure shear*, for which the velocity gradient is

$$\begin{pmatrix} \dot{\epsilon} & 0 & 0 \\ 0 & -\dot{\epsilon} & 0 \\ 0 & 0 & 0 \end{pmatrix} \tag{3-92}$$

Thus the material is stretched in direction 1, compressed in direction 2, and its third dimension is held fixed. After a steady planar extension for a time, t, the planar strain is $\epsilon = \dot{\epsilon}\, t$, and the Finger strain tensor is

$$\underset{\approx}{C}^{-1} = \begin{pmatrix} \exp(2\epsilon) & 0 & 0 \\ 0 & \exp(-2\epsilon) & 0 \\ 0 & 0 & 1 \end{pmatrix} \tag{3-93}$$

In a suitably rotated frame, this tensor is that of simple shear:

$$\underset{\approx}{C}^{-1} = \begin{pmatrix} 1+\gamma^2 & \gamma & 0 \\ \gamma & 1 & 0 \\ 0 & 0 & 1 \end{pmatrix} \tag{3-94}$$

where $\gamma^2 = \exp(2\epsilon) + \exp(-2\epsilon) - 2$. Thus a planar extension, or any other plane strain, is just a simple shear in a rotated frame. It is crucial to note, however, that the *velocity gradients* for steady shear and for steady planar extension are *not* equal, no matter what frame is chosen. Thus although the *elastic response* to a *step* shear and a *step* planar extension are identical, the *viscoelastic responses in steady flow* to these types of deformation are *not* identical. Nevertheless, the Kaye-BKZ equation allows one to predict the stresses in a given type of steady flow once the stresses after steps of the same type of strain are known. Thus the Kaye-BKZ equation allows one to predict the response to any planar flow, once the response to step shear is measured.

3.5.6.2 Non-Planar Deformations For nonplanar deformations, one must have u as a function of I_1 *and* I_2, with $I_1 \neq I_2$. This means, for example, that to predict the response in any uniaxial elongational deformation, one must measure the response of the material to step uniaxial elongations or some other set of uniaxial elongational deformations. Each geometrically different type of deformation requires a different set of measurements, typically requiring a different experimental apparatus. Only rarely are experiments on a material carried out in even two or three deformation types, let alone enough to determine u as a function of I_1 and I_2.

Thus the weakness of the Kaye-BKZ class of equations is that its generality demands that an almost impossible suite of experimental data be taken, if one wishes to predict the response to an arbitrary deformation history. As we found to be the case with the Oldroyd differential equations, continuum-mechanical principles, such as frame invariance, are not in themselves strong enough to restrict the range of possible constitutive equations to manageable proportions. So, in the next chapters we turn again to *molecular theories* for guidance in the choice of Kaye-BKZ kernel functions for integral equations, or nonlinear stress and strain rate terms to be included in differential equations.

3.6 OTHER STRAIN HISTORY INTEGRALS

3.6.1 Wagner's First Equation

One more difficulty of the Kaye-BKZ class of equations is that stresses are always calculated from *derivatives* of the potential function u. If sets of nonlinear stress data are available, these data allow the determination of $\partial u/\partial I_1$ and $\partial u/\partial I_2$, from which one must calculate $u(I_1, I_2, s)$. This is not trivial to do. Wagner avoided the problem by simply dropping the term involving the Cauchy tensor from the separable Kaye-BKZ equation, Equation 3-80. Wagner's equation[25] is

$$\underset{\approx}{\sigma} = \int_{-\infty}^{t} m(t-t')\,h(I_1, I_2)\,\underset{\approx}{C}^{-1}\,dt' \tag{3-95}$$

Wagner found a *damping function*, $h(I_1, I_2)$, that fit both shear and elongational data for Melt I.

The loss of the term involving the Cauchy tensor, $\underset{\approx}{C}$, means that Wagner's equation has no second normal stress difference in shear. More seriously, unless h is a function of I_1 only, the term $h(I_1, I_2)\underset{\approx}{C}^{-1}$ is *not* derivable from a potential function $U(I_1, I_2)$. This means that in deformations rapid enough that relaxation doesn't occur, the equation can predict a net *production* of work done *on the environment* by the material as it is rapidly deformed in a *cyclic strain* — a strain in which the final shape is the same as the initial.[26] To avoid this drawback both terms of the Kaye-BKZ equation should be retained.

Wagner[27] proposed a second equation that overcomes these difficulties of Equation 3-95. We shall refer to this second equation, which is discussed in Chapter 5, as the Wagner II equation; Equation 3-95 is the Wagner I equation.

3.6.2 Superposition Integral Equation

The Kaye-BKZ and Wagner equations are special cases of the *superposition integral equation*

$$\underset{\approx}{\sigma} = \int_{-\infty}^{t} [\phi_1(I_1, I_2, t-t')\underset{\approx}{C}^{-1} - \phi_2(I_1, I_2, t-t')\underset{\approx}{C}]\, dt' \tag{3-96}$$

Here ϕ_1 and ϕ_2 are in general independent functions. This equation, first proposed by Rivlin and Sawyers[28], can be derived from continuum mechanical assumptions discussed in Section 9.4.3.

If one requires that Equation 3-96 have an elastic limit at high strain rates, then it can be shown that

$$\phi_1 = 2\frac{\partial u}{\partial I_1} \quad \text{and} \quad \phi_2 = -2\frac{\partial u}{\partial I_2}$$

where u is a potential function. For this choice of ϕ_1 and ϕ_2, Equation 3-96 becomes the Kaye-BKZ equation. If ϕ_2 is set to zero and $\phi_1 = h(I_1, I_2)m(t-t')$, Equation 3-96 is the Wagner equation.

3.6.3 Tanner-Simmons Equation

Tanner and Simmons[29] proposed a simple modification of the Green and Tobolsky temporary network model or Lodge equation (Section 2.5), that alleviates its extremely elastic nonlinear behavior. The Tanner-Simmons equation is

$$\underset{\approx}{\sigma} = \int_{t-t_R}^{t} m(t-t')\underset{\approx}{C}^{-1}\, dt' \tag{3-97}$$

where t_R is a "rupture time" for the network. Thus contributions to the stress from network strands created before time t_R are cut off; such strands are the most strained and assumed to have "ruptured." t_R is a function of I_1 and I_2, chosen for example so that

$$\mathrm{tr}\,\underset{\approx}{C}^{-1}(t, t-t_R) - 3 = I_1(t, t-t_R) - 3 = B^2$$

where B is a parameter that describes the strength of the network. The Lodge equation is recovered if $B \rightarrow \infty$. In steady shear, $I_1(t,t') = (t-t')^2\dot{\gamma}^2 + 3$; thus $t_R = B/\dot{\gamma}$. As $\dot{\gamma}$ increases, t_R decreases and shear thinning occurs. If $m(t-t')$ is a sum of exponentials, then

$$m(t-t') = \sum_i \frac{G_i}{\lambda_i} \exp[(t'-t)/\lambda_i]$$

The steady-state shear viscosity and first and second normal stress coefficients of the Tanner-Simmons equation are

$$\eta = \sum_i G_i\lambda_i \{1 - [1 + (B/\lambda_i\dot{\gamma})]\, e^{-B/\lambda_i\dot{\gamma}}\}$$

$$\Psi_1 = \sum_i G_i\lambda_i^2 [2 - (2 + \frac{2B}{\lambda_i\dot{\gamma}} + \frac{B^2}{\lambda_i^2\dot{\gamma}^2})\, e^{-B/\lambda_i\dot{\gamma}}]$$

$$\Psi_2 = 0$$

3.7 SUMMARY

For small strains, no matter what the type of deformation, polymeric fluids obey the constitutive equation of linear viscoelasticity. Significant deviations from linear viscoelasticity only occur when the strain exceeds a threshold; if the strain rate is small enough, deviations don't occur at all. The equation of linear viscoelasticity cannot be valid for deformations of arbitrary magnitude and rate, because the equation violates the principle of frame invariance.

Oldroyd developed frame-invariant differential constitutive equations by defining time derivatives in frames that deform with the material elements. From these "convected" time derivatives, simple frame-invariant constitutive equations, the upper-convected and lower-convected Maxwell equations, are derived. If the time derivative is taken in a frame that rotates with the angular velocity of the material element, then the corotational Maxwell equation is obtained. This simple equation shows shear thinning. By including low-order, frame-invariant products of the velocity gradient tensor and the stress tensor, Oldroyd obtained a general differential constitutive equation with eight empirical constants based on continuum arguments alone, with no molecular assumptions. This equation makes useful qualitative predictions but is not quantitatively accurate.

Kaye and Bernstein, Kearsley, and Zapas proposed an integral constitutive equation based on general rubber elasticity theory, which also makes no molecular assumptions. The theory has an elastic free energy and stress tensor that relax with time. The stress is an integral over past time of two terms: one proportional to the Finger tensor, the other to the Cauchy tensor. The coefficients of these tensors are functions of elapsed time and of the invariants of the Finger strain tensor that can be determined in step-strain experiments. Once these functions are determined, the Kaye-BKZ equation accurately predicts important rheological responses such as stress growth and relaxation. The equation can be inaccurate, however, in strain histories involving strain reversal, such as reversing double-step strains and elastic recovery.

Often the stress after a step strain can be factored into a product of a time-dependent term multiplied by a strain-dependent term. When this occurs, the Kaye-BKZ equation can be considerably simplified. Even with this simplification, many different step-strain experiments must be carried out to specify completely the functions of the Kaye-BKZ equation unless one obtains some help from molecular theory.

After a step shear, the Kaye-BKZ and many other constitutive equations predict that the ratio of the first normal stress difference to the shear stress equals the applied shear strain. The validity of this result, known as the Lodge-

Meissner relationship, is supported experimentally; it can be derived from the assumption that the principal axes of stress and of strain rotate together in a step shear.

References

1. E.V. Menezes and W.W. Graessley, Rheol. Acta, *19*, 38 (1980).

2. J.C. Maxwell, Phil. Trans. Roy. Soc., *A157*, 49 (1867).

3. L. Boltzmann, Pogg. Ann. Phys., *7*, 624 (1876).

4. R.G. Larson, Rheol. Acta, to be submitted.

5. F.T. Trouton, Proc. Roy. Soc., *A77*, 426 (1906).

6. C. Truesdell and W. Noll, *The Non-linear Field Theories*, p. 46, Springer-Verlag, Berlin (1965).

7. G. Ryskin and J.M. Rallison, J. Fluid Mech., *99*, 513 (1980).

8. R. I. Tanner, *Engineering Rheology*, Oxford, New York (1985).

9. J.G. Oldroyd, Proc. Roy. Soc. *A200*, 523 (1950).

10. R.B. Bird, R.C. Armstrong, and O. Hassager, *Dynamics of Polymeric Liquids*, Vol. 1, John Wiley & Sons, New York (1977)

11. H. Jeffreys, *The Earth*, Cambridge University Press, p. 265 (1929).

12. J.D. Goddard and C. Miller, Rheol. Acta, *5*, 177 (1966).

13. J.G. Oldroyd, Proc. Roy. Soc., *A245*, 278 (1958).

14. H. Giesekus, Rheol. Acta, *5*, 29 (1966); J. Non-Newtonian Fluid Mech., *11*, 69 (1982).

15. A. Kaye, College of Aeronautics, Cranford, U.K., Note No. 134 (1962).

16. B. Bernstein, E.A. Kearsley, and L.J. Zapas, Trans. Soc. Rheol., *7*, 391 (1963).

17. K. Osaki, S. Kimura, and M. Kurata, J. Rheol., *25*, 549 (1981).

18. R.G. Larson and V.A. Valesano, J. Rheol., *30*, 1093 (1986).

19. H.M. Laun, Rheol. Acta, *17*, 1 (1978).

20. M.H. Wagner, Rheol. Acta, *15*, 136 (1976).

21. K. Osaki, Proc. Seventh International Congress on Rheology, Gothenburg, Sweden, 1976.

22. A.S. Lodge and J. Meissner, Rheol. Acta, *11*, 351 (1972).

23. R.S. Rivlin, Proc. Second Rubber Technology Conf., London, June 23, 1948.

24. R.G. Larson, S.A. Khan, and V.R. Raju, J. Rheol., submitted (1987).

25. M.H. Wagner, Rheol. Acta, *18*, 427 (1979).

26. R.G. Larson, J. Non-Newtonian Fluid Mech., *13*, 279 (1983).

27. M.H. Wagner, J. Non-Newtonian Fluid Mech., *4*, 39 (1978).

28. R.S. Rivlin and K.N. Sawyers, Ann. Rev. Fluid Mech., *3*, 117 (1971).

29. R.I. Tanner and J.M. Simmons, Chem. Eng. Sci., *22*, 1803 (1967).

CHAPTER 4

Reptation Theories For Melts and Concentrated Solutions

"*We see how quickly through the colander*
The wines will flow; on the other hand,
The sluggish olive-oil delays; no doubt,
Because 'tis wrought of elements more large,
Or else more crook'd and intertangled."

Lucretius: "Of the Nature of Things" (96-55 BC)*

4.1 INTRODUCTION

Recent progress in the development of molecular-based constitutive equations has been faster for polymer melts than for dilute solutions. At first sight this seems surprising, since the rheological properties of a dilute solution are controlled by the dynamics of an isolated polymer chain in its solvent, while in the melt the dynamics of each molecule must be affected by the many other molecules with which it is in intimate contact. The simplicity afforded by the isolation of the single coil is offset, however, by some serious complications that arise because of the presence of the solvent.

4.2 SIMPLIFYING FEATURES OF MELTS

4.2.1 Chains in Melts Are Ideal

One serious difficulty in dilute solution is that a coil dissolved in solvent does not follow random walk, Gaussian statistics except under "theta conditions." These theta conditions occur when the solvent is just poor enough that the polymer feels an effective repulsion toward the solvent that is just strong enough to balance the expansion of the coil caused by the excluded volume of the polymer molecule. For a given solvent, this perfect balance occurs at a special temperature called the *theta temperature.*[1]

---

* as cited by W.H. Herschel, J. Rheol. *1*, 440 (1930) and H. Markovitz, J. Rheol. *29*, 777 (1985).

Suppose the polymer-solvent interaction is neutral or athermal; that is, monomers of the polymer are no more attracted to the solvent than they are to other monomers. Then the polymer coil *swells*, and the end-to end distance, R, is not proportional to $M^{1/2}$, as it would be if it followed a random walk, but to $M^{3/5}$ (M is the molecular weight). The swelling can be explained as follows. Consider all random walk configurations of the chain. Except for self-intersecting conformations, which must be excluded, all random-walk configurations are equally likely, since there is no effective polymer-solvent interaction. The true configuration distribution is thus that of a random walker with self-intersecting walks removed. Now compact chain configurations with small R are more likely to be self-intersecting than are open configurations with large R, so a higher proportion of compact configurations must be excluded. Hence, the configuration distribution will be skewed toward the open configurations with large R. Thus the coil is, on the average, swollen.

In the melt, chains are ideal or Gaussian. To understand this, consider again all random walk configurations of all chains in the melt. If a chain intersects itself *or any other chain*, that configuration must be excluded. Now compact configurations are more likely to be self-intersecting. But the increased likelihood of self-intersection is exactly compensated by the decreased likelihood of intersection with a different chain, since the density of polymer is the same everywhere. Thus, equal fractions of compact and open configurations are excluded because equal fractions of configurations have intersections either of the chain with itself or with another chain. The excluded volume effects, which would otherwise swell the chain, are said to be *screened* by the surrounding chains. Thus *the distribution of configurations of a chain in the melt is Gaussian*. Flory[1] was the first to realize this, and it has since been confirmed by neutron scattering studies.[2]

This result, that polymer molecules in the melt are ideal at equilibrium, greatly simplifies theories of melt rheology. No matter how the configuration distribution is altered by flow, at least at equilibrium it is simple. The same is not true of dilute solutions.

4.2.2 No Hydrodynamic Interaction In Melts

A second complication that appears in dilute solutions but not in melts is *hydrodynamic interaction*. This occurs because the solvent transmits drag not only from the macroscopic boundaries to the chain but also from one part of the chain to another. This effect completely alters and drastically complicates the simple Rouse picture. Contrary to the Rouse picture, the dynamics of a real coil in a solvent are controlled not only by the *statistics* of polymer conformations (together with a simple drag law) but by an intimate coupling between the *statistics and the hydrodynamics*. Each conformation of the chain must be treated as a complex hydrodynamic entity containing many hydrodynamically interacting beads. Then the statistics of the many conformations, and the effect of flow on these statistics, must be accounted for. (An analogous problem exists in the mechanics of concentrated suspensions.) The dynamics of polymer coils with hydrodynamic interaction have not yet been analyzed adequately except when velocity gradients are small.[3] Computer calculations provide the most hopeful avenue of attack on this problem.

In the melt, experiments show that hydrodynamic interaction is screened out.[3] The drag on a chain may be complex because of the influence of neighboring chains, but the drag on one part of the chain does not influence the drag on a remote part of the same chain.

4.2.3 Stress-Optic Law For Melts

A final complication in dilute solutions under some conditions is that the equation for the stress tensor can involve not only a viscoelastic component from the stretching of chains, namely $\langle \underset{\sim\sim}{RR} \rangle$, but also purely dissipative components. These can come from several sources, as discussed in Chapter 8, including solvent-solvent friction, solvent-polymer friction, and *internal friction* within the polymer molecule. The purely dissipative contributions to the stress are not determined by the instantaneous chain conformation alone; in particular they are not proportional to $\langle \underset{\sim\sim}{RR} \rangle$. Instead, they depend directly on the velocity field and disappear as soon as the flow ceases. The solvent contribution to the stress tensor, $\underset{\approx}{\sigma}_s$, from Equation 2-49, is an example; its contribution to the stress tensor is $2\eta_s \underset{\approx}{D}$, which is zero when flow ceases. The viscoelastic term $\langle \underset{\sim\sim}{RR} \rangle$, on the other hand, does not drop to zero immediately when flow ceases but instead relaxes gradually.

Only the viscoelastic component contributes significantly to the stress tensor in a melt. This is shown by the nearly universal validity of the ***stress-optic law*** for melts. The stress-optic law states that the stress tensor is proportional to the ***refractive index tensor,*** also called the ***optical anisotropy tensor.*** The refractive index tensor defines how the index of refraction of a material (i.e., the relative velocity of a beam passing through the material), depends on the direction of the polarization of the beam. Light polarized in one direction travels at a different velocity through an anisotropic medium than does light polarized in a different direction. The difference is minute but still readily measurable. This is so because the difference in distance traveled by collinear beams polarized at right angles to each other need only be approximately the wavelength of the light; originally coherent beams will then interfere with each other after passing through the material.

Suppose, for example, that the beams travel along axis 3 in a Cartesian coordinate system. Since their polarization must be perpendicular to the direction of propagation, we let one beam be polarized in direction 1, the other in direction 2. The difference, $n_1 - n_2$, in refractive index between polarization directions 1 and 2 can be measured.

4.2.3.1 Theoretical Basis The index of refraction of a material is controlled by the ***polarizability*** of the molecules of which the material consists. Anisotropy in the index of refraction occurs when polarizable molecules, such as polymer molecules, are oriented. Polarizability is a tensor quantity; a directed electric field, $\underset{\sim}{E}$, produces a ***polarization,*** $\underset{\sim}{P}$:

$$\underset{\sim}{P} = \underset{\approx}{\alpha} \cdot \underset{\sim}{E} \tag{4-1}$$

where $\underset{\approx}{\alpha}$ is the polarizability tensor.

Now suppose the flexible polymer chain is modeled as a freely jointed chain (See Chapter 2). Consider a single link in the chain oriented, say along axis 1. We suppose that the polarizability of each link in the chain is axially symmetric. This is true if there is enough rotational freedom among the chemical units that make up the link that anisotropies in the *transverse* polarizability are averaged out. The polarizability tensor for link i, is then

$$\underset{\approx}{\alpha}_i = \begin{pmatrix} \alpha_1 & 0 & 0 \\ 0 & \alpha_2 & 0 \\ 0 & 0 & \alpha_2 \end{pmatrix} \tag{4-2}$$

where α_1 is the axial polarizability and α_2 is the transverse polarizability of the link. This tensor can be divided into anisotropic and isotropic pieces:

$$\underset{\approx}{\alpha}_i = \begin{pmatrix} \alpha_1 - \alpha_2 & 0 & 0 \\ 0 & 0 & 0 \\ 0 & 0 & 0 \end{pmatrix} + \begin{pmatrix} \alpha_2 & 0 & 0 \\ 0 & \alpha_2 & 0 \\ 0 & 0 & \alpha_2 \end{pmatrix} \tag{4-3}$$

The isotropic polarizability has no effect on the *anisotropy* of refractive index, and is dropped hereafter. Let $\underset{\sim}{u}_i$ be a unit vector pointing in the direction of the link (which is oriented along axis 1); that is, $\underset{\sim}{u}_i = (1,0,0)$. Then

$$\underset{\approx}{\alpha}_i = (\alpha_1 - \alpha_2)\, \underset{\sim}{u}_i \underset{\sim}{u}_i \tag{4-4}$$

This formula also gives the polarizability if the orientation, $\underset{\sim}{u}$, of the link is rotated away from the 1 direction.

The average polarizability of the whole chain is

$$\underset{\approx}{\alpha} = (\alpha_1 - \alpha_2) < \sum_i \underset{\sim}{u}_i \underset{\sim}{u}_i > \tag{4-5}$$

where $< \cdot >$ denotes an average over all conformations of the chain. The summation is over the N_K links, each of length b, of the chain. Since the links take on independent orientations,

$$\underset{\approx}{\alpha} = (\alpha_1 - \alpha_2) \sum_i < \underset{\sim}{u}_i \underset{\sim}{u}_i > \tag{4-6}$$

Suppose the chain is constrained to have an end-to-end vector $\underset{\sim}{R}$. This constraint biases the orientation of each of the links. As long as R is not so large that the limits of the Gaussian approximation are exceeded, Kuhn and Grün[4] showed that

$$\sum_i < \underset{\sim}{u}_i \underset{\sim}{u}_i > = N_K < \underset{\sim}{u}\underset{\sim}{u} > = \frac{3}{5} \frac{1}{N_K b^2} \underset{\sim}{R}\underset{\sim}{R} \tag{4-7}$$

If there are ν chains per unit volume, the anisotropic part of the *total polarizability per unit volume*, $\underset{\approx}{\overline{\alpha}}$, is

$$\underset{\approx}{\overline{\alpha}} = \nu(\alpha_1 - \alpha_2)\frac{2}{5}\beta^2 <\underset{\sim\sim}{RR}> \tag{4-8}$$

where $\beta^2 \equiv 3/(2N_K b^2)$. The polarizability is related to the index of refraction by the Lorentz-Lorenz formula. Since the anisotropy in the index of refraction is small compared to the average index of refraction, the Lorentz-Lorenz formula can be linearized to give a relation between the anisotropies in index of refraction and in polarizability:[5]

$$\begin{aligned} \underset{\approx}{n} &= \frac{(n^2+2)^2}{6n}\frac{4\pi}{3}\underset{\approx}{\overline{\alpha}} \\ &= (\alpha_1 - \alpha_2)\frac{2\pi}{45}\frac{(n^2+2)^2}{n}2\nu\beta^2 <\underset{\sim\sim}{RR}> \\ &= C\underset{\approx}{\sigma} \end{aligned} \tag{4-9}$$

where $\underset{\approx}{\sigma} = 2\nu\beta^2 kT<\underset{\sim\sim}{RR}>$ is the polymer contribution to the stress tensor; see Equation 2-9. The scalar, n, is the isotropic part of the index of refraction; the tensor $\underset{\approx}{n}$ is the anisotropic part. Thus the *index of refraction tensor is proportional to the stress tensor.* This is called the *stress-optic law.* The proportionality constant C, the *stress-optic coefficient,* has the value[5]

$$C = \frac{2\pi}{45kT}\frac{(n^2+2)^2}{n}(\alpha_1 - \alpha_2) \tag{4-10}$$

4.2.3.2 Validity Experimental studies have nearly universally supported the validity of the stress-optic law for melts and entangled solutions. Figure 4-1 shows an example for a silicon polymer in which a mechanical measurement of shear viscosity as a function of shear rate is compared with an optical measurement of the corresponding quantity. The ratio of n_{12} to σ_{12}, which is the stress-optic coefficient C, is independent of shear rate, in accord with the stress-optic law. C is also independent of molecular weight, which is consistent with Equation 4-10, wherein the stress-optic coefficient depends only on properties of a single link of the chain, and on the average index of refraction, but not on the molecular weight. The validity of the stress-optic law confirms the proportionality of the stress tensor to $<\underset{\sim\sim}{RR}>$. Thus the stress in the material at any instant in time is determined by the configurations of the polymer chains at that instant. *Hydrodynamic drag serves to orient polymer molecules,* that is, to influence the distribution of configurations, and therefore affect $<\underset{\sim\sim}{RR}>$, *but otherwise makes no significant contribution to the stress in melts.*

In the Rouse theory for dilute solutions, a direct contribution of hydrodynamic drag is acknowledged by the inclusion of a solvent term, $2\eta_s\underset{\approx}{D}$, in the

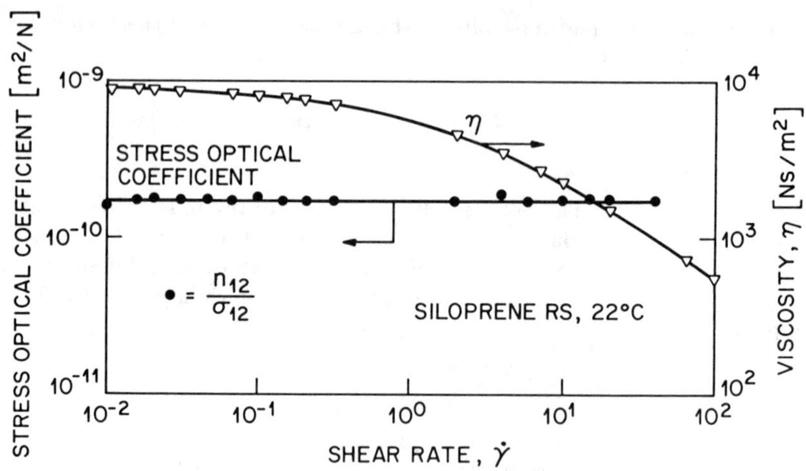

Fig. 4-1 Experimental demonstration of the stress-optic law; the component, n_{12}, of the index-of-refraction tensor is proportional to the corresponding component of the stress tensor, σ_{12}. (from Wales[7])

equation for the stress tensor, $\underset{\approx}{\sigma} = 2\nu kT\beta^2 <\underset{\sim}{R}\underset{\sim}{R}> + 2\eta_s \underset{\approx}{D}$. This term, being purely viscous in nature, is not proportional to $<\underset{\sim}{R}\underset{\sim}{R}>$, so it does not satisfy the stress-optic law. In the Rouse theory for dilute solutions, the *polymer* contribution *does* take the form $<\underset{\sim}{R}\underset{\sim}{R}>$; thus if the solvent term is subtracted from the stress tensor, the result, namely the polymer contribution, should be proportional to the index-of-refraction tensor. This is indeed found to be the case experimentally for *low-frequency* oscillatory shearing flows, that is, in the terminal region, but *at high frequencies the stress-optic law is violated,* even after the solvent contribution, $2\eta_s \underset{\approx}{D}$, has been subtracted.[3,8] This is discussed in more detail in Chapter 8.

In the melt (or in dilute solution), there will be a violation of the stress-optic law if the polymer chain gets extended beyond its Gaussian limit. This apparently occurs in uniaxial extensions of polystyrene melt,[9] when the stress exceeds 2 x 10^6 Pa.* Such high stresses rarely occur in rheological measurements.

We conclude that melts and concentrated solutions are simpler than dilute solutions in the sense that the stress tensor in the former is determined only by the instantaneous configuration of the polymer molecules; in the latter, various

* Deviations from the stress-optic law are also found at high frequencies in oscillatory deformations for melts near their glass transition temperatures.[6]

kinds of friction introduce a direct dependence of the stress tensor on *strain rate*. These friction contributions, which are purely dissipative, can come from friction in the solvent, from solvent-polymer friction, and from friction arising from conformation changes within the polymer molecule itself.

We have argued that the rheology of melts is in some ways simpler than that of dilute solutions. For melts the stress tensor is determined by the instantaneous distribution of configurations; this distribution is Gaussian at equilibrium, and hydrodynamic interaction is screened out. The same is not true of dilute solutions. The great complicating feature of melts is that the motion of each chain is affected by the presence of surrounding chains.

There is a limit, however, in which this complication disappears. If each polymer chain in the melt has a backbone of less than about 300 carbon atoms, then it turns out that the drag forces that neighboring chains exert on a given chain are like those of a simple low-molecular-weight solvent, except that hydrodynamic interaction is screened out. For these melts there are no special drag effects that emerge because the "solvent" for each chain consists of other polymer molecules. Thus low-molecular-weight melts avoid the complications of dilute solutions without incurring complications associated with the interpenetration of the polymer coils. These melts therefore obey the Rouse theory at low and moderate frequencies. At high frequencies, glassy modes not considered in the Rouse theory dominate the response.[3] Beyond the linear viscoelastic limit, low-molecular-weight melts ought to satisfy the upper-convected Maxwell or Lodge equations with the Rouse spectrum of relaxation times. One limitation that we expect, however, is that low molecular-weight chains can more easily be stretched beyond their Gaussian limit. Thus if deformations become large, non-Gaussian effects should be taken into consideration. Ways of accounting for deviations from Gaussian behavior are discussed in Chapter 8.

4.3 CROSSOVER TO ENTANGLEMENT EFFECTS

4.3.1 Appearance of a Plateau Modulus

The Rouse theory predicts a specific dependence of the relaxation spectrum on molecular weight, namely that each relaxation time is proportional to M^2. The zero-shear viscosity is proportional to a characteristic relaxation time, multiplied by the number of molecules per unit volume. Since the density of a melt remains constant as M increases, the number of chains decreases as 1/M. The zero-shear viscosity is therefore predicted by the Rouse theory to be proportional to M. Figure 4-2 is a plot from Berry and Fox[10] of log η_0 versus log M for a diversity of melts. There are two regimes. When M is below a critical value, M_c, the Rouse result, $\eta_0 \propto M$, is obtained; for $M > M_c$, η_0 increases much more steeply with increased molecular weight, namely $\eta_0 \propto M^{3.4}$. M_c varies from melt to melt. For polystyrene, $M_c \sim 38{,}000$. For highly flexible polymers, M_c usually corresponds to about 300–600 atoms in the backbone of the chain.[3] If the polymer melt is very *polydisperse* — that is, it has a broad distribution of molecular weights — then the transition from Rouse behavior is less abrupt than that depicted in Figure 4-2.

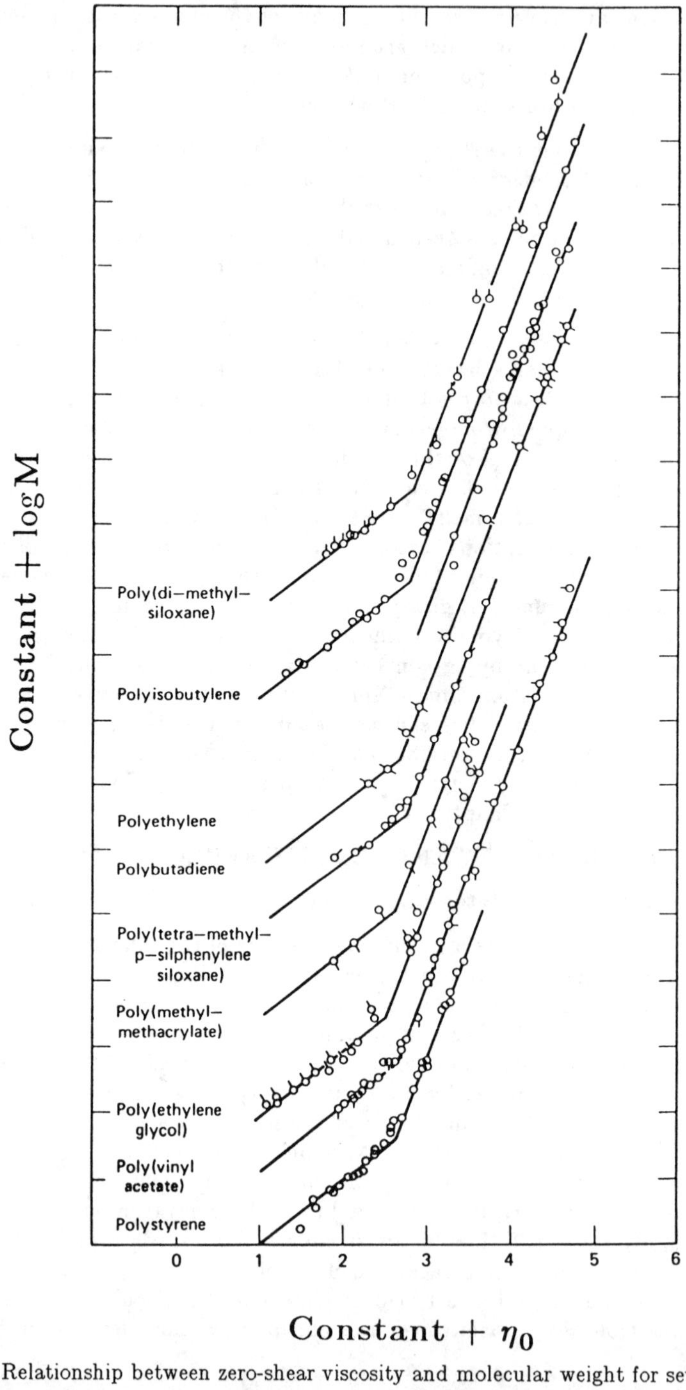

Fig. 4-2 Relationship between zero-shear viscosity and molecular weight for several melts of narrow polydispersity (from Berry and Fox,[10] reproduced with permission from Advances in Polymer Science).

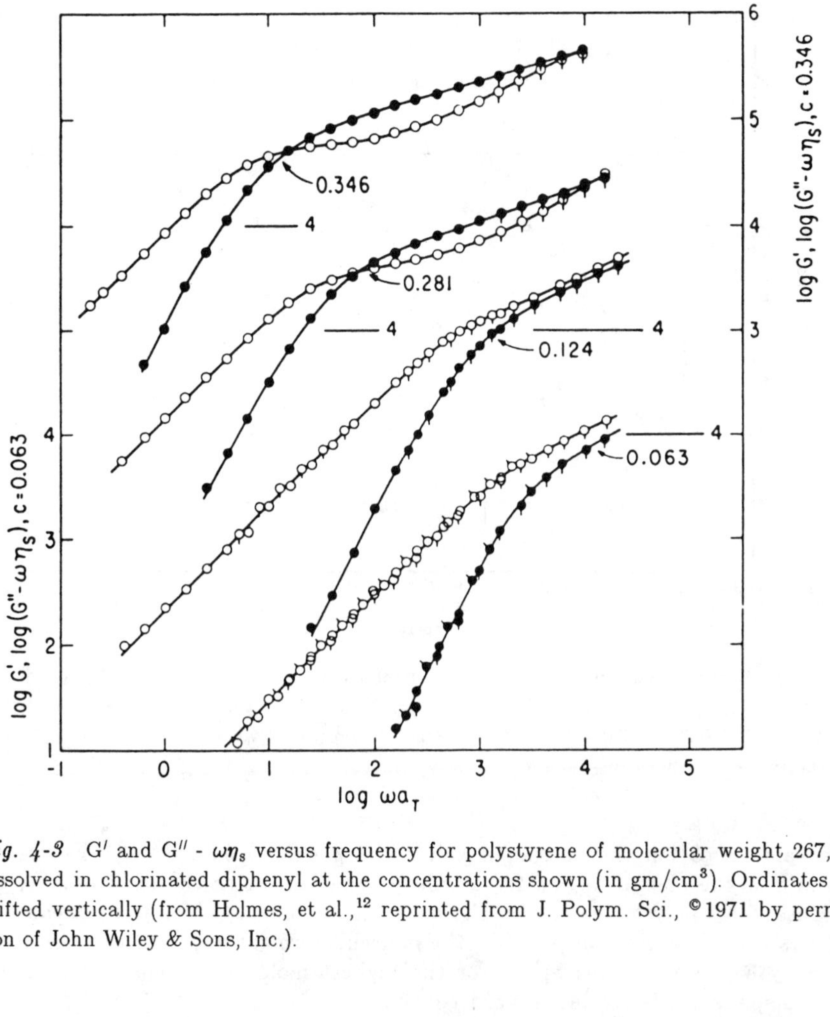

Fig. 4-3 G′ and G″ - $\omega\eta_s$ versus frequency for polystyrene of molecular weight 267,000 dissolved in chlorinated diphenyl at the concentrations shown (in gm/cm^3). Ordinates are shifted vertically (from Holmes, et al.,[12] reprinted from J. Polym. Sci., ©1971 by permission of John Wiley & Sons, Inc.).

If a solvent is added to the melt, the transition can be pushed to higher molecular weights. The dependence of M_c on concentration suggests that the transition is controlled by the degree of interpenetration of the polymer coils. If so much solvent is added that the coils do not overlap at all, then Rouse behavior gives way to the "Zimm" regime. In the Zimm[11], regime, discussed in Chapter 8, hydrodynamic interaction is important. Thus Rouse behavior occurs at intermediate concentrations. Figure 4-3 shows G′ and G″ for a polystyrene solution at four different concentrations. At the low concentration, c = 0.063 gm/cm^3, the coils do not overlap and Zimm behavior prevails. In this regime, $G'' = \sqrt{3} = G' \propto \omega^{2/3}$ at frequencies above the terminal regime. When c = 0.124, the coils overlap and the solution follows the Rouse theory, with $G' = G'' \propto \omega^{1/2}$ at high frequencies. At c = 0.281 and above, the coils are highly overlapping and a new regime appears at intermediate frequencies; in this regime, G′ is insensitive to ω, and G″ *crosses over* G′. This is called the *plateau region*.

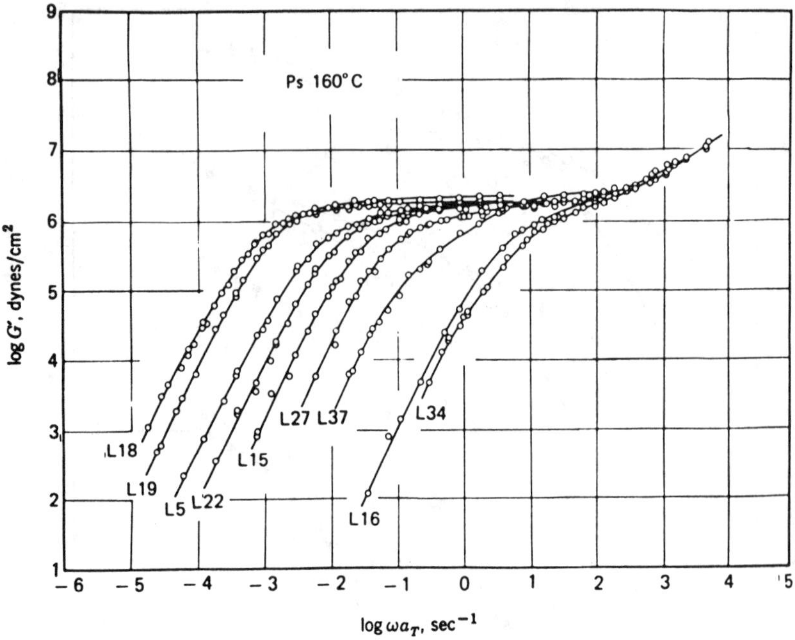

Fig. 4-4 Storage modulus G′, as a function of frequency reduced to 160°C for polystyrenes of narrow molecular-weight distribution. Molecular weight decreases from left to right from 580,000 to 47,000 (from Onogi et al.,[13] reprinted with permission from Macromolecules ©1970 American Chemical Society).

The value of G′ on the plateau is called the *plateau modulus*, G_N^0. Figure 4-4 shows how the plateau region for a bulk polymer grows as the molecular weight increases for a polystyrene melt. The molecular weight at which the plateau first appears roughly corresponds to M_c, the molecular weight at which the zero-shear viscosity begins to rise as $M^{3.4}$. For the highest molecular weight, 600,000, the plateau spans nearly six decades of frequency!

4.3.2 Meaning of the Plateau

The plateau in G′ implies that there is little relaxation occurring within this range of frequencies. This dearth of relaxation processes is also evidenced by the loss modulus, G″, which shows a minimum near the center of the plateau; see Figure 4-5. In this frequency range, the melt acts like an elastic rubber! For a perfect rubber, G′ is constant and G″ is zero. Unlike perfect rubbers, high-molecular-weight melts do relax finally, but λ_1, the longest relaxation time, increases rapidly with M. Both η_0 and λ_1 are proportional to $M^{3.4}$. Thus the slowest relaxation process — that is, the relaxation of the chain as a whole — is greatly retarded when M becomes large.

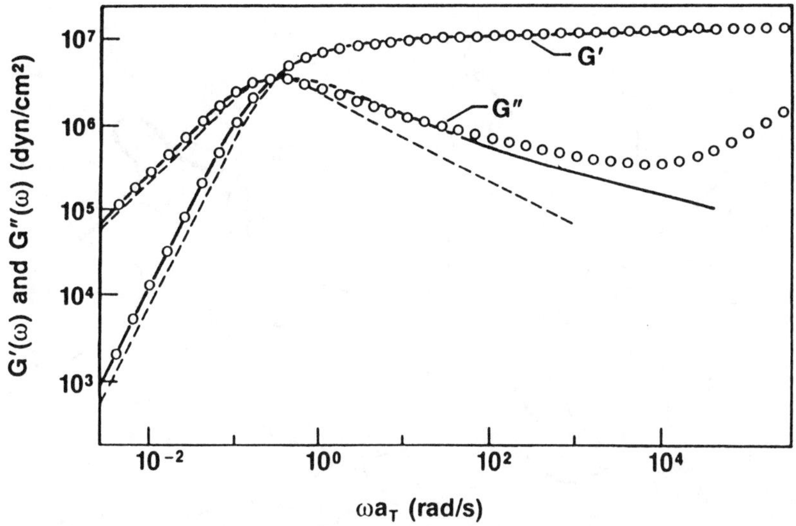

Fig. 4-5 G′ and G″ versus frequency shifted to 27°C for polybutadiene of molecular weight 360,000 and narrow polydispersity. The dashed line is the reptation theory according to Equation 4-13; the solid line includes fluctuations in the length of the primitive path (from Pearson[14])

It is now generally accepted that the chain's relaxation is greatly retarded because it becomes *entangled* with other chains. A precise definition of an *entanglement* remains elusive, but it is understood to be related to the *topological constraints* on a chain that prevent it from passing through another chain.[3,15] The idea of an entanglement was captured, in a crude way, by Green and Tobolsky's idea of a temporary cross-link; see Chapter 2. This simple model of Green and Tobolsky predicts a plateau in G′, but makes no prediction of the dependence of λ_1 on M. G′ for the Green and Tobolsky model is

$$G' = \nu kT \frac{1}{1 + \lambda^2\omega^2} \tag{4-11}$$

The plateau modulus, G_N^0, is here νkT, where ν is the number of entangled strands per unit volume. Figure 4-4 shows that G_N^0 is roughly independent of molecular weight, and from its magnitude one can estimate ν. From ν and the bulk density of the polymer, one can obtain the molecular weight of a typical entangled strand. This quantity is called the *molecular weight between entanglements*, M_e. It turns out that $M_e \sim M_c/2$.

The picture that emerges from these observations is as follows. Portions of the chain smaller than M_e relax rapidly by Rouse or glassy modes. Larger portions are trapped by entanglements, and take a much longer time, $\lambda_1 \propto M^{3.4}$, to relax.

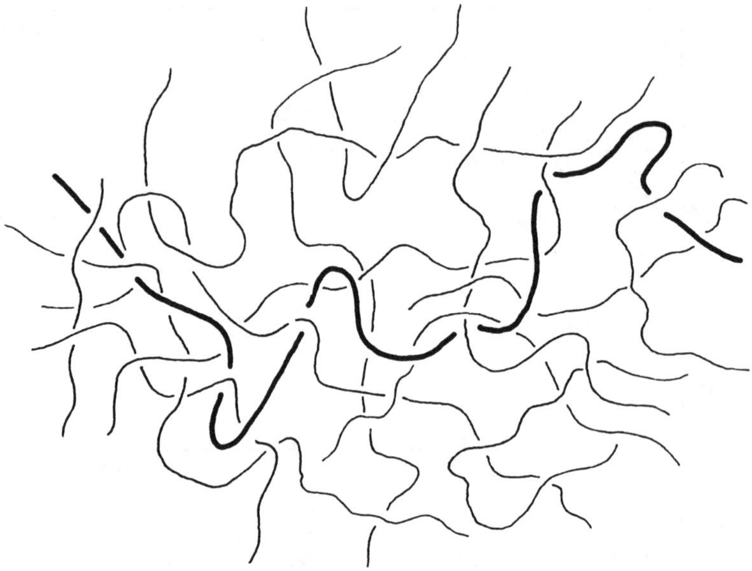

Fig. 4-6 Polymer molecule entangled in a mesh of other polymer chains (from Graessley,[18] reprinted with permission from Advances in Polymer Science).

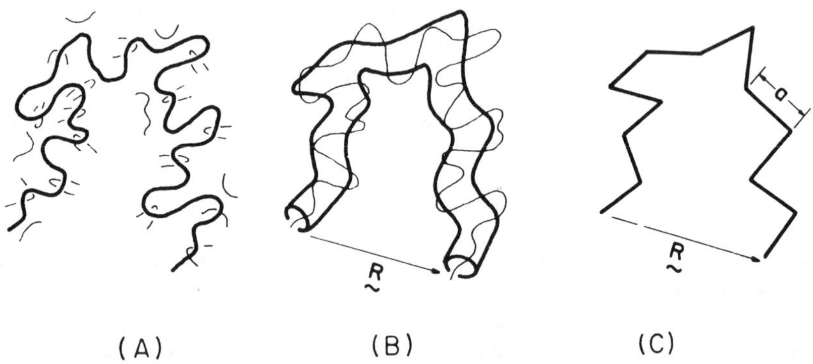

Fig. 4-7 Polymer molecules with which a given molecule is entangled (A) confine the given chain to a tube-like region (B). The tube contour (C) is roughly that of a random walk with step size equal to the tube diameter a. This random walk is called the primitive path; its contour length is much less than the contour length of the chain (From Graessley,[18] reprinted with permission from Advances in Polymer Science).

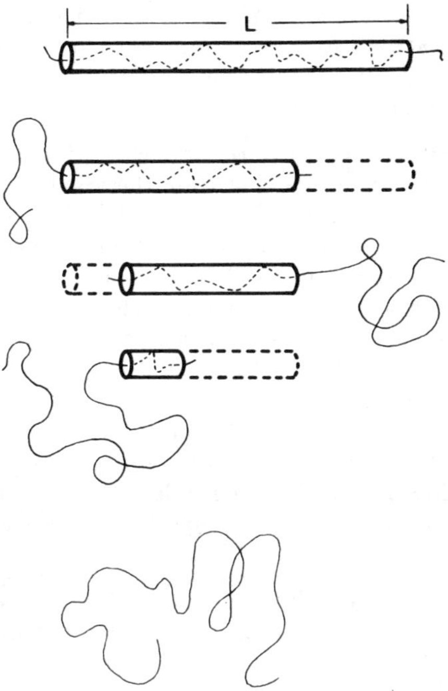

Fig. 4-8 Reptative diffusion of a polymer molecule out of its tube. For simplicity of visualization, the tube of Fig. 4-7 has been straightened out (From Graessley[18]).

4.4 THE DOI-EDWARDS CONSTITUTIVE EQUATION

4.4.1 Reptation

The simple model of Green and Tobolsky captures the idea of an entanglement spacing, but gives no explanation for the dependence, $\lambda_1 \sim M^{3.4}$. An explanation, at least a partial one, was provided by de Gennes[16] in 1971. He considered a simpler problem, that of a single long polymer chain lodged in a cross-linked rubber network; see Figure 4-6. Since the chain cannot cross the rubber strands, which are themselves held in place by chemical cross-links, the chain's lateral motion is limited. De Gennes called this motion *reptation*. The mesh of constraints confines the molecule laterally to a *tubelike region*,[17] Figure 4-7. The chain alters its overall conformation by sliding first one end out of the tube, then the other. When an end of the tube is vacated, it is forgotten; the chain end need not return to the same set of constraints when it reverses direction. By moving back and forth, the chain gradually forgets more and more of the original tube ends; a "new" conformation diffuses from the ends of the chain inward; see Figure 4-8. Since it is a diffusion process, the time required for the chain to vacate the original tube is proportional to the square of the *contour length* of the tube divided by the rate of the snaking motion. The square of the tube's contour length is proportional to M^2. The drag force resisting the snaking motion of the chain as a whole is proportional to M; the rate of snaking motion therefore goes as M^{-1}. Thus the *reptation time*, the time for escape from the tube, is proportional to $M^2/M^{-1} = M^3$. This is the time required for relaxation of an oriented configuration and hence is the largest relaxation time, λ_1.

The longest relaxation time measured in melts is found to scale as $\lambda_1 \propto M^{3.4}$. Although this result deviates somewhat from the scaling predicted for a polymer in a network, other related experiments, such as diffusivity measurements, (discussed briefly later) are very close to de Gennes' prediction for a single chain in a rubber network; thus the reptation process is thought to control relaxation in the melt. In the melt the molecules surrounding a chain form a "tube" of constraints that block lateral motion.[17] The diameter, a, of this tube must correspond to the entanglement spacing, M_e. That is, a strand of polymer having molecular weight M_e spans a random walk end-to-end distance a, Figure 4-7. The tube itself is a random walk, each step of which has length a. This random walk is called the "primitive path" of the chain. The contour length of the tube, or the primitive path, is therefore aM/M_e, which is much less than the contour length of the chain. The tube diameter, a, is large enough to allow the chain wide latitude for meandering about the primitive path, Figure 4-7.

De Gennes[16] worked out the dynamics of the reptation of the chain and found that the fraction of the original tube that remains unvacated, as a function of time, is

$$P(t) = \sum_{i\,\mathrm{odd}} \frac{8}{\pi^2 i^2} \exp[-i^2 t/\lambda^2] \qquad (4\text{-}12)$$

The relaxation modulus is just this fraction times G_N^0:

$$G(t) = \sum_{i\,\mathrm{odd}} G_i \exp[-t/\lambda_i]\,, \quad G_i = 8G_N^0/\pi^2 i^2\,, \quad \lambda_i = \lambda_1/i^2 \qquad (4\text{-}13)$$

From G(t), G' and G'' can be obtained. The result, shown in Figure 4-5, bears a resemblance to the measured G' and G'', except at high frequency, where processes other than reptation occur.

4.4.2 Nonlinear Modulus

The prediction of the linear relaxation spectrum for melts and concentrated solutions of high molecular weight is a major accomplishment. To create a complete constitutive equation, however, we must understand the nonlinear response. Crucial insights were provided by the step shear experiments of Osaki and coworkers.[19] Figure 4-9a shows the time-dependent modulus, $G(t,\gamma)$ — that is the stress divided by the strain — for various step strains, γ, imposed on a concentrated polystyrene solution. It therefore diffuses by snaking back and forth along its own contour. At times larger than a limiting time, λ_k, after the step, $G(t,\gamma)$ is factorable into time- and strain-dependent functions: $G(t,\gamma) = G(t)h(\gamma)$. This can be seen in Figure 4-9b, where vertical shifting brings the relaxation moduli for various γ into coincidence at times greater than $\lambda_k \sim 20$ sec. The damping function, $h(\gamma)$, shown in Figure 4-10 shows considerable *shear thinning*; that is, h decreases with increasing γ. The Lodge equation, based on the temporary network model of Green and Tobolsky, predicts that $h(\gamma) = 1$ for all γ (i.e., no shear

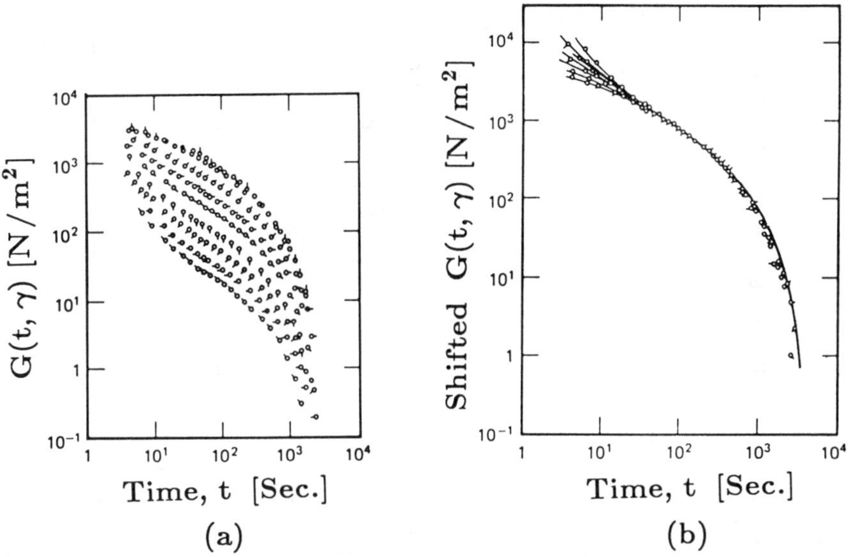

Fig. 4-9 (a) $G(t, \gamma) = \sigma_{12} / \gamma$ as a function of time for various strain magnitudes for concentrated polystyrene solutions. Each curve represents a different strain; the lower curves correspond to higher strains. (b) The curves can be superimposed at times longer than about 20 sec. by vertical shifting by an amount, $h(\gamma)$, that depends on strain (from Einaga et al.,[19] reprinted with permission from Polymer J.)

thinning). Osaki[21] speculated that since other melts and concentrated solutions with low polydispersity show the same dependence of h on γ as that depicted in Figure 4-10, this relationship must reflect material-independent "geometric factors."

The tube model suggests a mechanism to explain the shear thinning in $h(\gamma)$ observed by Osaki. Suppose the melt is rapidly deformed and that the tube and the chain it contains are stretched affinely. In the temporary network model, the strands would remain stretched until they escape their entanglements; see Figure 4-11 (top). In the tube model, Figure 4-11 (bottom), the tube prevents any rapid relaxation of the molecule by *lateral motion* but *does not inhibit* **retraction** *of the molecule* along the tube contour. Since retraction does not violate the tube constraints, it occurs quickly compared to reptation. The short-time nonfactorable relaxation seen in Figure 4-9 is interpreted as a manifestation of the retraction. At times larger than λ_k, retraction is complete, and only reptation can complete the relaxation. The reduction of stress brought about by the retraction process leads to the shear thinning evident in $h(\gamma)$.

The retraction mechanism was suggested by Doi and Edwards in a seminal series of papers[17,20,23,24] in which they derived a complete constitutive equation based on the reptation mechanism and the tube picture. Doi and Edwards

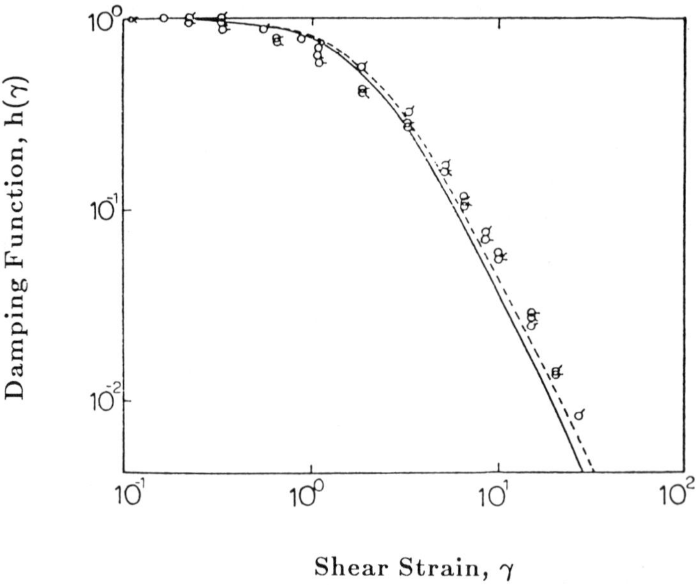

Fig. 4-10 Damping function h(γ) obtained by the vertical shifting depicted in Fig. 9. The data are from Fukuda et al.[21] The solid line is the prediction of Doi and Edwards with the independent alignment approximation; the dashed line is without this approximation (from Doi and Edwards,[20] reproduced with permission from J. Chem. Soc., Faraday Trans. II).

quantified the shear thinning produced by retraction by arguing that the chain *retracts enough to bring the contour length of its primitive path back to its equilibrium value* aM/M_e.

Because of retraction, when the sample is deformed, a given portion of chain does not remain associated with the same portion of tube. The deformation, however, is assumed to act on the constraints — that is, *on the tube* — and not directly on the chain. Since a strand moves from one part of the tube to another by retraction, the strand's orientation is determined not by the orientation of the part of the tube it originally occupied, but by the orientation of the part of the tube into which it moves. It is therefore very difficult in a general deformation history to determine how a given strand is oriented. Doi and Edwards avoided this difficulty by making a simplifying approximation, namely that each *strand deforms independently*. This is called the *independent alignment approximation*. Since the retraction process keeps the overall primitive path length constant, each step in the primitive path must return to the same length, a, after retraction. The effect of the deformation, therefore, is to *orient* strands *without stretching them*. Since rigid rods also respond to a deformation by rotating without stretching, the Doi-Edwards (DE) theory for melts bears a resemblance to theories for rigid-rod molecules.[25] The strand motion is therefore *nonaffine*.

One strand in the primitive path has end-to-end vector, $a\underset{\sim}{u}'$, with $\underset{\sim}{u}'$ a unit

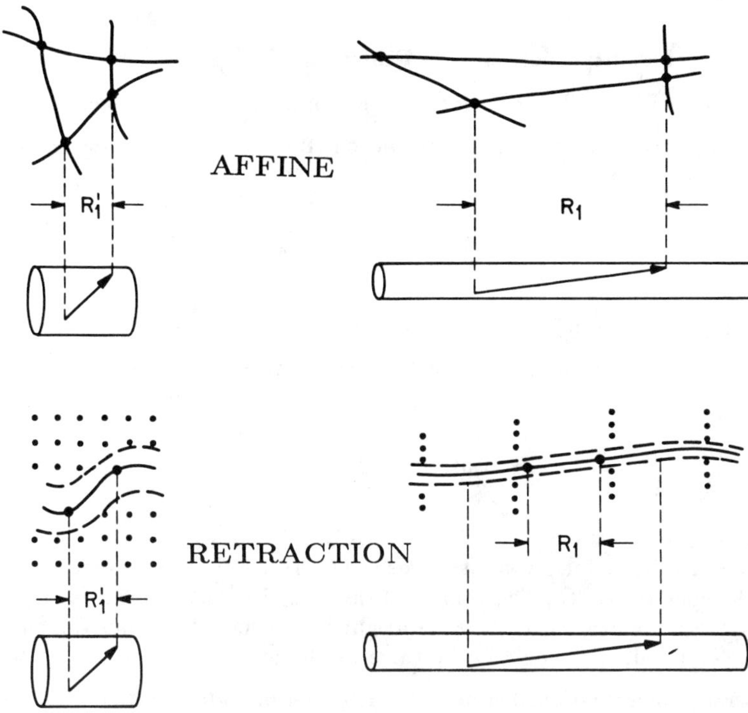

Fig. 4-11 *top*: Affine response of an entangled strand to a deformation according to the temporary network model. *bottom*: Polymer molecule is confined to a tubelike region by surrounding mesh of polymer molecules. After a deformation, the strand rapidly retracts within the deformed tube according to the Doi-Edwards model. The final strand end-to-end length equals its initial end-to-end length. (from Larson,[22] reprinted from the Journal of Rheology ©1984 Society of Rheology, by permission of John Wiley & Sons, Inc.).

vector. Following a step deformation, but before retraction, this end-to-end vector becomes

$$a\underset{\sim}{u}' \cdot \underset{\approx}{F}^{-1} \tag{4-14}$$

where $\underset{\approx}{F}^{-1}$ is the imposed deformation gradient. After retraction, the end-to-end distance must return to its original length, a. Therefore

$$a\underset{\sim}{u} = \frac{a\,\underset{\sim}{u}' \cdot \underset{\approx}{F}^{-1}}{|\underset{\sim}{u}' \cdot \underset{\approx}{F}^{-1}|} \tag{4-15}$$

where $a\underset{\sim}{u}$ is the end-to-end vector of the strand after the deformation; $\underset{\sim}{u}$ is a unit vector.

4.4.3 The Probability Distribution Function

If the melt is in equilibrium before the deformation, then the vector $\underset{\sim}{u}'$ takes on a random orientation. The distribution function, $\psi_o(\underset{\sim}{u}')$, for randomly oriented vectors is

$$\psi_0(\underset{\sim}{u}')\,d\Omega' = \frac{1}{4\pi}d\Omega' \tag{4-16}$$

where $\underset{\sim}{u}'$ resides in the infinitesimal solid angle $d\Omega'$. As a result of the deformation, all vectors between $\underset{\sim}{u}'$ and $\underset{\sim}{u}' + d\Omega'$ are rotated to new orientations between $\underset{\sim}{u}$ and $\underset{\sim}{u} + d\Omega$:

$$\psi(\underset{\sim}{u})d\Omega = \psi_0(\underset{\sim}{u}')\,d\Omega' \tag{4-17}$$

Consider a cone-shaped volume element of the continuum, where the solid angle at the cone tip is $d\Omega'$, The volume of this cone is $r'^3 d\Omega'$, where r' is the cone height. After the deformation, the cone height is deformed to r, and the angle to $d\Omega$. This deformed cone angle is the same as the angle in which the vectors $\underset{\sim}{u}$ reside after the deformation, because the deformation *rotates* strand end-to-end vectors affinely. Because the volume of the continuum is unchanged by the deformation, the cone volume remains constant; therefore

$$\frac{1}{3}r^3 d\Omega = \frac{1}{3}r'^3 d\Omega' \tag{4-18}$$

Since the continuum is deformed affinely, r is related to r′ by

$$r = |\underset{\sim}{r}' \cdot \underset{\approx}{F}^{-1}| = r'|\underset{\sim}{u}' \cdot \underset{\approx}{F}^{-1}| \tag{4-19}$$

From Equations 4-17—4-19,

$$\psi(\underset{\sim}{u}) = \psi_0(\underset{\sim}{u}')\frac{d\Omega'}{d\Omega} = |\underset{\sim}{u}' \cdot \underset{\approx}{F}^{-1}|^{-3} \tag{4-20}$$

This simple derivation of $\psi(\underset{\sim}{u})$ was first presented by Marrucci and Grizzuti.[26] An equivalent expression was obtained earlier by Curtiss and Bird.[27]

4.4.4 The Free Energy and the Stress Tensor

The expression for the free energy, $W = 2kT\beta^2 \langle R^2 \rangle$, used in the temporary network model is a constant for the DE model, because the strands do not stretch; R^2 is not affected by the deformation. This term, which arises from "spring" forces in the molecules, will therefore not contribute to the stress tensor. For the DE model, the *combinatoric term*, Equation 2-14, gives the only contribution[26,28] to $\underset{\approx}{\sigma}$:

$$W = \nu kT <\log\psi> = -3\nu kT <\log|\underset{\sim}{u}'\cdot\underset{\approx}{F}^{-1}|>_0 + \text{constant} \tag{4-21}$$

Here the average over configuration space, $<\cdot>$, reduces to an average over the orientation distribution for $\psi(\underset{\sim}{u})$:

$$<\cdot> = \int \psi(\underset{\sim}{u}) \cdot d\Omega \tag{4-22}$$

The "·" denotes the quantity being averaged. Because of Equation 4-17, the above is equal to an average over the equilibrium distribution function, $<\cdot>_0$,

$$<\cdot> = <\cdot>_0 = \int \psi_0 \cdot d\Omega_0 = \frac{1}{4\pi}\int_0^{2\pi}\int_0^{\pi} \cdot \sin\theta\, d\theta\, d\phi \tag{4-23}$$

where θ is the polar angle and ϕ is the azimuthal angle.

W in Equation 4-21 is a function of the deformation. Because of arguments presented in Chapter 3, we know that it must be a function only of the strain invariants I_1 and I_2 or of the principal stretch ratios, λ_1, λ_2, and λ_3. By considering an extensional deformation, and using arguments analogous to those of Chapter 3, one can show that the stress tensor corresponding to this free energy is

$$\underset{\approx}{\sigma} = 3\nu kT<\underset{\sim}{u}\underset{\sim}{u}> = 3\nu kT a^{-2} <\underset{\sim}{R}\underset{\sim}{R}> \tag{4-24}$$

where $\underset{\sim}{R} = a\underset{\sim}{u}$ is the end-to-end vector of a strand. Note that this expression for $\underset{\approx}{\sigma}$ obeys the stress-optic law, since $\underset{\approx}{n}$ in Equation 4-9 is also proportional to $<\underset{\sim}{R}\underset{\sim}{R}>$. This expression is the same as that of the temporary network model except for the coefficient $3\nu kTa^{-2}$. The *dynamics* of the response of a strand to an "instantaneous" deformation history are different, however. (Here "instantaneous" means fast compared to reptation but slow compared to retraction.) In the temporary network model, $\underset{\sim}{R} = \underset{\sim}{R}'\cdot\underset{\approx}{F}^{-1}$, where $\underset{\sim}{R}$ and $\underset{\sim}{R}'$ are the strand's end-to-end vectors before and after the deformation. In the DE theory, $\underset{\sim}{R} = \underset{\sim}{R}'\cdot\underset{\approx}{F}^{-1}/|\underset{\sim}{u}'\cdot\underset{\approx}{F}^{-1}|$; the strand's length remains constant because of retraction. Thus in the DE theory

$$\underset{\approx}{\sigma} = 3\nu kT <\underset{\sim}{u}\underset{\sim}{u}> = 3\nu kT < \frac{\underset{\sim}{u}'\cdot\underset{\approx}{F}^{-1}\,\underset{\sim}{u}'\cdot\underset{\approx}{F}^{-1}}{|\underset{\sim}{u}'\cdot\underset{\approx}{F}^{-1}|^2} >_0$$

$$= \frac{3}{5}\nu kT\underset{\approx}{Q} \tag{4-25}$$

where

$$\underset{\approx}{Q} = 5 < \frac{\underset{\sim}{u'} \cdot \underset{\approx}{F}^{-1} \underset{\sim}{u'} \cdot \underset{\approx}{F}^{-1}}{|\underset{\sim}{u'} \cdot \underset{\approx}{F}^{-1}|^2} >_0 \tag{4-26}$$

$\underset{\approx}{Q}$ is the "geometric" ***universal tensor*** sought by Osaki. The factor of 5 is introduced into it so that $\underset{\approx}{Q}$ approaches the linear viscoelastic strain, $\underset{\approx}{\gamma}$, when the strain is small. $G_N^0 = \frac{3}{5}\nu kT$ is then the plateau modulus.

For a given deformation, one can obtain $\underset{\approx}{Q}$ by inserting the appropriate tensor $\underset{\approx}{F}^{-1}$ and carrying out the angle averages by numerical integration.[24] In shear, Q_{12}, the shear component of $\underset{\approx}{Q}$ can be expressed as $\gamma h(\gamma)$, where the "damping function" $h(\gamma)$ is the nonlinearity in the relationship between shear stress and strain. It is plotted in Figure 4-10. The astounding agreement with Osaki's experimental $h(\gamma)$ is powerful evidence for the correctness of the DE model. Figure 4-10 also shows the damping function obtained when the independent alignment approximation is not made. For single-step strains, the affect of this approximation is small.

4.4.5 The Constitutive Equation

Equation 4-25 gives the stress following the step deformation and the retraction of the chains, but before the final relaxation — that is before reptation. The effects of reptation after or during deformation can be taken into account using an argument analogous to that in Chapter 2 for the Green and Tobolsky model. Let us calculate the fraction, $P_{t',t}$, of strands that last vacated their "old" tubes at a past time, t′, to take on a "new" initially random orientation, which has since been oriented by the deformation. First consider the fraction of strands that entered their ***present*** tubes at time t′ or before; this is equal to the fraction that survive without escaping their tubes for an interval of time t − t′ or longer. From Equation 4-12, this fraction is just $P(t-t')$. Of these, only $P_{t',t} = (dP/dt')\,dt'$ entered the present tube at time t′ to within dt′. From Equation 4-12,

$$P_{t',t} = \sum_{i\,\mathrm{odd}} \frac{G_i}{\lambda_i} \exp[-(t-t')/\lambda_i] \equiv m(t-t') \tag{4-27}$$

According to Equation 4-25, these strands contribute to the present stress the increment

$$d\underset{\approx}{\sigma} = 3\nu kT P_{t't} < \frac{\underset{\sim}{u'} \cdot \underset{\approx}{F}^{-1}(t',t)\, \underset{\sim}{u'} \cdot \underset{\approx}{F}^{-1}(t',t)}{|\underset{\sim}{u'} \cdot \underset{\approx}{F}^{-1}(t',t)|^2} >_0 = G_N^0 P_{t't}\, \underset{\approx}{Q}(t',t) \tag{4-28}$$

Adding up the contributions from all strands gives the complete stress tensor

$$\underset{\approx}{\sigma} = G_N^0 \int_{-\infty}^{t} m(t-t')\, \underset{\approx}{Q}(t',t)\, dt' \tag{4-29}$$

Since $\underset{\approx}{Q}$ was derived from a free energy, Equation 4-29 is a *separable Kaye-BKZ equation.*[23] W in Equation 4-21 is the potential function $U(I_1,I_2)$. Thus

$$\underset{\approx}{\sigma} = \int_{-\infty}^{t} m(t-t')\underset{\approx}{E}(t,t')\, dt' \tag{4-30}$$

where

$$\underset{\approx}{E} = 2\frac{\partial U}{\partial I_1}\underset{\approx}{C}^{-1} - 2\frac{\partial U}{\partial I_2}\underset{\approx}{C} = \underset{\approx}{Q} \tag{4-31}$$

4.4.6 Premises of the Doi-Edwards Model

The crucial assumptions of the model of Doi and Edwards are:

1. Surrounding chains form a mesh of entanglements that constrain a given chain to remain within an average distance, a, of a primitive path. The constraints relax slowly compared to the relaxation time of the chain.

2. The chain rapidly retracts during deformation so that the contour length of its primitive path remains constant.

3. Relaxation occurs by reptation only.

In addition, there is a simplification, the independent alignment approximation, that makes possible the derivation of a closed form for the constitutive equation:

4. Each step in the primitive path of the chain orients independently of the other steps.

4.5 APPROXIMATIONS TO THE DOI-EDWARDS EQUATION

4.5.1 Currie's Potential

As it stands, the DE equation is unwieldy because $U(I_1,I_2)$ must be obtained numerically for each deformation. Currie, however, found an accurate analytic approximation,[28]

$$U(I_1,I_2) \approx \frac{5}{2}\log\left(\frac{J-1}{7}\right) \tag{4-32}$$

with

$$J \equiv I_1 + 2\sqrt{I_2 + 13/4} \tag{4-33}$$

From this potential function, one obtains

$$\underset{\approx}{Q} \approx \left[\frac{5}{J-1}\right]\underset{\approx}{C}^{-1} - \left[\frac{5}{(J-1)(I_2+13/4)^{1/2}}\right]\underset{\approx}{C} \tag{4-34}$$

4.5.2 Larson's Potential

A simpler and cruder approximation is[29]

$$U \sim \frac{5}{2}\log(1+\frac{1}{5}(I_1-3)) \tag{4-35}$$

$$\underset{\approx}{Q} \sim \frac{\underset{\approx}{C}^{-1}}{1+\frac{1}{5}(I_1-3)} \tag{4-36}$$

This latter approximation highlights the difference between the DE equation and the temporary network model of Green and Tobolsky. The strain dependence of the DE equation is "softer" than that of the temporary network model, roughly by the factor $1+3(I_1-3)/5$. The Doi-Edwards model also makes a specific prediction for the relaxation spectrum, Equation 4-13; the Green and Tobolsky (GT) model does not. Doi and Edwards noted, however, that the longest relaxation mode accounts for most of the relaxation of the DE equation; it accounts for 96% of the zero-shear viscosity.[24] Thus the *Doi-Edwards model is nearly a single-relaxation-time model,* as was the original temporary network model of Green and Tobolsky. The models of Doi and Edwards and of Green and Tobolsky are also much alike in that both envision strands trapped by entanglements either between temporary cross-links (GT), or in a tube (DE). When escape occurs, the strand in either model *suddenly* takes on a configuration typical of equilibrium.

The approximation, Equation 4-36, to the DE $\underset{\approx}{Q}$ tensor is reasonably accurate for most purposes. Its most significant failing is that it predicts a zero second normal stress difference in shear, whereas $N_2/N_1 = -2/7$ for the DE equation at low shear rates. Currie's approximation gives the correct prediction for N_2/N_1.

4.5.3 Approximation Based on the Seth Elastic Strain Measure

Another approximation to $\underset{\approx}{Q}$ that gives the correct normal stress ratio, N_2/N_1, at low shear rates is

$$\underset{\approx}{Q} \approx \frac{\frac{7}{3}\underset{\approx}{C}^{-3/7}}{1+\frac{7}{15}(\mathrm{tr}\underset{\approx}{C}^{-3/7}-3)} \tag{4-37}$$

Here $\underset{\approx}{C}^{-3/7}$ is the Finger tensor, $\underset{\approx}{C}^{-1}$, raised to the 3/7 power. A nonintegral power, $\underset{\approx}{C}^{-a}$, of a symmetric tensor is obtained as follows. First rotate the principal axes of $\underset{\approx}{C}^{-1}$ so that $\underset{\approx}{C}^{-1}$ is diagonalized (Chapter 3 explains how this is done.):

$$\hat{\underset{\approx}{C}}^{-1} = \begin{pmatrix} \lambda_1^2 & 0 & 0 \\ 0 & \lambda_2^2 & 0 \\ 0 & 0 & \lambda_3^2 \end{pmatrix} \tag{4-38}$$

λ_i is a stretch ratio. Next raise each diagonal element to the power a:

$$\hat{\underset{\approx}{C}}^{-1} = \begin{pmatrix} \lambda_1^{2a} & 0 & 0 \\ 0 & \lambda_2^{2a} & 0 \\ 0 & 0 & \lambda_3^{2a} \end{pmatrix} \tag{4-39}$$

Finally rotate back to the original frame to obtain $\underset{\approx}{C}^{-a}$. $a^{-1}\underset{\approx}{C}^{-a}$ is called the *Seth* tensor;[30] the potential function corresponding to it is

$$U = \frac{1}{2a} \mathrm{tr} \underset{\approx}{C}^{-a} \tag{4-40}$$

The Seth tensor has been used to describe rubbers, and Chang, Bloch, and Tschoegl[31] based an integral constitutive equation for viscoelastic liquids on it. Their equation is

$$\underset{\approx}{\sigma} = \int_{-\infty}^{t} m(t-t') \frac{1}{a} \underset{\approx}{C}^{-a}(t,t')\, dt' \tag{4-41}$$

The ratio N_2/N_1 for this equation at low shear rates is $-(1-a)/2$. Equation 4-41 can be considered a simple generalization of the Lodge equation. Like the Lodge equation, Equation 4-41 has steady extensional viscosities that become infinite at finite extension rates, although these singularities are pushed out to higher strain rates as a is decreased. The tensor in Equation 4-37, like the DE $\underset{\approx}{Q}$ tensor, avoids extensional singularities.

4.5.4 Differential Approximation

Marucci[32] and Larson[22] independently proposed a differential approximation to the Doi-Edwards equation. It is based on the observation that the end-to-end vector, $\underset{\sim}{R}$, of a strand that rotates in a deformation but does not extend follows the differential equation

$$\dot{\underset{\sim}{R}} = \underset{\sim}{R} \cdot \underset{\sim}{\nabla} \underset{\sim}{v} - \underset{\approx}{D} : \underset{\sim}{u}\underset{\sim}{u}\underset{\sim}{R} \tag{4-42}$$

where $\underset{\sim}{u} = \underset{\sim}{R}/R$. The first term on the right side of Equation 4-42 represents affine motion. The second subtracts a component parallel to $\underset{\sim}{R}$; the magnitude of this component is just large enough to cancel the strand's extension that would otherwise occur because of the affine term $\underset{\sim}{R} \cdot \underset{\sim}{\nabla} \underset{\sim}{v}$. Equation 4-42 can be integrated to give

$$\underset{\sim}{R} = \frac{\underset{\sim}{R}' \cdot \underset{\approx}{F}^{-1}}{|\underset{\sim}{u}' \cdot \underset{\approx}{F}^{-1}|} \tag{4-43}$$

where $\underset{\approx}{F}^{-1}$ is the cumulative deformation gradient. Here $\underset{\sim}{R}'$ is the initial strand

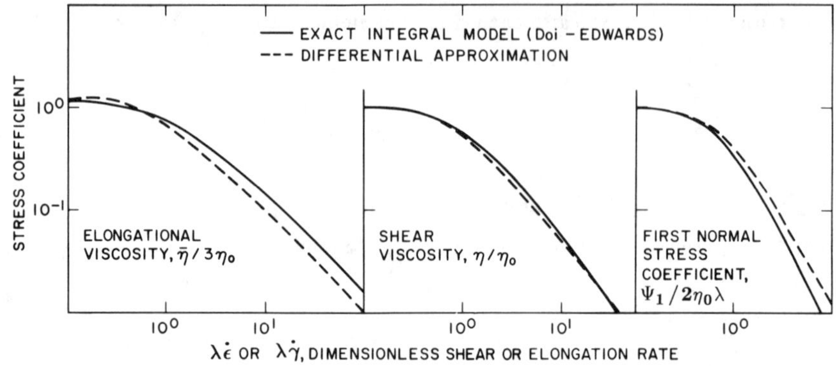

Fig. 4-12 Comparison of the Doi-Edwards integral equation with a differential approximation to it for steady-state elongational viscosity, shear viscosity, and first normal stress coefficient (from Larson,[22] reprinted from the Journal of Rheology © 1984 Society of Rheology, by permission of John Wiley & Sons, Inc.).

end-to-end vector; $\underset{\sim}{u}' \equiv \underset{\sim}{R}'/R'$. Equation 4-43 leads to the Doi-Edwards integral constitutive equation as discussed earlier. Equation 4-42 is the differential equivalent to Equation 4-43. The time derivative $d(R^2)/dt = \dot{\underset{\sim}{R}}\cdot\underset{\sim}{R} + \underset{\sim}{R}\cdot\dot{\underset{\sim}{R}}$ averaged over ψ, the configuration distribution function, is obtained from Equation 4-42: Thus

$$\frac{\partial}{\partial t}\langle \underset{\sim\sim}{RR}\rangle = \underset{\sim\sim}{\nabla v}^T\cdot\langle \underset{\sim\sim}{RR}\rangle + \langle \underset{\sim\sim}{RR}\rangle\cdot\underset{\sim\sim}{\nabla v} - 2\underset{\approx}{D}\langle \underset{\sim\sim\sim\sim}{uuRR}\rangle \tag{4-44}$$

The first three of these terms can be grouped together to form the upper-convected time derivative of $\langle \underset{\sim\sim}{RR}\rangle$. In the last term, approximating the average of the product $\underset{\sim\sim}{uu}$ times $\underset{\sim\sim}{RR}$ by the product of the averages, we obtain

$$\langle \underset{\sim\sim\sim\sim}{uuRR}\rangle \approx \langle \underset{\sim\sim}{uu}\rangle\langle \underset{\sim\sim}{RR}\rangle = a^{-2}\langle \underset{\sim\sim}{RR}\rangle\langle \underset{\sim\sim}{RR}\rangle \tag{4-45}$$

where a is here the length of the strand. $3\nu kTa^{-2}\langle \underset{\sim\sim}{RR}\rangle$ is the stress according to Equation 4-24. Thus Equation 4-44, when multiplied by the $3\nu kTa^{-2}$, gives

$$\dot{\underset{\approx}{\sigma}} - \underset{\sim\sim}{\nabla v}^T\cdot\underset{\approx}{\sigma} - \underset{\approx}{\sigma}\cdot\underset{\sim\sim}{\nabla v} + \frac{2}{3G}\underset{\approx}{D}:\underset{\approx\approx}{\sigma\sigma} = \overset{\nabla}{\underset{\approx}{\sigma}} + \frac{2}{3G}\underset{\approx}{D}:\underset{\approx\approx}{\sigma\sigma} = \underset{\approx}{0} \tag{4-46}$$

where $G \equiv 3\nu kT$. Equation 4-46 describes the stress in the material as a result of deformation *without reptational relaxation.* It describes the *instantaneous response to a sudden strain or a step strain.*

When the effect of relaxation is included, the constitutive equation is

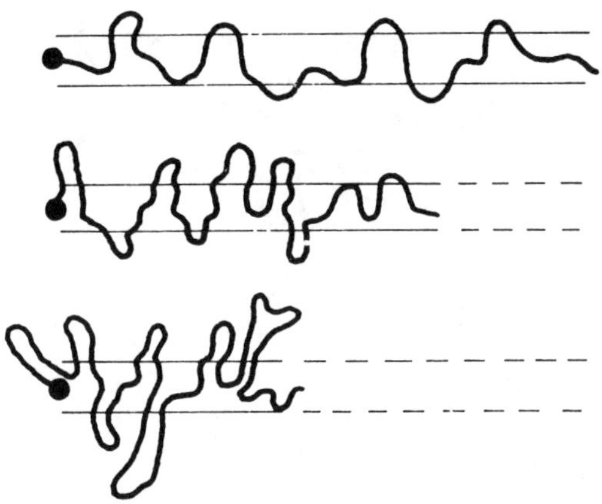

Fig. 4-13 Fluctuation of primitive path length occurs when a chain randomly pulls its end away from the end of the tube. Large fluctuations are unlikely, and relaxation by this process is much slower than by reptation. (from Graessley,[18] reprinted with permission from Advances in Polymer Science).

$$\overset{\nabla}{\underset{\approx}{\sigma}} + \frac{2}{3G}\underset{\approx}{D} : \underset{\approx}{\sigma}\underset{\approx}{\sigma} + \frac{1}{\lambda}(\underset{\approx}{\sigma} - G\underset{\approx}{\delta}) = \underset{\approx}{0} \tag{4-47}$$

This equation differs from the UCM equation only by the term $(2/3G)\underset{\approx}{D} : \underset{\approx}{\sigma}\underset{\approx}{\sigma}$. Its predictions are not far from those of the "exact" DE equation. Figure 4-12, for example, compares the elongational viscosity, shear viscosity, and first normal stress coefficient as functions of shear rate for the two equations. The difference between the two comes from the approximation made in Equation 4-45, which was necessary to obtain a differential equation in which all averages involving ψ are replaced by terms involving $\underset{\approx}{\sigma}$. This equation is discussed further in Chapters 5 and 7.

4.6 PREDICTIONS OF REPTATION THEORIES

The reptation model has been enormously popular, not only because of the details of the predictions that it makes but also because of the *scope* of the phenomena that it addresses. It offers an appealing framework not only for constitutive theories but for entangled polymer dynamics in general. Furthermore, the basic theory readily admits refinements that can often improve its predictions.

4.6.1 Molecular-Weight Dependence

4.6.1.1 Zero-Shear Viscosity The most crucial prediction of the theory is its prediction that the zero-shear viscosity and longest relaxation time scale as M^3. This scaling behavior is close enough to the measured $M^{3.4}$ dependence to lend credence to the reptation concept, but the deviation is significant and has attracted much attention. Additional mechanisms that could affect the shear viscosity include the following:

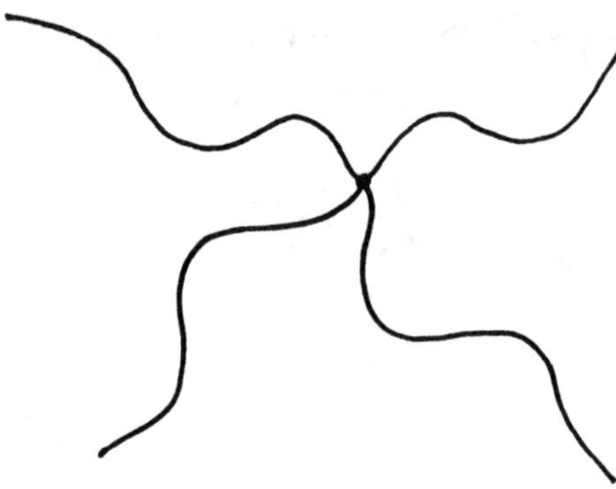

Fig. 4-14 A star molecule with four arms.

1. *tube renewal,* i.e, the release of topological constraints on a molecule by the movement of the *surrounding* molecules.[18,14] Although affecting the magnitude of the shear viscosity, this mechanism would apparently not change the M^3 dependence.[18]

2. *fluctuations of primitive path length,* i.e., escape from entanglements by a random shrinkage of the primitive path, Figure 4-13, followed by expansion into a new set of constraints.[33] These fluctuations effect not only the dependence of η_0 on M but also lead to predictions of $G'(\omega)$ and $G''(\omega)$ that are in better accord with experiment than is the simple Doi-Edwards model; see Figure 4-5.

4.6.1.2 Diffusivity The prediction that the center-of-mass diffusivity, D, of a polymer chain in a melt of like chains should be proportional to M^{-2} has been borne out experimentally.[34] The simultaneous prediction of reasonable dependencies of both η_0 and D on M is a major triumph of de Gennes' reptation model.

4.6.2 Relaxation Spectrum

4.6.2.1 Linear Molecules The reptation model predicts that for a nearly monodisperse melt the relaxation spectrum is dominated by a single relaxation time, λ_1. This is in reasonable accord with experimental data at low and moderate frequencies; see Figure 4-5. As the frequency increases, however, there is ever more deviation from the Doi-Edwards model. In the DE model, reptation is the only relaxation process considered. Any additional relaxation processes that might be present will tend to broaden the spectrum, thereby bringing $G'(\omega)$ and $G''(\omega)$ closer to the experimental curves. Inclusion of the process of tube length fluctuation, for example, removes some of the discrepancy between experiment and theory but does not account for the upturn in G'' at high ω. This upturn is apparently caused in part by Rouse-like motions of parts of the molecule that lie within a primitive path step. Sample polydispersity also broadens the relaxation spectrum.[35] Commercial melts are often so polydisperse that the maximum in G'' versus ω doesn't appear at all; see Figure 2-8 for example.

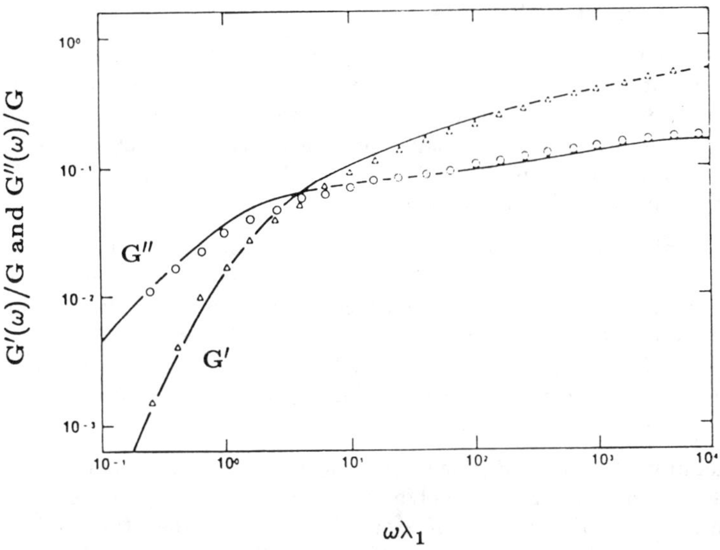

Fig. 4-15 Dimensionless storage modulus G'/G (triangles) and loss modulus G'' / G, (circles), versus dimensionless frequency for an 8-arm star, where λ_1 is the longest relaxation time. The modulus G was obtained by fitting the theory (lines) to the data (from Pearson and Helfand,[39] reprinted with permission from Macromolecules ©1984 American Chemical Society).

4.6.2.2 Star Molecules Ideas related to reptation can successfully predict the mechanical properties of molecules with unusual geometry. An example is the star polymer depicted in Figure 4-14. De Gennes[36] recognized that the arms of the star cannot reptate because they are anchored at the branch point. Relaxation thus occurs by slower processes. Doi and Kuzuu[37] proposed that stars relax by primitive path fluctuations, depicted in Figure 4-13. For complete relaxation to occur, the end of a branch must diffuse to the branch point, from whence it can re-extend into a new tube. This process takes a time that grows *exponentially* with the branch length. Thus Doi and Kuzuu predicted that $\log \eta_0 \underset{\sim}{\propto} M$. This prediction has been confirmed experimentally.[38] Pearson and Helfand analyzed the relaxation spectrum of stars that relax by primitive path fluctuations and found excellent agreement with experiment[38]; see Figure 4-15. These predictions, particularly the exponential growth of η_0 with M, are profoundly different from the corresponding predictions for linear molecules. The agreement obtained with experiment is therefore a striking confirmation of reptation ideas.

4.6.2.3 Rings Ring-shaped polymer molecules have also been studied to test the reptation model. Although they should not be able to relax by reptation or even by primitive path fluctuation, since they have no loose ends, rings have a dependence of η_0 on M that is similar to linear molecules.[39,40] This behavior is not yet understood and could conceivably demand a major reworking of reptation theory.

4.6.3 Nonlinear Viscoelasticity

An impressive feature of the Doi-Edwards model is that it has only two parameters that must be obtained by fits to rheological experiments. These parameters are the entanglement spacing a — which is determined by the plateau modulus G_N^o — and the longest relaxation time λ_1. Both parameters are fixed by linear viscoelastic measurements. Thus the nonlinear predictions of the DE equation involve *no adjustable parameters.*

4.6.3.1 Nonlinear Modulus As already noted, the nonlinear shear relaxation modulus, for linear molecules with little polydispersity, is in excellent agreement with the Doi-Edwards model at long times. There are molecular weight limitations, however. At very high M (e.g., M > 1,000,000 for polystyrene) the measured damping function, $h(\gamma)$, is drastically lower than the DE prediction.[19,41,42] It has been speculated that a mechanical instability is responsible for this deviation.[26,42]

The damping function, $h(\gamma)$, measured on star polymers follows the DE prediction, just as does $h(\gamma)$ for linear polymers (except at very high molecular weight). Although one end of each arm of a star is anchored, the other end is free to retract, leading to the same strain softening as in linear polymers.[14]

Damping functions have been measured for other kinds of step strains. In elongation of filaments of melt, large step strains usually rupture the filament, so comparisons are restricted to modest strains. In biaxial extension, however, large step strains can be executed. The DE model seems to describe the measured biaxial damping functions reasonably well, as is discussed in Chapter 7.

4.6.3.2 Steady-State Shear The Doi-Edwards model for monodisperse melts predicts extreme shear thinning, namely $\eta \propto \dot{\gamma}^{-1.5}$, at high shear rates. Thus as the shear rate increases the shear stress $\eta\dot{\gamma}$ passes through a maximum and then *decreases* with further increases in $\dot{\gamma}$. This means that there are *two* shear rates for each shear stress. Such circumstances would lead to mechanical instabilities in steady shearing that are not observed. Experimental results for melts of low polydispersity[43] show $\eta \propto \dot{\gamma}^{-0.82}$ at large $\dot{\gamma}$ rather than the predicted $\dot{\gamma}^{-1.5}$ dependence. At least some of the discrepancy certainly arises from the narrowness of the Doi-Edwards relaxation spectrum. As we saw in our discussion of the corotational Maxwell model in Chapter 3, any broadening of the relaxation spectrum will reduce the level of shear thinning. The presence of polydispersity and non-reptative relaxations will tend to bring the predicted shear viscosity into closer agreement with experiment. The same can be said for the predicted normal stress coefficients Ψ_1 and Ψ_2, which are also too shear thinning, $\Psi_1 \propto \dot{\gamma}^{-2}$, $\Psi_2 \propto \dot{\gamma}^{-2.5}$, at large $\dot{\gamma}$. Experimentally, Ψ_1 goes as $\dot{\gamma}^{-1.5}$ at large shear rates.[43]

4.6.3.3 Second Normal Stress Difference The DE equation gives an a priori prediction of the ratio Ψ_2/Ψ_1 that is apparently within a factor of 2 or so of experimental data. The predicted value of Ψ_2/Ψ_1 at low shear rates is $-2/7 = -0.29$. This particular prediction, however, is sensitive to the independent alignment approximation. When this approximation is removed, the DE model predicts[44] $\Psi_2/\Psi_1 = -1/7 = -0.14$ at low $\dot{\gamma}$. Most experimental measurements[6] of this ratio yield values in the range –0.05 to –0.14. The experiments in which the values were obtained, however, involved for the most part very polydisperse

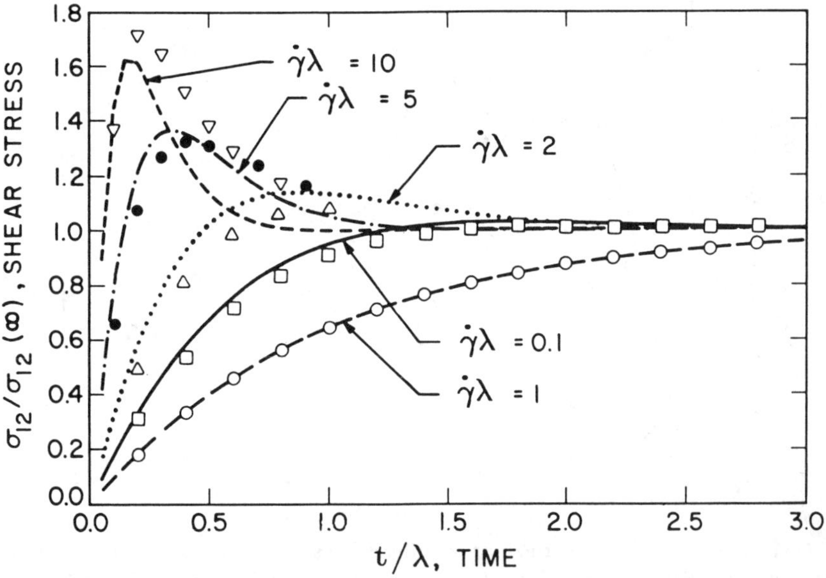

Fig. 4-16 (a) Normalized shear stress as a function of dimensionless time, t/λ, where λ is the relaxation time, for various dimensionless shear rates. Depicted are the Doi-Edwards model (symbols) and the differential approximation to the DE model (lines) (from Larson,[22] reprinted from the Journal of Rheology ©1984 Society of Rheology, by permission of John Wiley & Sons, Inc.).

polymers and were not measured at shear rates low enough for N_2 to be in the low shear-rate range where it has a quadratic dependence on $\dot{\gamma}$. Since the Doi-Edwards equation (and most other constitutive equations that give a non-zero N_2) predicts that Ψ_2/Ψ_1 *decreases* as $\dot{\gamma}$ increases, the true low-shear-rate value of Ψ_2/Ψ_1 could be significantly higher than the range of values typically seen. It is unfortunate that there are so few measurements of Ψ_2/Ψ_1, especially for melts or concentrated solutions of narrow polydispersity, for the prediction of this quantity is a sensitive test of constitutive equations.

4.6.3.4 Stress Overshoot The Doi-Edwards equation predicts an overshoot in shear stress as a function of time after inception of steady shearing; see Figure 4-16. This behavior accords with experimental measurements, as can be seen by referring to Figure 3-1a. The DE equation predicts that the maximum occurs at a roughly constant value of the imposed strain, $\dot{\gamma}t$, of about 2. Experimentally for modest strain rates the maximum is indeed seen at $\dot{\gamma}t \sim 2$; but at high strain rates it shifts to higher strains.[45]

Experiments also show an overshoot in the first normal stress difference, Figure 3-1b. This overshoot is *not* predicted by the DE equation, Figure 4-17. Pearson and Kiss[46] have suggested that the overshoot in Ψ_1 occurs because polymer strands are stretched during a brief period after start-up of shearing but before the strands are nearly completely aligned in the shearing plane. The Doi-

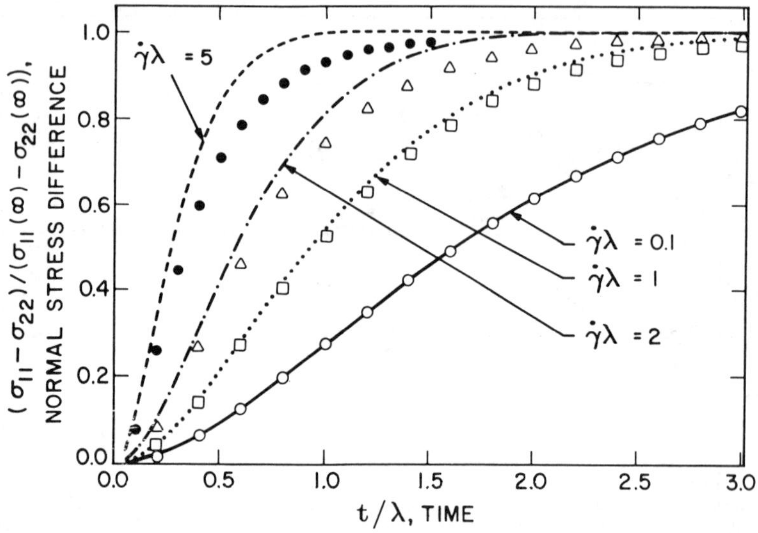

Fig. 4-17 Normalized first normal stress difference as a function of dimensionless time for various dimensionless shear rates. Depicted are the Doi-Edwards model (symbols) and the differential approximation to the DE model (lines) (from Larson,[22] reprinted from the Journal of Rheology ©1984 Society of Rheology, by permission of John Wiley & Sons, Inc.).

Edwards model assumes that the strands remain at their equilibrium length throughout the deformation; that is retraction is assumed to occur fast compared to the rate at which strands are stretched by the deformation. Pearson and Kiss were led to their hypothesis by experiments of Menezes and Graessley[45] which showed that when the imposed steady shear rate equals or exceeds the reciprocal of the *retraction time*, λ_k, the maxima in the first normal stress difference and in the shear stress occur at values of the imposed strain, $\dot{\gamma}t$, that become ever higher, the higher the value of the imposed steady shear rate. At low shear rates the maxima occur at fixed values of the strain, roughly $\dot{\gamma}t = 2$ and 5, respectively. Models have been proposed by Menezes and Graessley and by Pearson and Kiss (see Section 6.3.5.2) that account for incomplete retraction; these models give qualitative agreement with the experimental overshoots.

4.6.3.5 Double-Step Strains Like Kaye-BKZ equations in general, the Doi-Edwards equation often poorly predicts shear stress relaxation after a reversing double-step strain. Some improvement is possible if the independent alignment approximation is removed. The DE equation is also likely to give poor predictions of recoverable strains.[22]

4.6.3.6 Effects of Polydispersity Since most polymer fluids that are of commercial importance have high polydispersity (values of M_w/M_n of 2 or more) it is important that a prospective constitutive equation be capable of describing polydisperse materials. The simple-minded way of amending the Doi-Edwards

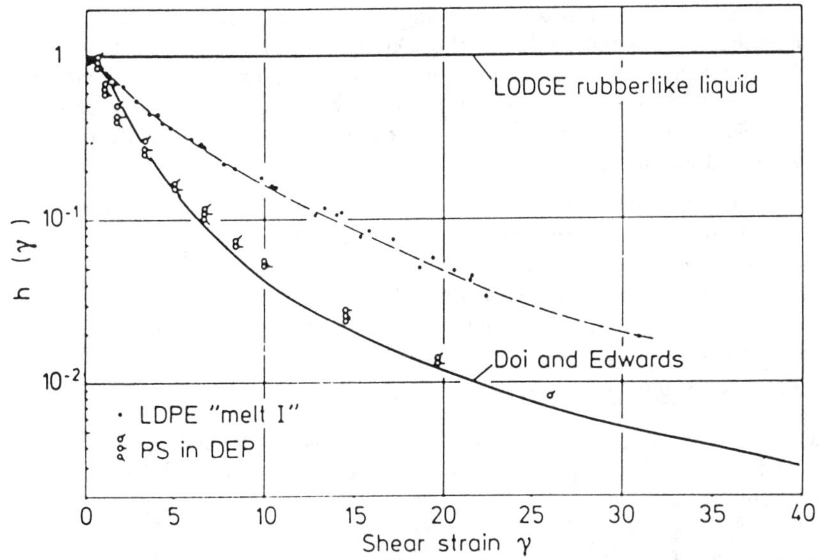

Fig. 4-18 Damping function, h(γ), versus shear strain for a concentrated nearly monodisperse linear polystyrene solution and a branched polydisperse low density polyethylene, compared to the prediction of the Doi-Edwards model and of the Lodge equation (from Wagner[47], reproduced with permission from *Rheology*).

equation for polydispersity is to ***add*** together contributions to the stress from each component in the molecular-weight distribution. Since the time-dependent memory function, $m(t-t')$, depends on the molecular weight, but $\underset{\approx}{Q}$ does not, this change does not affect the predicted damping function, $h(\gamma)$, for example. Yet experimentally, $h(\gamma)$ *is* affected by polydispersity. Both polydispersity and the presence of long-chain side branches shift upwards for a given γ. "Long" side branches are branches longer than the entanglement spacing; branches shorter than this influence the friction coefficient but otherwise don't affect reptation. Figure 4-18, for example, shows that $h(\gamma)$ for Melt I is significantly greater than that predicted by the Doi-Edwards theory. Melt I is polydisperse with $M_w/M_n = 28$ and has long side branches. It is not surprising that long side branches should effect $h(\gamma)$, since it is easy to image them interfering with the retraction process. (The interference would occur if there are more than one long side branch spaced along the backbone; star-branching doesn't hinder retraction, as noted earlier.) Polydispersity alone however, also affects $h(\gamma)$, probably because the various relaxation mechanisms are dependent on molecular weight. In a highly polydisperse sample, some of the longer molecules might require as long to retract as the shorter ones require to reptate. Thus *mixing of relaxation modes* could produce an *effective* damping function that corresponds to a less relaxed condition than the complete retraction assumed by Doi and Edwards.

It has also been found in linear viscoelastic measurements[48] that the relaxation of long chains in a polydisperse melt or concentrated solution is assisted by the reptation of surrounding short chains. Since the reptation time of each chain in the melt is proportional to $M^{3.4}$, short chains in the melt can reptate out of the way of long chains more quickly than the long chains themselves can reptate. This process of relaxation by reptation of the surrounding chains is called "tube renewal." It is apparently important even in monodisperse melts of linear or star-branched polymers.

Accounting for arbitrary levels of polydispersity or of long-chain branching at a level of rigor similar to that of the Doi-Edwards model would likely be difficult and tedious, and even if successful would yield a constitutive equation too complicated to be very useful. Therefore a simple ad hoc generalization of the DE equation is presented in Chapter 5. This equation can successfully describe some commercial melts.[22]

4.7 CURTISS-BIRD THEORY

Motivated by Doi and Edward's work, Curtiss and Bird in 1981 presented their own version of a reptation-based constitutive equation.[27] The Curtiss-Bird constitutive equation is

$$\underset{\approx}{\sigma} = \nu k T \left[\int_{-\infty}^{t} m(t-t') \langle \underset{\sim}{u}\underset{\sim}{u} \rangle_{t't} \, dt' + 2\epsilon \underset{\approx}{D} : \int_{-\infty}^{t} G(t-t') \langle \underset{\sim}{u}\underset{\sim}{u}\underset{\sim}{u}\underset{\sim}{u} \rangle_{t't} \, dt' \right] \quad (4\text{-}48)$$

The term $\langle \cdot \rangle_{t't}$ is an average over oriented unit vectors, $\underset{\sim}{u} = \underset{\sim}{u}' \cdot \underset{\approx}{F}^{-1}(t',t) / | \underset{\sim}{u}' \cdot \underset{\approx}{F}^{-1}(t',t) |$, where $\underset{\sim}{u}'$ is randomly oriented at time t'. With this definition, the Doi-Edwards $\underset{\approx}{Q}$ tensor equals $\frac{1}{5} \langle \underset{\sim}{u}\underset{\sim}{u} \rangle_{t't}$. Thus $\langle \underset{\sim}{u}\underset{\sim}{u}\underset{\sim}{u}\underset{\sim}{u} \rangle_{t't}$ is proportional to the fourth moment of the same distribution function of which $\underset{\approx}{Q}$ is the second moment.

This Curtiss-Bird constitutive equation was derived using the Curtiss-Bird phase-space analysis. This approach is a formal one that eschews the freewheeling intuitive arguments of Doi and Edwards. It eliminates the Doi-Edwards "tube" and the retraction mechanism and instead treats the polymer chain as a freely jointed chain of beads and rods that diffuses predominantly along its own contour. Each rod in this idealization is not a Kuhn step, but is a portion of chain with end-to-end length a, the entanglement spacing.

This description of the chain as a sequence of rigid rods is precisely the result that emerges from Doi and Edwards' approach. While Curtiss and Bird assumed this description at the outset, Doi and Edwards arrived at it through a series of intuitive arguments involving retraction and the independent alignment approximation. Thus it is not surprising that the first term of Equation 4-48 is identical to the DE equation except for a factor of 3 difference in modulus.

The second term arises from an assumed frictional resistance that surrounding chains exert on the reptating chain. It is analogous to the frictional term that appears in theories for dilute solutions of rodlike polymers as discussed in Chapter 8. Its empirical coefficient ϵ is the "link tension coefficient." This term can offset the extreme shear thinning of the first term and lead to shear-rate dependencies of the steady-state shear viscosity and first normal stress coefficient that compare well with experiment.[49]

This second term, however, has an explicit dependence on the rate of strain, $\underset{\approx}{D}$, and is therefore a purely dissipative term that does not satisfy the stress-optic law. This term also manifests itself in overly large stresses immediately after start-up of steady straining and in values of $\eta' = G''/\omega$ that, at high frequency, are a couple of orders of magnitude too large.[18] The second term would therefore seem to be unwelcome, despite the improved fit that it gives to steady-state stresses in shear.

4.8 SUMMARY

There are several features of the dynamics of flexible macromolecules in the melt or in concentrated solution that simplify theories of viscoelasticity for these fluids in comparison to theories for dilute solutions. These features are the Gaussian equilibrium distribution of configurations, the absence of hydrodynamic interactions, and the absence of contributions to the stress tensor other than those from configurational entropy. This latter feature is responsible for the validity of the stress-optic law. There are complications, however, because long, flexible polymer molecules at high concentration entangle with each other.

Entanglements lead to a plateau in G' versus ω; for deformations at plateau frequencies, the material behaves almost elastically. Because of entanglements, long-range polymer diffusion occurs by a snakelike motion called reptation by de Gennes. Reptation accounts for the plateau in G'. The molecule can be thought of as being confined within a tube that prohibits significant lateral motion, but permits reptation and, if the polymer is deformed, the tube permits a rapid retraction process to occur. Retraction accounts for the nonlinearity in the shear modulus as a function of shear strain.

Based on reptation and retraction, Doi and Edwards developed a constitutive model for concentrated solutions and melts. Adding an additional mathematical approximation, the independent alignment approximation, they obtained a constitutive equation of Kaye-BKZ class that requires only two parameters to define the linear viscoelastic response and none at all for the nonlinear response. If desired, additional approximations allow one to further simplify the integral equation or even convert it to differential form.

The prediction that the Doi-Edwards model makes for the dependence of zero-shear viscosity on molecular weight is in reasonable agreement with experimental data, and its prediction of molecular diffusivity is excellent. A related model for star polymers gives excellent predictions for linear viscoelastic properties, including an exponential dependence of the zero-shear viscosity on molecular weight. The Doi-Edwards model for linear molecules has flaws in some of its predictions of nonlinear properties, such as an exaggerated shear thinning, no overshoot in the growth of the first normal stress difference after start-up of steady shearing, and poor predictions in reversing double-step strains. The constitutive equation is also inadequate for polydisperse commercial materials.

The problem of excessive shear thinning is avoided in the constitutive equation of Curtiss and Bird, which is like that of Doi and Edwards except that it has an additional term with an empirical coefficient. The additional term leads to other difficulties, however, such as a violation of the stress-optic law.

References

1. P.J. Flory, *Principles of Polymer Chemistry,* Cornell University Press, Ithaca, N.Y. (1953).
2. For references and discussion see P.-G. de Gennes, *Scaling Concepts in Polymer Physics,* Chap. 2, Cornell University Press, Ithaca, 1979.
3. J.D. Ferry, *Viscoelastic Properties of Polymers,* 3rd ed., John Wiley & Sons, New York (1980).
4. W. Kuhn and F. Grün, Kolloid Z., *101,* 248 (1942).
5. L.R.G. Treloar, *The Physics of Rubber Elasticity,* 3rd ed., Clarendon Press, Oxford (1975).
6. H. Janeschitz-Kriegl, *Polymer Melt Rheology and Flow Birefringence,* Springer-Verlag, New York (1983).
7. J.L.S. Wales, *The Application of Flow Birefringence to Rheological Studies of Polymer Melts,* Delft University Press (1976).
8. C.J. Landry, Ph.D. Thesis, University of Wisconsin, 1985.
9. R. Muller and D. Froelich, Polymer, *26,* 1477 (1985).
10. G.C. Berry and T.G. Fox, Adv. Polym. Sci., *5,* 261 (1968).
11. B.H. Zimm, J. Chem. Phys., *24,* 269 (1956).
12. L.A. Holmes, S. Kusamizu, K. Osaki, and J.D. Ferry, J. Polym. Sci., A-2, *9,* 2009 (1971).
13. S. Onogi, T. Masuda, and K. Kitagawa, Macromolecules, *3,* 109 (1970).
14. D.S. Pearson, Rubber Chem. and Technol., Rubber Reviews, 1987.
15. S.F. Edwards, Proc. Phys. Soc. *92,* 9 (1967).
16. P.G. de Gennes, J. Chem. Phys., *55,* 572 (1971).
17. M. Doi and S.F. Edwards, J. Chem. Soc., Faraday Trans. II, *74,* 1789 (1978).
18. W.W. Graessley, Adv. Polym. Sci., *47,* 68 (1982).
19. Y. Einaga, K. Osaki, and M. Kurata, S. Kimura, and M. Tamura, Polym. J., *2,* 550 (1971).
20. M. Doi and S.F. Edwards, J. Chem. Soc., Faraday Trans. II, *74,* 1802 (1978).
21. M. Fukuda, K. Osaki, and M. Kurata, J. Polym. Sci., Polym. Phys. Ed., *13,* 1563 (1975).
22. R.G. Larson, J. Rheol., *28,* 545 (1984).
23. M. Doi and S.F. Edwards, J. Chem. Soc., Faraday Trans. II, *74,* 1818 (1978).
24. M. Doi and S.F. Edwards, J. Chem. Soc., Faraday Trans. II, *75,* 32 (1979).
25. M. Doi and S.F. Edwards, J. Chem. Soc., Faraday Trans. II, *74,* 560, 918 (1978).
26. G. Marrucci and N. Grizzuti, J. Rheol., *27,* 433 (1983).

27. C.F. Curtiss and R.B. Bird, J. Chem. Phys., *74*, 2016; 2026 (1981).

28. P.K. Currie in "Rheology," the Proc. Eighth International Congress on Rheology, Naples, Italy, 1980.

29. R.G. Larson in "Rheology," the Proc. Ninth International Congress on Rheology, Acapulco, Mexico, 1984.

30. B.R. Seth in M. Reiner and D. Abir, Eds., *Second-Order Effects in Elasticity, Plasticity, and Fluid Mechanics*, Pergamon, Oxford, 162 (1964).

31. W.V. Chang, R. Bloch, and N.W. Tschoegl, Rheol. Acta, *15*, 367 (1976).

32. G. Marrucci in *Advances in Transport Process*, Vol. V, A.S. Mujamdar and R.A. Mashelkar, Eds., John Wiley & Sons, 1984.

33. M. Doi, J. Polym. Sci., Poly. Phys. Ed., *21*, 667 (1983).

34. M. Tirrell, Rubber Chem. Technol., *57*, 523 (1984).

35. Y.-H. Lin, J. Rheol., *28*, 1 (1984).

36. P.-G. de Gennes, J. Phys. (Paris), *36*, 1199 (1975).

37. M. Doi and N.Y. Kuzuu, J. Polym. Sci., Polym. Lett. Ed., *18*, 775 (1980).

38. D.S. Pearson and E. Helfand, Macromolecules, *17*, 888 (1984).

39. G.B. McKenna, Symp. Faraday Soc., *18*, 210 (1983).

40. J. Roovers, Macromolecules, *18*, 1359 (1985).

41. C.M. Vrentas and W.W. Graessley, J. Rheol., *26*, 359 (1982).

42. R.G. Larson, V.R. Raju, and S.A. Khan, J. Rheol. in press (1987). 1987.

43. E.V. Menezes and W.W. Graessley, Rheol. Acta, *19*, 38 (1980).

44. K. Osaki, S. Kimura, and M. Kurata, J. Polym. Sci., Polm. Phys. Ed., *19*, 517 (1981).

45. E.V. Menezes and W.W. Graessley, J. Polym. Sci: Polym. Phys. Ed., *20*, 1817 (1982).

46. D.S. Pearson, and A.D. Kiss, presentation at the March Meeting of the Amer. Phys. Soc., New York, 1987.

47. M.H. Wagner in "Rheology," Vol. 2, the Proceedings of the Eighth International Congress on Rheology, Naples, Italy 1980.

48. M. Rubinstein, E. Helfand, and D.S. Pearson, Macromolecules, *20*, 822 (1987).

49. H.H. Saab, R.B. Bird, and C.F. Curtiss, J. Chem. Phys., *77*, 4758 (1982).

CHAPTER 5

Constitutive Models with Nonaffine Motion

"for who can make that straight, which he hath made crooked?"

Ecclesiastes 7:13

5.1 INTRODUCTION

The Doi-Edwards model presented in Chapter 4 gives a relatively complete description of the dynamics of entangled polymer molecules during deformation, relaxation, and flow. In this chapter and the next, we discuss models that have much less complete molecular underpinnings. Though based on rough molecular or thermodynamic ideas, the connection between each such *semi-empirical* constitutive equation and the postulated molecular processes is loose, and the form of the equation is determined more by mathematical convenience than by the demands of the molecular picture. These semi-empirical constitutive equations stand or fall based on their stress predictions and not on the accuracy of their description of molecular processes.

Why resort to semi-empirical constitutive models when a more precise description, that of Doi and Edwards, is available? There are two major reasons. First, the DE model has several failings in its predictions of stress, even if the material is an ideal nearly monodisperse melt. Its prediction of the relaxation spectrum is too narrow, it predicts no first normal stress overshoot in start-up of steady shearing, and it predicts excessive shear thinning in the steady-state shear viscosity. Second, the DE model does not adequately describe *commercial* polymer melts, which are usually very polydisperse and sometimes contain long side branches. In principle, the failings of the DE model for both ideal and commercial materials can be alleviated by refining the model, using concepts of the DE picture, such as retraction, together with recently studied ideas such as tube renewal and tube length fluctuations. These concepts will hopefully allow the development of a more accurate constitutive equation for ideal materials, as well as constitutive equations for polydisperse branched materials that are as rigorous as the DE equation is for monodisperse unbranched melts. Such equations, however, will be complex and unwieldy, hardly tractable for calculation of simple steady flows, let alone complex unsteady processing flows. The "simple" DE model is already complex enough, especially without the independent alignment approximation that reduces it to Kaye-BKZ form.

This is not to say that more rigorous constitutive equations should not be derived. By deriving more precise and hence more complex constitutive equations and comparing their predictions with experimental data, we test our molecular understanding. But such complex equations are not likely to be used in numerical simulation of complex fluid flow. Limitations on both computer time and programmer patience imply that *the more complex the flow is, the less room there is for complexity in the constitutive equation.*

The need for simplicity in the constitutive equation often restricts the allowable equations to simple integral or differential forms. A semi-empirical equation is typically derived from a simple model of the molecular behavior by making whatever mathematical approximations are required to force the constitutive equation into a simple form. We have already seen an example of this in the derivation of the DE equation. The DE model with no mathematical approximations leads to a complex constitutive equation that is barely tractable even for simple flows. Doi and Edwards invoked the independent alignment approximation because it allowed them to reduce the equation to a convenient single integral of Kaye-BKZ form. But Kaye, as well as Bernstein, Kearsley, and Zapas had already arrived at this form by a semi-empirical approach; the Kaye-BKZ equation is a mathematically simple expression that models a viscoelastic liquid as a relaxing rubber. The demands of tractability, therefore, led Doi and Edwards to make the mathematical approximations necessary to reduce their model to the well-known Kaye-BKZ form.

Thus, while more rigorous constitutive equations for commercial melts will no doubt eventually be found, we expect that after these more rigorous equations are simplified, they will bear a resemblance to the equations presented in this chapter and the next.

In this chapter, we consider models in which the assumption of affine motion is dropped. We concern ourselves mainly with models for concentrated solutions and melts, although some of the equations discussed here have been used to describe dilute solutions. The simplest nonaffine model is that of Gordon and Schowalter,[1] in which strands of polymer slip with respect to the deformation of the macroscopic continuum. The formula for the stress tensor is changed when the motion becomes nonaffine. We then present the Gordon-Schowalter convected derivative, which generalizes the upper-convected derivative of the Maxwell equation. We discuss the strengths and weaknesses of this model, particularly the problems it has in step shear strains. These problems are overcome in a type of nonaffine motion proposed by Larson, which is a generalization of the nonaffine motion in the DE model; Larson's model with nonaffine motion realistically describes the behavior of commercial melts in step strains. We then argue that nonaffine motion is actually irreversible, whereas most constitutive equations assume it to be reversible. Fortunately, there is a method proposed by Wagner for incorporating irreversibility into the damping function and into integral constitutive equations. Irreversible nonaffine motion can also be included in differential constitutive equations. Finally the White-Metzner equation is presented; it too gives an irreversible description of step strains but violates the Lodge-Meissner relationship.

5.2 GORDON-SCHOWALTER CONVECTED DERIVATIVE

In chapter 4, we saw that because of the retraction process, the DE model involves a *nonaffine* response of polymer strands to an imposed deformation. Gordon and Schowalter[1] had earlier proposed a simple model for nonaffine motion, based on Ericksen's theory for anisotropic fluids. Gordon and Schowalter's proposal was that a strand end-to-end vector, $\underset{\sim}{R}$, responds to a deformation according to

$$\dot{\underset{\sim}{R}} = \underset{\sim}{R} \cdot \underset{\sim}{\nabla} \underset{\sim}{v} - \xi \underset{\sim}{R} \cdot \underset{\approx}{D} \tag{5-1}$$

where ξ is an empirical coefficient. For $\xi = 0$, the motion is affine. The term $\xi \underset{\sim}{R} \cdot \underset{\approx}{D}$ represents a "slippage" of the strand in the continuum. It is simpler than the nonaffine motion asserted by Doi and Edwards,

$$\dot{\underset{\sim}{R}} = \underset{\sim}{R} \cdot \underset{\sim}{\nabla} \underset{\sim}{v} - \underset{\sim}{u}\underset{\sim}{u} : \underset{\approx}{D} \underset{\sim}{R} \tag{5-2}$$

where $\underset{\sim}{u} \equiv \underset{\sim}{R}/|\underset{\sim}{R}|$, in that Gordon and Schowalter's nonaffine motion is *linear* in $\underset{\sim}{R}$. Although Gordon and Schowalter suggested Equation 5-1 for dilute solutions, Johnson and Segalman[2] and Phan-Thien and Tanner[3] applied it to polymer melts. We discuss the Johnson-Segalman model here; discussion of the Phan-Thien/Tanner model is deferred to Chapter 6.

5.2.1 The Stress Tensor

The Gordon-Schowalter (GS) nonaffine motion is *not consistent* with the usual form for the stress tensor

$$\underset{\approx}{\sigma} = 2\nu k T \beta^2 \langle \underset{\sim}{R}\underset{\sim}{R} \rangle \qquad \text{(affine motion)} \tag{5-3}$$

The correct formula[4] is found by equating the time rate of change of the stored elastic energy, $\dot{A}$, with the power required to carry out the deformation, $\dot{W} = \underset{\approx}{\sigma} : \underset{\approx}{D}$. The stored elastic energy is $A = \nu k T \beta^2 \langle R^2 \rangle$. Thus using Equation 5-1,

$$\underset{\approx}{\sigma} : \underset{\approx}{D} = \dot{A} = 2\nu k T \beta^2 \langle \dot{\underset{\sim}{R}} \cdot R \rangle = 2a\nu k T \underset{\approx}{D} : \beta^2 \langle \underset{\sim}{R}\underset{\sim}{R} \rangle \tag{5-4}$$

where $a \equiv 1 - \xi$. Equation 5-4 implies that

$$\underset{\approx}{\sigma} = 2a\nu k T \beta^2 \langle \underset{\sim}{R}\underset{\sim}{R} \rangle \qquad \text{(nonaffine motion)} \tag{5-5}$$

Thus for the GS nonaffine motion, the equation for the stress tensor is reduced from that for an affine motion by the factor a. If Equation 5-3 is used instead of Equation 5-5, then the absurd conclusion follows that a stretched material, by releasing its elastic energy, *can do an amount of work on its surroundings that is greater than the elastic energy that is stored in the stretched strands of the material.* Equation 5-5, rather than 5-3, is therefore the correct formula for the stress tensor.

The physical explanation for the introduction of the factor, "a," into the equation for the stress tensor is that since the deformation of the continuum, that is the deformation at the macroscopic level, is by definition affine, the strands "slip" with respect to motion of the continuum. Because the continuum "slips" past the strands, the continuum doesn't "feel" all the tension in the strands. Thus each strand transmits only a fraction of its tension to the continuum. If one attempts to stretch an elastic cord with oily hands, for example, the slippage of one's hands would reduce the felt force.

5.2.2 The Convected Derivative

We obtain the *Gordon-Schowalter convected derivative* by taking the time derivative of the stress tensor:

$$\dot{\underset{\approx}{\sigma}} = 2a\nu kT\beta^2 \langle \underset{\sim}{\dot{R}}\underset{\sim}{R} \rangle + 2a\nu kT\beta^2 \langle \underset{\sim}{R}\underset{\sim}{\dot{R}} \rangle$$

$$= \underset{\sim}{\nabla}\underset{\sim}{v}^T \cdot \underset{\approx}{\sigma} + \underset{\approx}{\sigma} \cdot \underset{\sim}{\nabla}\underset{\sim}{v} - \xi(\underset{\approx}{\sigma} \cdot \underset{\approx}{D} + \underset{\approx}{D} \cdot \underset{\approx}{\sigma}) \tag{5-6}$$

where we have used Equation 5-1 to eliminate $\dot{\underset{\sim}{R}}$ from the above.

Thus

$$\overset{\square}{\underset{\approx}{\sigma}}_a = \underset{\approx}{0}\ , \tag{5-7}$$

where $\overset{\square}{\underset{\approx}{\sigma}}_a$, the GS convected derivative, is defined by

$$\overset{\square}{\underset{\approx}{\sigma}}_a \equiv \dot{\underset{\approx}{\sigma}} - \underset{\sim}{\nabla}\underset{\sim}{v}^T \cdot \underset{\approx}{\sigma} - \underset{\approx}{\sigma} \cdot \underset{\sim}{\nabla}\underset{\sim}{v} + \xi(\underset{\approx}{\sigma} \cdot \underset{\approx}{D} + \underset{\approx}{D} \cdot \underset{\approx}{\sigma})$$

$$\equiv \dot{\underset{\approx}{\sigma}} - \underset{\approx}{\omega}^T \cdot \underset{\approx}{\sigma} - \underset{\approx}{\sigma} \cdot \underset{\approx}{\omega} - a(\underset{\approx}{D} \cdot \underset{\approx}{\sigma} + \underset{\approx}{\sigma} \cdot \underset{\approx}{D}) \tag{5-8}$$

Equation 5-7 gives the stress during sudden deformations, with no time allowed for any relaxation. Special cases of this convected derivative are

$a = 1$	upper-convected derivative
$a = 0$	corotational derivative
$a = -1$	lower-convected derivative

5.3 JOHNSON-SEGALMAN MODEL

Johnson and Segalman[2] proposed that both the Rouse model for dilute solutions, Section 2.7, and the Green and Tobolsky model for melts, Section 2.5, can be modified by allowing slippage of the kind discussed above. Thus Johnson and Segalman replaced the upper-convected derivative in the upper-convected Maxwell model by the GS convected derivative, thereby obtaining

$$\lambda \overset{\square}{\underset{\approx}{\sigma}}_{a} + \underset{\approx}{\sigma} = \frac{G}{a} \underset{\approx}{\delta} \tag{5-9}$$

with $G = a^2 \nu kT$.

5.3.1 Elastic Strain Measure

An equivalent result in integral form[2,5] is obtained by replacing the Finger tensor in the Lodge equation with a more general tensor, $\underset{\approx}{E}$:

$$\underset{\approx}{\sigma} = \int_{-\infty}^{t} m(t-t') \underset{\approx}{E}(t,t') \, dt' \tag{5-10}$$

where $\underset{\approx}{E}(t,t')$ is the solution of the differential equation

$$\overset{\square}{\underset{\approx}{E}}_{a} = \dot{\underset{\approx}{E}} - \underset{\approx}{\omega}^{T} \cdot \underset{\approx}{E} - \underset{\approx}{E} \cdot \underset{\approx}{\omega} - a(\underset{\approx}{D} \cdot \underset{\approx}{E} + \underset{\approx}{E} \cdot \underset{\approx}{D}) = \underset{\approx}{0} \tag{5-11}$$

with $\underset{\approx}{E}(t,t) = a^{-1} \underset{\approx}{\delta}$. When $m(t-t')$ is a single exponential decay, one can show that Equation 5-10 is equivalent to Equation 5-9 using an argument analogous to that given in Section 1.6.2. It is the linearity in $\underset{\approx}{\sigma}$ of the convected derivative, Equation 5-8, that allows one to express the same constitutive equation in both integral and differential forms. If $m(t-t')$ is a sum of exponentials, then the integral equation is equivalent to a series of differential equations; see Section 3.4.4.1. Thus Equation 5-10 *is the integral form of the Johnson-Segalman model* with a distribution of relaxation times.

The tensor $\underset{\approx}{E}$, the *elastic strain measure*, represents the instantaneous response of the material to a rapid or *sudden deformation*. Suppose a finite strain is suddenly applied to a relaxed material at time t=0. Then Equation 5-10 gives for the stress tensor immediately after the step strain:

$$\underset{\approx}{\sigma} = G(t) \underset{\approx}{E} \tag{5-12}$$

where $G(t)$ is the time-dependent linear relaxation modulus,

$$G(t) = \int_{-\infty}^{0} m(t-t') \, dt' \tag{5-13}$$

If $m(t-t')$ is a single exponential decay, then $G(t) = G \exp(-t/\lambda)$.

Consider the differential form of the JS model, Equation 5-9, in a sudden deformation. In a sudden deformation, $\underset{\sim}{\nabla} \underset{\sim}{v}$ is very large and so is the rate of change of the stress $\dot{\underset{\approx}{\sigma}}$. The terms $\underset{\approx}{\sigma}$ and $G \underset{\approx}{\delta}$ are therefore eclipsed by the convected derivative, which contains $\underset{\sim}{\nabla} \underset{\sim}{v}$ and $\dot{\underset{\approx}{\sigma}}$. Hence

$$\overset{\square}{\underset{\approx}{\sigma}}_a \equiv \dot{\underset{\approx}{\sigma}} - \underset{\approx}{\omega}^T \cdot \underset{\approx}{\sigma} - \underset{\approx}{\sigma} \cdot \underset{\approx}{\omega} - a(\underset{\approx}{D} \cdot \underset{\approx}{\sigma} + \underset{\approx}{\sigma} \cdot \underset{\approx}{D}) = \underset{\approx}{0} \tag{5-14}$$

Thus *the convected derivative of the stress tensor $\underset{\approx}{\sigma}$ is zero in a rapid deformation.*

From the definition of $\underset{\approx}{E}$, Equation 5-11, we find from Equation 5-14 that $\underset{\approx}{\sigma} = G\underset{\approx}{E}$ immediately following a rapid deformation. This, of course, is the same result as we obtained for the equivalent integral equation, Equation 5-10. Thus *the convected derivative, or equivalently the elastic strain measure, describes the material's elastic response to a sudden deformation.* In Chapter 1, an identity analogous to Equation 5-14 was found to hold for the Finger tensor:

$$\overset{\nabla}{\underset{\approx}{C}^{-1}} = \underset{\approx}{0} \tag{5-15}$$

5.3.1.1 Extension Consider, for example, an extensional deformation with strain rates $\dot{\epsilon}_1$, $\dot{\epsilon}_2$, and $\dot{\epsilon}_3$, along the principal directions. Equation 5-11 gives

$$\dot{E}_{ii} = 2a\dot{\epsilon}_i \tag{5-16}$$

Integrating this result subject to the initial condition $E_{ii} = a^{-1}$ at $t=0$ gives

$$E_{ii} = \frac{1}{a}\exp(2a\epsilon_i) \tag{5-17}$$

where

$$\epsilon_i = \int_0^{0+} \dot{\epsilon}_i \, dt \tag{5-18}$$

is the total Hencky strain applied at time $t=0$. Thus in a sudden extensional deformation,

$$\underset{\approx}{E} = \frac{1}{a}\begin{pmatrix} e^{2a\epsilon_1} & 0 & 0 \\ 0 & e^{2a\epsilon_2} & 0 \\ 0 & 0 & e^{2a\epsilon_3} \end{pmatrix} \tag{5-19}$$

In extensional deformations, $\underset{\approx}{E}$ is the same as the *Seth* elastic strain measure,[6] $\underset{\approx}{C}^{-a}$; see Section 4.5.3.

5.3.1.2 Shear One can likewise work out $\underset{\approx}{E}$ for a sudden shear. It has the components[5]

$$\begin{aligned} E_{12} &= \sin(\gamma\sqrt{1-a^2})/\sqrt{1-a^2} \\ E_{11}-E_{22} &= 2\,[1-\cos(\gamma\sqrt{1-a^2})]/(1-a^2) \\ E_{22}-E_{33} &= -(1-a)(E_{11}-E_{22})/2 \end{aligned} \tag{5-20}$$

The components of $\underset{\approx}{E}$ are therefore *oscillatory functions of shear strain.*

5.3.2 Predictions of the Johnson-Segalman Model

In terms of the stress tensor, $\underset{\approx}{\tau} \equiv \underset{\approx}{\sigma} - G\,\underset{\approx}{\delta}$, Equation 5-9 takes the form

$$\lambda \overset{\square}{\underset{\approx}{\tau}}_a + \underset{\approx}{\tau} = 2\eta_0 \underset{\approx}{D} \tag{5-21}$$

where $\eta_0 \equiv G\lambda = a^2\nu kT\lambda$. Because of incompressibility, $\underset{\approx}{\sigma}$ and $\underset{\approx}{\tau}$, which differ by an isotropic constant, cannot be distinguished experimentally, and so neither can Equations 5-9 and 5-21; see Section 1.6.2. At equilibrium, the stress tensor $\underset{\approx}{\tau}$ equals $\underset{\approx}{0}$; $\underset{\approx}{\sigma}$ equals $G\underset{\approx}{\delta}$. In sudden deformations, the convected derivative of $\underset{\approx}{\sigma}$ equals zero; the convected derivative of $\underset{\approx}{\tau}$ equals $2\eta_0\underset{\approx}{D}/\lambda$.

The JS differential equation is a special case of the Oldroyd eight-constant equation (set $\mu_1 = a$, and all other μ's and ν's and λ_r to zero in Equation 3-46). The corotational Maxwell (CRM) equation, Equation 3-37, is in turn, a special case of the JS equation obtained when $a = 0$. In the corotational limit, $a = 0$, however, the stress tensor, given by Equation 5-5, goes to zero. In this limit, Equation 5-1 yields a rotation of the strands with no stretching:

$$\dot{\underset{\sim}{R}} = \underset{\sim}{R}\cdot\underset{\approx}{\omega} \tag{5-22}$$

The corotational limit differs from the stretch-free rotation assumed by Doi and Edwards, Equation 5-2, in that Equation 5-2 produces a *concentration* of strands along the major stretch direction. In Equation 5-22, however, there is no concentration of strands, because all strands are rotated through the same angle. Thus the isotropic strand distribution function characteristic of equilibrium is *not changed* by the deformation; all strands are merely rotated equally through some angle. Hence no stress is produced by the deformation. The corotational limit is therefore sensible only if "a" is thought of as nonzero but small, with νkT large, so that $\eta_0 = a^2\nu kT\lambda$ is finite. Equation 5-21 then applies, and $\overset{\square}{\underset{\approx}{\tau}}_a$ is approximately $\overset{0}{\underset{\approx}{\tau}}$, the corotational derivative.

5.3.2.1 Steady-State Shearing In steady-state shearing, the JS equation predicts shear thinning of a type similar to that of the CRM equation:[7]

$$\eta = \frac{\eta_0}{1+(1-a^2)\lambda^2\dot{\gamma}^2} \tag{5-23}$$

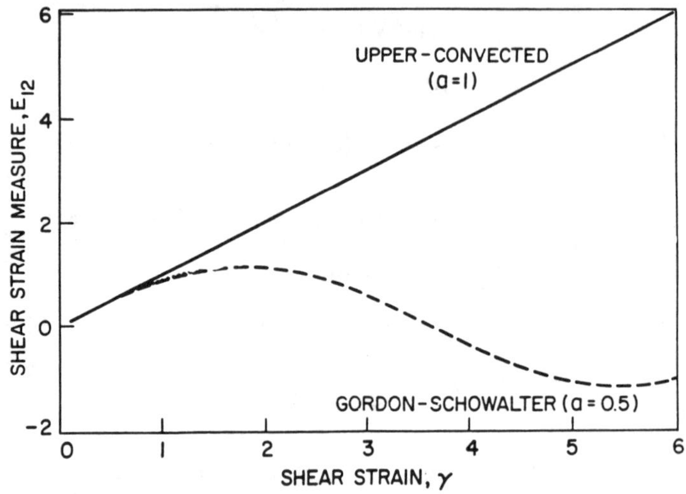

Fig. 5-1 Elastic strain measure in step shearing for the Gordon-Schowalter convected derivative compared to the upper-convected derivative.

When a = 1, the UCM result (no shear thinning) is obtained. As "a" decreases, the shear thinning occurs at progressively smaller strain rates, $\dot{\gamma}$.

The first and second normal stress differences, respectively, are

$$\Psi_1 = \frac{2\lambda\eta_0}{1+(1-a^2)\lambda^2\dot{\gamma}^2} \; ; \quad \Psi_2 = \frac{-(1-a)\lambda\eta_0}{1+(1-a^2)\lambda^2\dot{\gamma}^2} \tag{5-24}$$

Thus the ratio Ψ_2/Ψ_1 equals $-(1-a)/2$. A value Ψ_2/Ψ_1 between –0.05 and –0.30, which is the range usually found in experiments,[8–10] is obtained when "a" lies in the range 0.9 to 0.4. A value of "a" in this range usually also yields reasonable values of $\eta(\dot{\gamma})$, if the distribution of relaxation times is taken into account as described in Section 3.4.4.

5.3.2.2 Steady-State Extension In steady-state extension with $0<a\leq1$, the JS equation predicts that

$$\sigma_{ii} = \frac{\eta_0}{1-2a\dot{\epsilon}_i} \tag{5-25}$$

where $\dot{\epsilon}_i$ is the extension rate in direction i. Thus by decreasing "a", the singularity in σ_{ii} that occurs when $\dot{\epsilon}_i = 1/2a$ can be pushed out to higher values of $\dot{\epsilon}_i$.

In shear or extension, decreasing "a" makes the material response "softer."

5.3.2.3 Step Shear Strains Following a step shear, the stress is $\underset{\approx}{\tau} = G(t)(\underset{\approx}{E} - a^{-1}\underset{\approx}{\delta})$, where $\underset{\approx}{E}$ is given by Equation 5-20. Thus *the shear and normal stress differences are oscillatory functions of shear strain;* see Figure 5-1. The

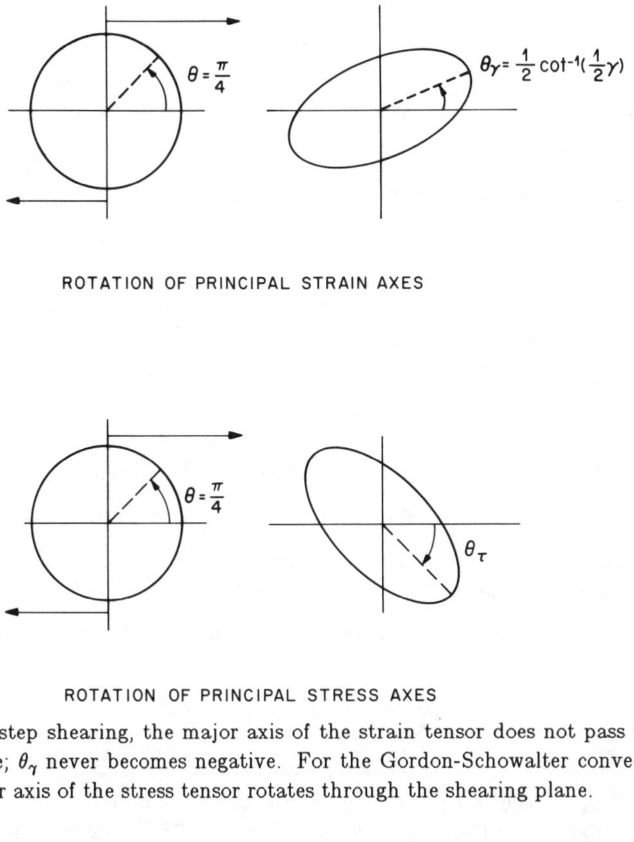

Fig. 5-2 In step shearing, the major axis of the strain tensor does not pass through the shearing plane; θ_γ never becomes negative. For the Gordon-Schowalter convected derivative, the major axis of the stress tensor rotates through the shearing plane.

shear stress and elastic recovery oscillate between positive and ***negative*** values. An oscillatory response occurs not only in step shears but also in start-up of steady shearing, if the dimensionless shear rate, $\dot{\gamma}\lambda$, becomes large.[11] The smaller the value of "a," the larger $\dot{\gamma}\lambda$ must become before the stress becomes oscillatory.

The oscillatory response is ***not*** seen in extensional deformations; see Equation 5-19. It occurs only in deformations with vorticity. In a shearing deformation, for example, the major principal axis of the Finger strain tensor, $\underset{\approx}{C}^{-1}$, rotates from an initial angle of 45° with respect to the shearing plane to an angle, θ_γ, given by

$$2\cot(2\theta_\gamma) = \gamma \tag{5-26}$$

which approaches 0 as $\gamma \rightarrow \infty$, but ***never crosses the shearing plane;*** see Figure 5-2. The major axis of $\underset{\approx}{E}$, however, ***rotates through the shearing plane,*** when $-1 <$ a < 1. Thus, for example, when a $= 0$, the angle θ that the major axis of $\underset{\approx}{E}$ makes with the shearing plane is

$$\theta = \frac{\pi}{4} - 4\gamma \tag{5-27}$$

In steady shear, the major axis of $\underset{\approx}{E}$ therefore rotates* at a constant rate $\dot{\theta} = -4\dot{\gamma}$.

Thus *the principal axes of* $\underset{\approx}{C}^{-1}$ *and* $\underset{\approx}{E}$ *are not coincident.* Neither are the principal axes of $\underset{\approx}{\sigma}$ and $\underset{\approx}{C}^{-1}$ coincident after a step shear. Since the Lodge-Meissner relationship, $\tau_{11} - \tau_{22} = \gamma\tau_{12}$, is derived from a coincidence of the principle axes of stress and strain,** as discussed in Section 3.5.5, *the Johnson-Segalman model does not obey the Lodge-Meissner relationship.*[12]

All the unreasonable results discussed above – oscillation and negative stresses in step strains and stress growth, negative elastic recovery, and the violation of the Lodge-Meissner relationship – stem from the failure of the principal axes of stress to corotate with the principal axes of strain in sudden deformations.

5.3.3 Forcing Corotation of Principal Stress and Strain Axes

The problems of the Johnson-Segalman equation discussed above disappear if $\underset{\approx}{E}$ is forced to be coaxial with $\underset{\approx}{C}^{-1}$. This can be done by replacing $\underset{\approx}{E}$ in Equation 5-15 with $\underset{\approx}{C}^{-a}$, the *Seth* elastic strain measure,[6] introduced in Section 4.5.3. Thus

$$\underset{\approx}{\sigma} = \int_{-\infty}^{t} m(t-t')\,\underset{\approx}{C}^{-a}\,dt' \tag{5-28}$$

predicts the same results as the JS model in extensional deformations, but avoids spurious oscillations in flows with vorticity, since $\underset{\approx}{C}^{-a}$ is coaxial with $\underset{\approx}{C}^{-1}$.

Forcing corotation of principal stress and strain axes in the differential version of the JS equation is not so easy. In general it can be done by adding an appropriate function of $\underset{\approx}{\sigma}$ and $\underset{\approx}{D}$ to the convected derivative:[13]

---

* The GS nonaffine motion with small "a" approximates the rotation of a rotund particle in a flow field, as is discussed in Chapter 8. In a shear field such a particle tumbles in a "Jeffery orbit," and the principal axes of stress therefore tumble as well. The GS convected derivative thus has utility in describing stress in particle suspensions or *dilute* solutions of rigid or possibly even flexible polymers; see Section 8.7.2.1. For melts and concentrated solutions, however, the molecules do not undergo tumbling motions, and the GS convected derivative is not appropriate.

** Tensors with the same principal axes are said to be *coaxial.* Since the tensor $\underset{\approx}{C}$ is coaxial with $\underset{\approx}{C}^{-1}$, the Kaye-BKZ elastic strain measure, $\underset{\approx}{E}_{KBKZ} = 2(\partial U/\partial I_1)\underset{\approx}{C}^{-1} - 2(\partial U/\partial I_2)\underset{\approx}{C}$, is coaxial with $\underset{\approx}{C}^{-1}$. Thus the integral version of the JS equation, Equation 5-10, is *not* in the Kaye-BKZ class. When $-1 < a < 1$, $\underset{\approx}{E}$ is not even a *function* of the cumulative strain $\underset{\approx}{C}^{-1}$, but is instead a *functional* of the deformation *path.* Thus the JS equation is not viscoelastic simple fluid as defined by Coleman and Noll; see Chapter 9.

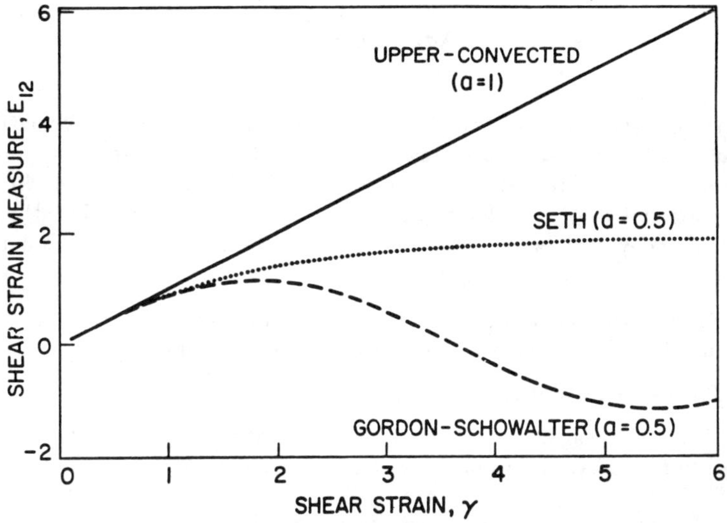

Fig. 5-3 Elastic strain measure in step shear showing that the oscillations present in the Gordon-Schowalter convected derivative are eliminated when the principal axes of stress and strain are forced to corotate, as in the Seth convected derivative (from Larson[13], reproduced, with permission, from the Jour. of Non-Newt. Fluid Mech.).

$$\overset{\blacksquare}{\underset{\approx}{\sigma}}_{a} \equiv \overset{\square}{\underset{\approx}{\sigma}}_{a} + \underset{\approx}{f}^{T} \cdot \underset{\approx}{\sigma} + \underset{\approx}{\sigma} \cdot \underset{\approx}{f} \tag{5-29}$$

If $\underset{\approx}{f}$ is chosen so that the principal axes of stress and strain corotate, then the solution to the equation $\overset{\blacksquare}{\underset{\approx}{\sigma}}_{a} = 0$ is proportional to $a^{-1}\,\underset{\approx}{C}^{-a}$, the Seth elastic strain measure, not only in extensional deformations but also in *any* deformation. We therefore call $\overset{\blacksquare}{\underset{\approx}{\sigma}}_{a}$ the *Seth convected derivative.* The required function $\underset{\approx}{f}$ is, in general, complex but takes a particularly simple form in *planar* deformations *when* $a = 1/2$:

$$2\underset{\approx}{f} = (\underset{\approx}{D} \cdot \underset{\approx}{\sigma} - \underset{\approx}{\sigma} \cdot \underset{\approx}{D})/(\mathrm{tr}\underset{\approx}{\sigma} - 1) \tag{5-30}$$

Here $\mathrm{tr}\underset{\approx}{\sigma} \equiv \sigma_{11} + \sigma_{22} + \sigma_{33}$. Figure 5-3 shows the shear component, E_{12}, of the solution to $\overset{\blacksquare}{\underset{\approx}{\sigma}}_{a} = 0$, in a shear deformation. Notice that when the principal axes of stress and strain are forced to corotate in sudden deformations, the oscillatory dependence of E_{12} on γ is removed.

The Seth convected derivative can be incorporated into a simple Maxwell differential equation,

$$\lambda \overset{\blacksquare}{\underset{\approx}{\sigma}}_a + \underset{\approx}{\sigma} = \frac{G}{a}\underset{\approx}{\delta} \tag{5-31}$$

which has behavior identical to the JS equation in extensional deformations, but in shear is different and more reasonable in that it avoids spurious oscillations and satisfies the Lodge-Meissner relationship. For Equation 5-31, η, Ψ_1, and Ψ_2 are proportional to $\dot{\gamma}^{-1}$, at large shear rate, $\dot{\gamma}$. For the JS equation, these coefficients scale as $\dot{\gamma}^{-2}$ at large $\dot{\gamma}$; see Equations 5-23 and 5-24. Thus Equation 5-31 shows less severe shear thinning than does the JS equation.

A general expression for the Seth convected derivative for $a \neq 1/2$ and for three-dimensional deformations has not been found, however.

5.3.4 Time-Strain Separability

As discussed earlier, the convected derivative describes the immediate response to a step strain. Consider a constitutive equation of the form

$$\lambda_i \overset{\square}{\underset{\approx}{\tau}}_i + \underset{\approx}{\tau}_i = 2\eta_{0,i} \underset{\approx}{D} \tag{5-32}$$

where $\overset{\square}{\underset{\approx}{\tau}}_i$ is some convected derivative. Here the total stress is the sum of contributions from a set of relaxation modes; $\underset{\approx}{\tau} = \sum_i \underset{\approx}{\tau}_i$. The relaxation after a step strain is governed by

$$\lambda_i \dot{\underset{\approx}{\tau}}_i + \underset{\approx}{\tau}_i = \underset{\approx}{0} \tag{5-33}$$

This equation has the solution

$$\underset{\approx}{\tau}_i / G_i = \exp(-t/\lambda_i)(\underset{\approx}{E} - \underset{\approx}{\delta}) \tag{5-34}$$

where $\underset{\approx}{E}$ satisfies $\overset{\square}{\underset{\approx}{E}} = \underset{\approx}{0}$. The stress tensor is therefore *factorable* into a time-dependent part, $G\exp(-t/\lambda)$, and a strain-dependent part, $\underset{\approx}{E} - \underset{\approx}{\delta}$. This *time-strain separability* comes about because the part of the constitutive equation that controls the relaxation, namely Equation 5-33, *is linear in* $\underset{\approx}{\tau}$. The strain-dependent part of the response is determined by the elastic strain measure $\underset{\approx}{E}$.

Approximate time-strain factorability is often seen at long times in both commercial melts[14–16] and concentrated solutions of low polydispersity.[17] As discussed in Chapter 4, time-strain separability implies that after a step shear, the shear stress is given by

$$\tau_{12} = G(t)\,h(\gamma)\,\gamma\,, \tag{5-35}$$

where $h(\gamma)$ is called the *damping function*. $h(\gamma)$ measured on polystyrene of narrow polydispersity[18] agrees with the Doi-Edwards equation, as Figure 5-4 shows. For commercial melts that are polydisperse and may also have molecules with

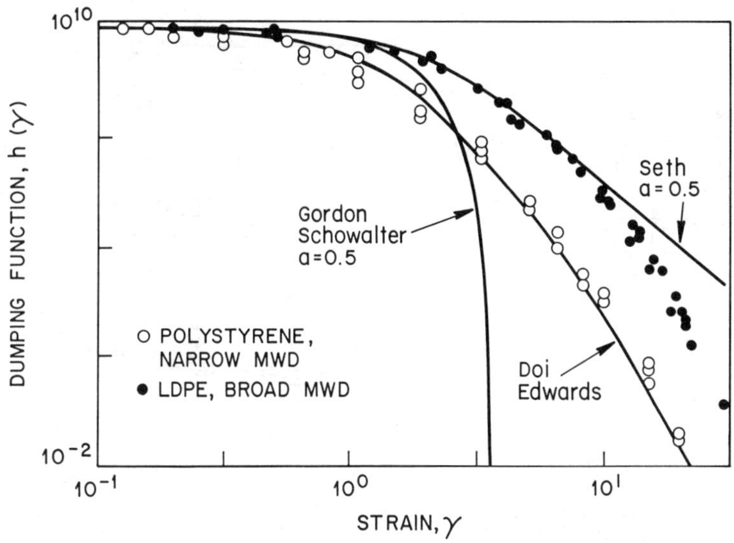

Fig. 5-4 Damping functions for polystyrene of narrow molecular weight distribution (MWD), and Melt I LDPE with a broad MWD compared to the damping functions of the Doi-Edwards theory, and those corresponding to the Gordon-Schowalter and Seth convected derivatives.

long side branches, the damping function is not as strain-softening as predicted by the Doi-Edwards theory. Figure 5-4, for example, shows that the damping function for Melt I is not as strain softening as the damping function for the nearly monodisperse polystyrene studied by Osaki and co-workers. The damping function of DE theory does not agree with the Melt I data.

The shear damping function predicted by a constitutive equation is determined by the elastic strain measure or the convected derivative, that is, by $h(\gamma) = E_{12}/\gamma$. Better agreement with experimental data for commercial melts can therefore be obtained by finding an elastic strain measure that is not as "soft" as the DE elastic strain measure, $\underset{\approx}{Q}$. At low strains, the damping function corresponding to the GS convected derivative with $a = 1/2$ is less strain softening than that of the DE model, but at high strains it becomes negative, as discussed earlier. The shear damping function corresponding to the Seth tensor or Seth convected derivative, Equations 5-29 — 5-30, however, is nonnegative and agrees with the commercial Melt I data out to strains of about 10. Many elastic strain measures or damping functions for shear deformations have been proposed; most of these are tabulated in an excellent review by Petrie.[5]

5.4 PARTIALLY-EXTENDING CONVECTED DERIVATIVE

Larson has proposed that $h(\gamma)$ for commercial melts deviates from the DE theory because the retraction process proposed by Doi and Edwards may not come to completion before the reptation process begins.[19] For example, a pair of side branches long enough to entangle with surrounding chains might limit the

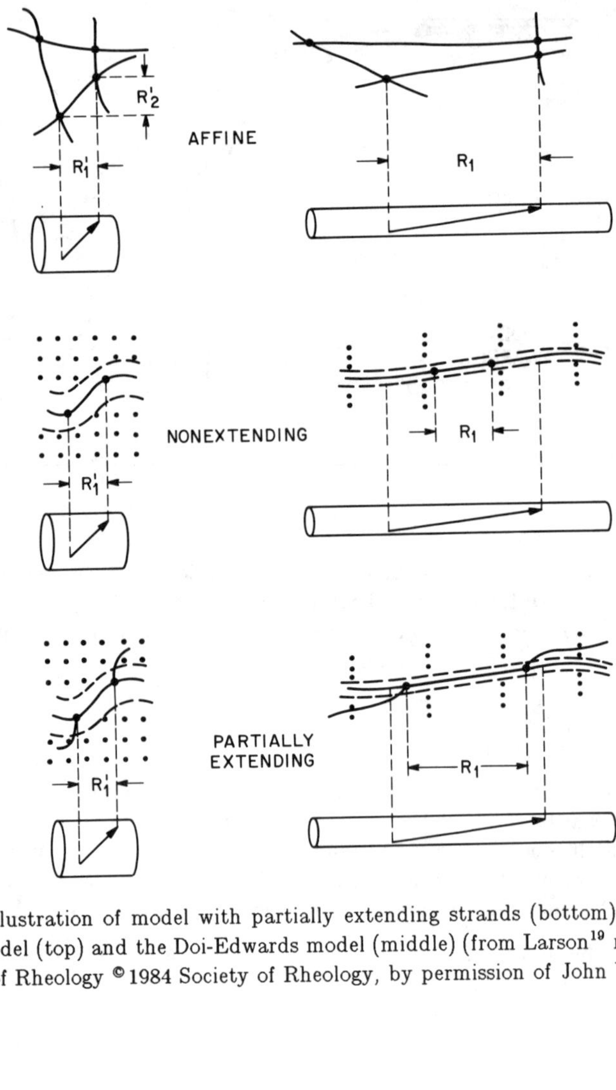

Fig. 5-5 Illustration of model with partially extending strands (bottom), compared to the affine model (top) and the Doi-Edwards model (middle) (from Larson[19] reprinted from the Journal of Rheology ©1984 Society of Rheology, by permission of John Wiley & Sons, Inc.).

retraction of the part of the backbone chain lying between the two side branches; see Figure 5-5. Polydispersity may have a similar effect, because long chains in the melt could take as long to retract as short chains require to reptate. Thus, on the average, chains are only partially retracted when reptation begins. Incomplete retraction would reduce the softness of the damping function; if no retraction occurs, for example, $h(\gamma)$ is unity for all γ.

A careful analysis of these effects would be complex and would likely yield an unwieldy constitutive equation. Larson proposed the following simple empirical expression for the instantaneous response to a sudden deformation of a partially retracting strand:

$$\dot{\underset{\sim}{R}} = \underset{\sim}{R} \cdot \underset{\sim}{\nabla} \underset{\sim}{v} - \xi \underset{\sim}{R} \underset{\sim}{u} \underset{\sim}{u} : \underset{\approx}{D} \tag{5-36}$$

When $\xi = 1$, the retraction is complete and Equation 5-36 is the DE expression. When $\xi = 0$, the response is affine. ξ is a measure of the *extent of retraction.*

An analysis similar to that of Section 4.4.3 yields for the configuration distribution function in step strains,

$$\psi(\underset{\sim}{R}) = \psi_0(\underset{\sim}{R'}) \mid \underset{\sim}{u'} \cdot \underset{\approx}{F}^{-1} \mid^{3\xi} \tag{5-37}$$

where $\underset{\sim}{R}$ and $\underset{\sim}{R'}$ are strand end-to-end vectors before and after the step strain, respectively. ψ and ψ_0 are the distribution functions before and after the deformation. $\underset{\approx}{F}^{-1}$ is the reciprocal of the deformation gradient. The stress tensor corresponding to this nonaffine motion is

$$\underset{\approx}{\sigma}/G = 2a\langle \beta^2 \underset{\sim}{R}\underset{\sim}{R} \rangle + 3\xi \langle \underset{\sim}{u}\underset{\sim}{u} \rangle \tag{5-38}$$

An especially simple and useful convected derivative arises from this expression if ξ is assumed not to be constant but to *increase* as the stress increases such that $\xi = \xi' \mathrm{tr}\underset{\approx}{\sigma}/3G$. This means that the retraction becomes more complete as the stress gets larger. The "partially-extending" convected derivative that emerges, after some approximations, is

$$\overset{\square}{\underset{\approx}{\sigma}}_{PE} \equiv \overset{\nabla}{\underset{\approx}{\sigma}} + \frac{2\xi'}{3G} \underset{\approx}{D} : \underset{\approx}{\sigma}\underset{\approx}{\sigma} \tag{5-39}$$

The elastic strain measure, $\underset{\approx}{E}$, that satisfies the equation $\overset{\square}{\underset{\approx}{E}}_{PE} = \underset{\approx}{0}$ is

$$\underset{\approx}{E}_{PE} = \frac{\underset{\approx}{C}^{-1}}{1 + \frac{\xi'}{3}(I_1 - 3)} \tag{5-40}$$

$\underset{\approx}{C}^{-1}$ and $\underset{\approx}{E}_{PE}$ are coaxial, so the Lodge-Meissner relationship holds and spurious oscillations are avoided.

5.4.1 Shear Damping Function

In shear, the damping function that corresponds to Equation 5-40 is

$$h(\gamma) = \frac{1}{1 + \frac{\xi'}{3}\gamma^2} \tag{5-41}$$

This simple form for the damping function, which has been used by several authors,[5] fits step-strain data for commercial melts, including Melt I, as shown in Figure 5-6. Similarly good fits are obtained for other commercial melts,[16] at least

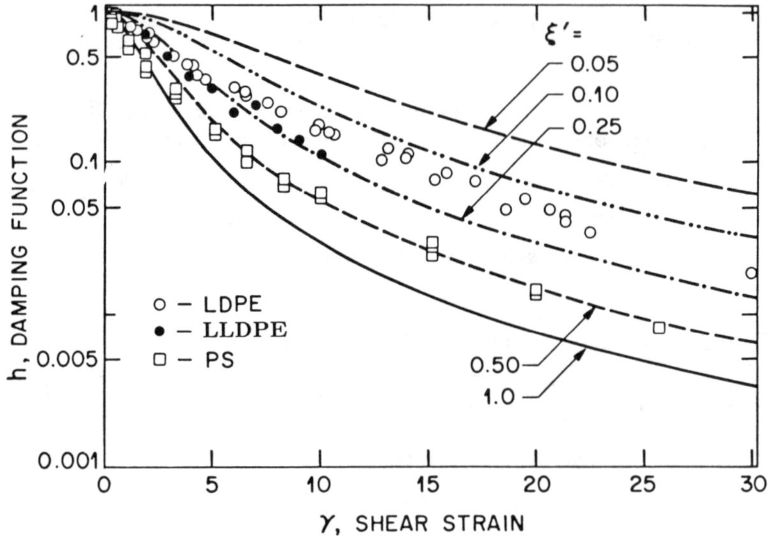

Fig. 5-6 Damping functions for polystyrene of narrow polydispersity, and polyethylenes of broad polydispersity with long branches (LDPE) and without (LLDPE) (from Larson[17], reprinted from the Journal of Rheology ©1985 Soc. of Rheol., by permission of John Wiley & Sons, Inc.).

up to shear strains of 10 or so.* The fitted value of ξ' ranges from 0.13 to 0.6. In general, as the polydispersity increases, ξ' decreases. The lowest values of ξ', with the least strain-softening, are obtained with low-density polyethylenes, which are polydisperse and have long side branches.

The differential constitutive equation of Larson,

$$\overset{\square}{\underset{\approx}{\sigma}}_{PE} \equiv \overset{\nabla}{\underset{\approx}{\sigma}} + \frac{2\xi'}{3G}\underset{\approx}{D}:\underset{\approx}{\sigma}\underset{\approx}{\sigma} = \frac{1}{\lambda}(\underset{\approx}{\sigma} - G\underset{\approx}{\delta}) \tag{5-42}$$

approximates the DE integral equation when $\xi' = 3/5$ and reduces to the UCM equation when $\xi' = 0$. (The Doi-Edwards limit is better approximated by a value $\xi' = 3/5$ than by $\xi' = 1$, because error introduced by a preaveraging approximation used to obtain Equation 5-39 is nearly canceled out when ξ' is reduced from unity to 3/5.)

5.4.2 Predictions in Steady Flows

At high rates of strain, Equation 5-42 with one relaxation mode predicts that $\eta \propto \Psi_1 \propto \dot{\gamma}^{-2}$, and $\Psi_2 = 0$ for all shear rates. The steady-state elongational

* At strains higher than 10 better fits are obtained with a damping function proposed by Soskey and Winter,[15] $h = (1 + \gamma^m)^{-1}$ with $m = 1.4 - 1.6$.

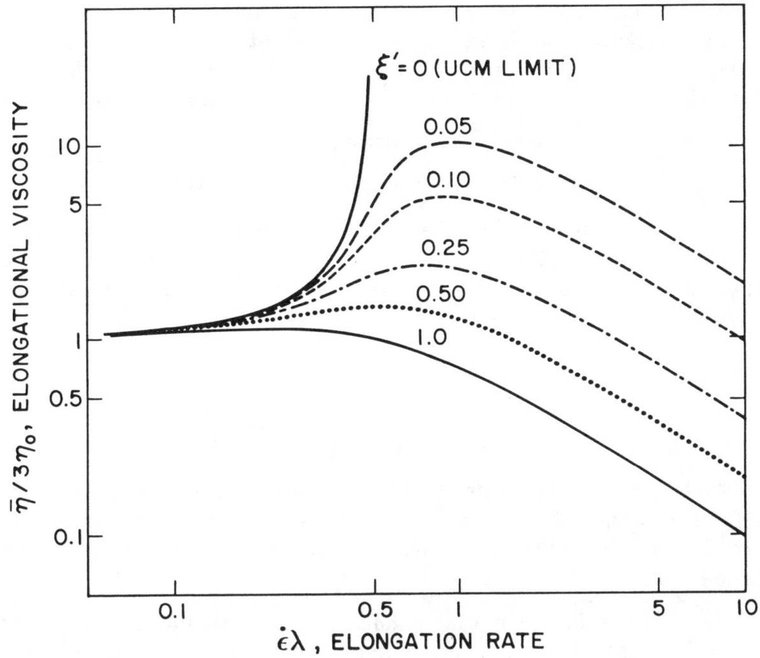

Fig. 5-7 Steady-state elongational viscosity predicted by Larson's model for various values of the parameter, ξ'. The Doi-Edwards model is approximated by $\xi' = 0.6$ (from Larson[19], reprinted from the Journal of Rheology ©1984 Soc. of Rheol., by permission of John Wiley & Sons, Inc.).

viscosity passes through a maximum as $\dot{\epsilon}$ increases, before the asymptotic behavior, $\bar{\eta} = (3G\lambda/\xi')\dot{\epsilon}^{-1}$, is attained; see Figure 5-7. This maximum in elongational viscosity is characteristic of commercial melts. A detailed comparison of a multi-mode version of Equation 5-42 with commercial melt data is presented in Chapter 7.

5.4.3 Integral Equation

A separable Kaye-BKZ integral constitutive equation can also be formed from the elastic strain measure of Equation 5-40:

$$\underset{\approx}{\sigma} = \int_{-\infty}^{t} m(t-t')\, \underset{\approx}{E}_{PE}(t,t')\, dt' \qquad (5\text{-}43)$$

The Kaye-BKZ potential function from which $\underset{\approx}{E}_{PE}$ can be derived is

$$U = \frac{3}{2\xi'} \log[1 + \frac{\xi'}{3}(I_1 - 3)] \qquad (5\text{-}44)$$

If $m(t-t')$ is a single decaying exponential or if the differential equation is generalized to include a spectrum of relaxation times (as described in Section 3.4.4.1), the differential and integral equations, Equations 5-42 and 5-43 are *identical* in step strains and relaxation after step strains, but because the convected derivative, Equation 5-42, is nonlinear in $\underset{\approx}{\sigma}$, the differential and integral equations are *not identical* for other deformation histories.

5.5 IRREVERSIBILITY OF NONAFFINE MOTION

The convected derivative describes the elastic response of the material and is therefore thermodynamically ***reversible***, in principle, if the deformation and its reversal occur rapidly enough. Thus in the limit of infinitely rapid deformations the stress tensor is

$$\underset{\approx}{\sigma} = G\underset{\approx}{E} \tag{5-45}$$

where $\underset{\approx}{E}$ is a function of the strain or the strain path, but is *independent of the rate at which the material is deformed*, assuming that the deformation is fast enough for Equation 5-45 to be valid. All work exerted to deform the material can, according to the constitutive equation, be recovered in the reversal of the deformation.

There is evidence, however, that nonaffine motion should not be regarded as reversible. The evidence is of necessity indirect because the prediction of perfect reversibility for infinitely rapid deformations can never be tested rigorously since no rate of deformation is rapid enough to render negligible the gradual relaxation processes that are included in the linear relaxation spectrum. Indirect evidence for irreversibility of nonaffine motion can be adduced however, by seeing if more irreversibility than is already included in the linear relaxation spectrum is required to obtain agreement with experimental data.

5.5.1 Reversing Deformations

One line of evidence is the experimental behavior of melts in reversing deformations. When the relaxation processes that contribute to the linear viscoelastic spectrum are included in the constitutive equation along with the nonlinear elastic strain measure, stresses in reversing double-step strains or recoverable strains are usually *overpredicted*.[20,21]

5.5.1.1 Reversing Double Step Strain Consider, for example, a double-step shear strain where the second strain is opposite in direction to the first step. The shear stress following the second step can easily be predicted from the Kaye-BKZ equation

$$\sigma = \int_{-\infty}^{t} m(t-t')\,h(\gamma)\,\gamma\,dt' \tag{5-46}$$

where $m(t-t')$ and $h(\gamma)$ are obtained from the single-step-strain experiment. If the second strain, γ_2, is separated from the first, γ_1, by an interval of time, t_1, then the shear stress at a time, t, after the second strain is predicted by the separable Kaye-BKZ equation to be

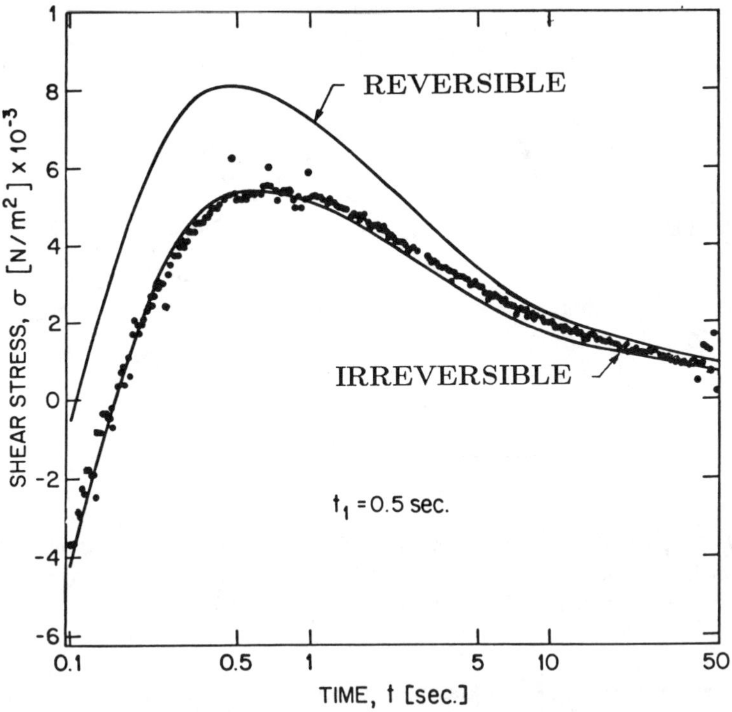

Fig. 5-8 Response of a commercial linear low density polyethylene to a double-step shear strain. The first step is 5 shear units, followed after 0.5 sec. by a second step of -1 shear units. The lines are from Eqns. 5-47 for the reversible model and 5-50 for the irreversible model (from Larson and Valesano[21], reprinted from the Journal of Rheology ©1986 Soc. of Rheol., by permission of John Wiley & Sons, Inc.).

$$\sigma(t,t_1,\gamma_1,\gamma_2) = [G(t) - G(t+t_1)]\, h(\gamma_2)\gamma_2 + G(t+t_1)\, h(\gamma_1+\gamma_2)(\gamma_1+\gamma_2) \quad (5\text{-}47)$$

All terms on the right side are measurable in single-step-strain experiments. Thus the behavior of the Kaye-BKZ equation in double-step strains is predictable using experimental data from single step strains.

Figure 5-8 shows the measured stress for a linear low density polyethylene (LLDPE) after a double-step strain.[21] The first shear strain, γ_1, is 5. The second shear strain is –1 and is therefore in a direction opposite to the first strain. $m(t-t')$ and $h(\gamma)$ for this LLDPE were obtained from single-step-strain experiments; $h(\gamma)$ obeys Equation 5-41 with $\xi' = 0.25$. Note in Figure 5-8 that the stress is initially ***negative***, because the material at first remembers the newer, though weaker, negative strain, $\gamma_2 = -1$. Soon the influence of the older, but larger, strain prevails, and the stress becomes positive. Finally the stress reaches a maximum and relaxes to zero. This behavior is predicted qualitatively by the Kaye-BKZ equation, but the predicted positive stresses are too large.

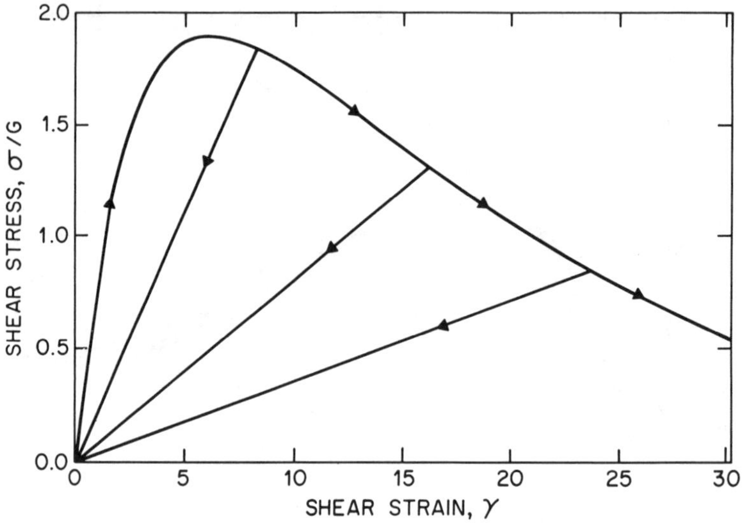

Fig. 5-9 Response to a reversing step strain showing hysteresis in the absence of relaxation (from Larson and Valesano[21], reprinted from the Journal of Rheology ©1986 Soc. of Rheol., by permission of John Wiley & Sons, Inc.).

Wagner proposed that the damping function represents a *breakage* or *loss* of network strands and therefore ought to decrease *irreversibly*.[22] Since the number of strands contributing to the stress is proportional to $h(\gamma)$, Wagner proposed that $h(\gamma)$, which is less than unity after a large deformation, should not increase back to unity when the strain history is reversed.

Consider a hypothetical deformation with no relaxation. A typical plot of shear stress versus shear strain is shown in Figure 5-9. It has a stress maximum at a finite shear strain. According to the Kaye-BKZ equation, if this strain history is reversed rapidly enough to avoid the relaxation processes included in $m(t-t')$, this stress-strain curve ought to be *retraced* in the opposite direction. This means that if the forward strain is large enough, when the strain is reversed the stress will initially *increase* as strain is *unloaded* from the material. Wagner's suggestion avoids this anomaly; it implies that when reversal occurs, $h(\gamma)$ remains at its minimum value, $h(\gamma_1)$, achieved just before the strain history is reversed. As a result, Wagner's stress-strain curves have *hysteresis* but no maximum; see Figure 5-9.

Wagner incorporated this irreversibility into the single integral equation:

$$\sigma = \int_{-\infty}^{t} m(t-t')\, H\{I_1, I_2\}\, \underset{\approx}{C}^{-1}(t,t')\, dt' \tag{5-48}$$

where

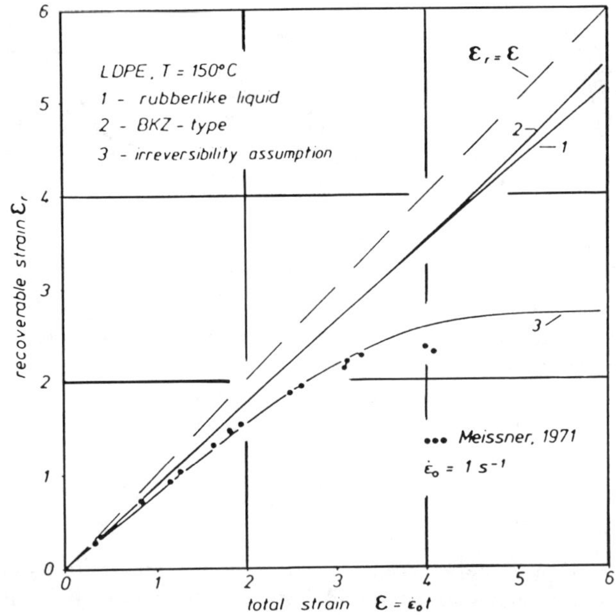

Fig. 5-10 Recoverable elongational strain after prior steady elongation for Melt I compared to: 1) the Lodge rubberlike liquid equation, 2) the reversible Wagner I equation, and 3) the irreversible Wagner II equation (from Wagner and Meissner[28], reprinted, with permission, from Makromolekular Chemie).

$$H\{I_1, I_2\} \equiv \min_{t''=t'}^{t''=t} h[I_1(t'',t'), I_2(t'',t')] \tag{5-49}$$

We call this Wagner's second equation (Wagner II) to distinguish it from his first (Wagner I), in which the damping function is reversible; see Section 3.5.6. Wagner's second equation is obtained from his first by replacing $h(I_1,I_2)$ with $H\{I_1,I_2\}$. If the strain history has no reversal of direction, Wagner's second equation makes stress predictions identical to those of his first. For the reversing double-step strain experiment, however, the prediction is different. For Wagner's second equation, Equation 5-48, the shear stress after the second step is given by

$$\sigma(t,t_1,\gamma_1,\gamma_2) = [G(t) - G(t+t_1)]\, h(\gamma_2)\,\gamma_2 + G(t+t_1)\, h(\gamma_1)\,(\gamma_1 + \gamma_2) \tag{5-50}$$

which differs in the last term from the prediction of Wagner's first equation or of the Kaye-BKZ equation, Equation 5-47. (The Kaye-BKZ equation and Wagner I equations are identical in their predictions of shear stress in shearing deformations.)

Figure 5-8 shows that excellent agreement with experiments on the LLDPE are obtained when Wagner's irreversibility is included in the integral equation.

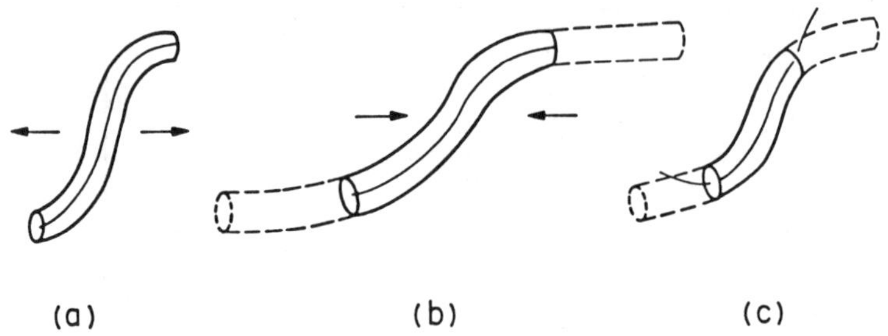

Fig. 5-11 Source of irreversibility in the tube model: (a) when a randomly oriented tube and the molecule in it are deformed, (b), the polymer retracts away from the tube ends, so that when the deformation is partially reversed, (c), the ends of the molecule are forced out of the tube and take on random configurations (from Larson and Valesano[21], reprinted from the Journal of Rheology © 1986 Soc. of Rheol., by permission of John Wiley & Sons, Inc.).

5.5.1.2 Recoverable Strain Wagner proposed his irreversibility assumption because of large discrepancies he found between the Kaye-BKZ equation and measurements of ***elongational strain recovery***. Figure 5-10 shows the recoverable elongational strain, ϵ_r, as a function of total imposed strain, $\epsilon = \dot{\epsilon}t$ Hencky units, measured by Meissner[23] on Melt I. The strain was imposed in a steady elongational deformation at a rate, $\dot{\epsilon}$, of 1 $\sec^{-1}$. As the prior strain, $\dot{\epsilon}t$, increases, the measured recoverable strain reaches a plateau. Thus the recoverable strain seems to be limited to about 2–3 Hencky units. The Wagner I or Kaye-BKZ equation, however, predicts no such limit. It predicts that the recoverable strain continues to increase as $\dot{\epsilon}t$ increases.

Now strain recovery is a reversing deformation, and the recoverable strain, ϵ_r, is reduced if Wagner's irreversibility hypothesis is introduced. Figure 5-10 shows that Wagner's irreversible damping function, Equation 5-49, applied to an elongational deformation, brings the predictions of the integral equation into agreement with the experimental data.

For recoverable shear, the difference between the Kaye-BKZ equation (or the Wagner I equation) and the Wagner II equation is not nearly so dramatic.

5.5.2 The Tube Picture

Another line of evidence that suggests irreversibility in nonaffine motion is the tube model of Doi. The DE model yields a Kaye-BKZ equation only after the independent alignment approximation is invoked;[24] see Chapter 4. The independent alignment approximation artificially forces each strand to remain associated with the same portion of tube as the deformation progresses. If the independent alignment approximation is *not* invoked, however, the deformed polymer strand shrinks away from the tube ends during a deformation, Figure 5-11. The ends of the initial tube are therefore "forgotten." When the deformation is reversed, the chain ends emerge from the still-remembered part of the tube and are free to take on random orientations. As a result, strand orientation, free energy, and stress are lost. This process occurs even if the deformation and its reversal are faster than the reptation process. Thus, because of the ***nonaffine process of retraction***, dissipation or irreversibility is introduced, which is not included in the linear relaxation spectrum $m(t-t')$. When this effect is accounted for — that is, when independent alignment is not assumed — the stress-strain relationship predicted by the DE model in the absence of reptation has hysteresis similar to that postulated by Wagner. The agreement of the DE equation with double-step strain experiments is improved when the independent alignment approximation is dropped, if the interval of time between the steps is greater than the retraction time.[20]

5.5.3 Differential Formulation of Irreversibility

The nonaffine motion proposed by Gordon and Schowaler cannot be made irreversible, but the nonaffine motion proposed by Larson can be. We take

$$\dot{\underset{\sim}{R}} = \underset{\sim}{R}\cdot\underset{\sim}{\nabla}\underset{\sim}{v} - \xi\underset{\sim}{R}\,[\underset{\sim}{u}\underset{\sim}{u}:\underset{\approx}{D}]_+ \tag{5-51}$$

The term in brackets, $[\cdot]_+$, is taken to be the value of the enclosed quantity if it is positive; if the enclosed quantity is negative, the term is set to zero. With this change, the retraction process is no longer reversed when the rate of deformation changes sign. The resulting convected derivative

$$\overset{\nabla}{\underset{\approx}{\sigma}} + \frac{2\xi'}{3G}[\underset{\approx}{\sigma}:\underset{\approx}{D}]_+\underset{\approx}{\sigma} \tag{5-52}$$

when set to zero for double-step strains with no relaxation between the steps has the solution

$$\underset{\approx}{\sigma} = H(I_1)\underset{\approx}{C}^{-1} \tag{5-53}$$

where

$$H\{I_1\} \equiv \min_{t''=t'}^{t''=t} h[I_1(t'',t')] \tag{5-54}$$

and $h = (1+\xi'(I_1-3)/3)^{-1}$. Here t' is an instant of time before the first deformation. This irreversibility, Equation 5-54, is the type proposed by Wagner and is depicted in Figure 5-9. Equation 5-52 can be incorporated into a simple differential equation

$$\overset{\nabla}{\underset{\approx}{\sigma}} + \frac{2\xi'}{3G}[\underset{\approx}{\sigma}:\underset{\approx}{D}]_+\underset{\approx}{\sigma} = -\left(\frac{1}{\lambda}\underset{\approx}{\sigma} - G\underset{\approx}{\delta}\right) \tag{5-55}$$

which is then an analog of Wagner's integral representation of irreversibility.

If we want to fit data in reversing double-step strain and recoverable strain experiments, it is not as necessary to introduce irreversibility into the differential equations as it is to introduce it into the integral equations, because the reversible differential equations are in better agreement with the data than are the reversible integral equations.[21]

5.6 WHITE-METZNER EQUATION

We have seen that nonaffine motion, which produces shear thinning and strain softening, should be thought of as a *relaxation* process that occurs faster than the deformation. In 1963, White and Metzner[25] introduced a modification of the UCM differential equation that, like irreversible nonaffine motion, allows relaxation that is faster than any deformation. The White-Metzner (WM) constitutive equation is

$$\lambda(\mathrm{II_D})\,\overset{\nabla}{\underset{\approx}{\sigma}} + \underset{\approx}{\sigma} = \mathrm{G}\,\underset{\approx}{\delta} \tag{5-56}$$

where λ *depends on the rate of strain*, particularly on the second invariant of the rate of strain tensor,

$$\mathrm{II_D} \equiv 2\,\underset{\approx}{\mathrm{D}}:\underset{\approx}{\mathrm{D}} \tag{5-57}$$

In simple shearing, $\mathrm{II_D} = \dot{\gamma}^2$. This equation can be transformed to a more familiar form by defining a new stress tensor, $\underset{\approx}{\tau} \equiv \underset{\approx}{\sigma} - \mathrm{G}\underset{\approx}{\delta}$. Then

$$\lambda(\mathrm{II_D})\,\overset{\nabla}{\underset{\approx}{\tau}} + \underset{\approx}{\tau} = 2\eta(\mathrm{II_D})\,\underset{\approx}{\mathrm{D}} \tag{5-58}$$

with $\eta(\mathrm{II_D}) \equiv \mathrm{G}\lambda(\mathrm{II_D})$.

5.6.1 Steady-State Flows

The dependence of λ on $\mathrm{II_D}$ was proposed as a simple means of introducing shear thinning. Stresses for the WM equation in steady flows are obtained from those of the UCM equation merely by replacing λ in the latter with $\lambda(\mathrm{II_D})$, and η with $\eta(\mathrm{II_D})$. The shear viscosity decreases with increased $\dot{\gamma}$ if λ is a *decreasing* function of $\mathrm{II_D}$. Thus high strain rates *reduce* the relaxation time. One form for $\lambda(\mathrm{II_D})$ proposed by Ide and White[26] is

$$\lambda(\mathrm{II_D}) = \frac{\lambda_0}{1 + \mathrm{a}\lambda_0(2\underset{\approx}{\mathrm{D}}:\underset{\approx}{\mathrm{D}})^{1/2}} \tag{5-59}$$

where "a" is an empirical parameter. With this form for $\lambda(\mathrm{II_D})$, the steady-state viscosity and normal stress coefficients are

$$\eta = \mathrm{G}\,\lambda(\mathrm{II_D}) = \frac{\mathrm{G}\lambda_0}{1 + \mathrm{a}\lambda_0\dot{\gamma}}$$

$$\Psi_1 = 2\mathrm{G}\,[\lambda(\mathrm{II_D})]^2 = \frac{2\mathrm{G}\lambda_0^2}{(1 + \mathrm{a}\lambda_0\dot{\gamma})^2}$$

$$\Psi_2 = 0 \tag{5-60}$$

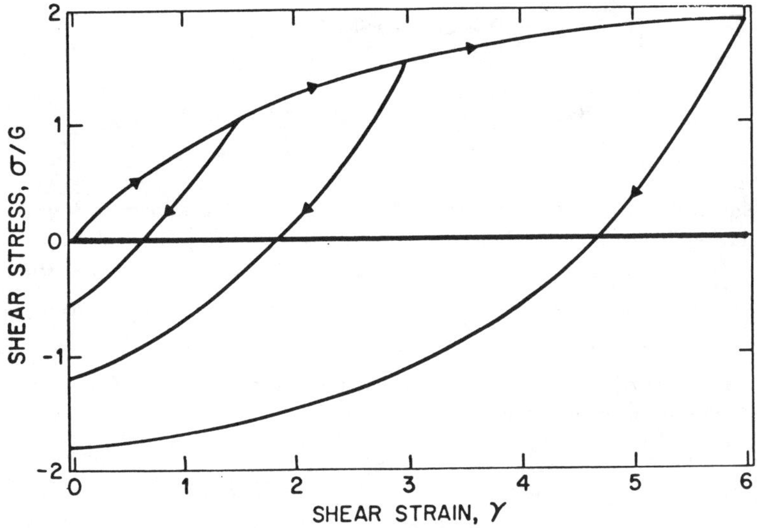

Fig. 5-12 Irreversibility of the White-Metzner equation, Eqns. 5-58 to 5-59, in step shearing.

In steady-state elongation,

$$\bar{\eta} = \frac{G(1 + a\sqrt{3}\lambda_0\dot{\epsilon})}{1 + (a\sqrt{3} - 2)\lambda_0\dot{\epsilon}} - \frac{G(1 + a\sqrt{3}\lambda_0\dot{\epsilon})}{1 + (a\sqrt{3} + 1)\lambda_0\dot{\epsilon}} \tag{5-61}$$

Thus the elongational singularity of the UCM equation is eliminated if $a > 2/\sqrt{3}$.

5.6.2 Sudden Deformations

The WM equation with the above form for $\lambda(II_D)$ bears a resemblance to irreversible nonaffine motion. Let the rate of deformation become arbitrarily large. Then $\lambda(II_D)$ becomes arbitrarily small, so the upper-convected derivative in Equation 5-56 does *not* dominate the other terms as it does for the UCM equation in sudden deformations. Instead $1/\lambda \rightarrow a(2\underset{\approx}{D}:\underset{\approx}{D})^{1/2}$. Thus for sudden deformations, the WM equation becomes

$$\overset{\nabla}{\underset{\approx}{\sigma}} + a(2\underset{\approx}{D}:\underset{\approx}{D})^{1/2}(\underset{\approx}{\sigma} - G\underset{\approx}{\delta}) = \underset{\approx}{0} \tag{5-62}$$

This result resembles a nonaffine convected derivative such as Equation 5-39. It is irreversible, since $(2\underset{\approx}{D}:\underset{\approx}{D})^{1/2}$ is always positive. After a step shear the stress components are

$$\sigma_{12} = \frac{G}{a}[1 - \exp(-a\gamma)]\, e^{-t/\lambda}$$

$$\sigma_{11} - \sigma_{22} = \frac{G}{a}[\frac{2}{a}(1 - (1 + a\gamma)\exp(-a\gamma))]e^{-t/\lambda} \tag{5-63}$$

The shear stress, σ_{12}, as a function of shear strain, γ, is shown in Figure 5-12. The deviation from linearity (from an affine response) is produced by the term involving $\underset{\approx}{D}:\underset{\approx}{D}$. This term is irreversible and yields hysteresis loops when the strain is reversed, just as does irreversible nonaffine motion; compare Figures 5-12 and 5-9.

The ratio of $(\sigma_{11} - \sigma_{22})/\sigma_{12}$ for the WM equation deviates from the shear strain γ, however, as γ increases,

$$\frac{\sigma_{11} - \sigma_{22}}{\sigma_{12}} = \frac{2}{a} - \frac{2\gamma}{e^{a\gamma} - 1} \tag{5-64}$$

Thus the Lodge-Meissner relationship is broken.[21]

5.7 SUMMARY

This chapter presented semi-empirical constitutive models in which the stretching of polymer strands in the limit of infinitely rapid deformation is not affine. Nonaffine motion is represented in differential constitutive equations by a convected derivative and in integral equations by an elastic strain measure. For affine motion, the convected derivative is the upper-convected derivative, and the elastic strain measure is the Finger tensor.

The simplest nonaffine motion is the "slippage" proposed by Gordon and Schowalter for polymers in dilute solution. Although others have applied the Gordon-Schowalter model of nonaffine motion to melts, it predicts behavior in transient shear flows that is uncharacteristic of melts, such as stress oscillations and a violation of the Lodge-Meissner relationship. This unrealistic behavior can be traced to the lack of corotation of the principal axes of stress and strain, predicted by the model in step shearing. Corotation of principal stress and strain axes is achieved in the convected derivatives or elastic strain measures of Seth and of Larson. Larson's nonaffine model is a semi-empirical generalization of the Doi-Edwards model that accounts for partial retraction of strands. It provides a reasonable description of commercial polymers in step shears.

Reversing strain histories such as double-step strains and elastic recovery, as well as tube theory, indicate that nonaffine motion should really be considered irreversible. Irreversibility can be accounted for in a relatively simple way in either the integral equations or differential constitutive equations. The White-Metzner equation is an early empirical constitutive equation that is irreversible in step strains.

References

1. R.J. Gordon and W.R. Schowalter, Trans. Soc. Rheol., *16*, 79 (1972).

2. M.W. Johnson, Jr. and D. Segalman, J. Non-Newtonian Fluid Mech., *2*, 225 (1977).
3. N. Phan Thien and R.I. Tanner, J. Non-Newtonian Fluid Mech., *2*, 353 (1977).
4. R.G. Larson, J. Non-Newtonian Fluid Mech., *13*, 279 (1983).
5. C.J.S. Petrie, J. Non-Newtonian Fluid Mech., *5*, 147 (1979).
6. B.R. Seth in M. Reiner and D. Abir Ed., *Second-Order Effects in Elasticity, Plasticity, and Fluid Mechanics*, Pergamon, Oxford, 162 (1964).
7. C.J.S. Petrie, *Elongational Flows*, Pitman, London, 1979.
8. M. Keentok, A.G. Georgescu, A.A. Sherwood, and R.I. Tanner, J. Non-Newtonian Fluid Mech., *6*, 303 (1980).
9. G.A. Alvarez, A.S. Lodge, and H.-J. Cantow, Rheol. Acta, *24*, 368 (1985).
10. S. Ramachandran, H.W. Gao, and E.B. Christiansen, Macromolecules, *18*, 695 (1985).
11. R.B. Bird, R.C. Armstrong, and O. Hassager, *Dynamics of Polymeric Liquids*, Vol. 1, John Wiley & Sons, New York (1977).
12. O. Hassager and S. Pedersen, J. Non-Newtonian Fluid Mech., *4*, 261 (1978).
13. R.G. Larson, J. Non-Newtonian Fluid Mech., *24*, 331 (1987).
14. H.M. Laun, Rheol. Acta, *17*, 1 (1978).
15. P.R. Soskey and H.H. Winter, J. Rheol., *28*, 625 (1984).
16. Y. Einaga, K. Osaki, and M. Kurata, S. Kimura, and M. Tamura, Polym. J., *2*, 550 (1971).
17. R.G. Larson, J. Rheol., *29*, 823 (1985).
18. M. Doi and S.F. Edwards, J. Chem. Soc., Faraday Trans. II, *75*, 32 (1979).
19. R.G. Larson, J. Rheol., *28*, 545 (1984).
20. K. Osaki, S. Kimura, and M. Kurata, J. Rheol, *25*, 549 (1981).
21. R.G. Larson and V.A. Valesano, J. Rheol., *30*, 1093 (1986).
22. M.H. Wagner, J. Non-Newtonian Fluid Mech., *4*, 39 (1978).
23. M.H. Wagner and J. Meissner, Makromol. Chem. *181*, 1533 (1980).
24. M. Doi and S.F. Edwards, J. Chem. Soc., Faraday Trans. II, *74*, 1802 (1978).
25. J.L. White and A.B. Metzner, J. Appl. Polym. Sci., *8*, 1367 (1963).
26. Y. Ide and J.L. White, J. Non-Newtonian Fluid Mech., *2*, 281 (1977).

CHAPTER 6

Nonseparable Constitutive Models

"the race is not to the swift,
nor the battle to the strong,
neither yet bread to the wise,
nor yet riches to men of understanding,
nor yet favour to men of skill;
but time and chance happeneth to them all."

Ecclesiastes 9:11

6.1 INTRODUCTION

In the molecular models of Chapter 5, strain softening occurs because of *nonaffine motion*. The White-Metzner (WM) model considered at the end of Chapter 5 is an exception to this; in the WM model strain softening occurs because *the relaxation is accelerated at high strain rates*. In this chapter, we consider other models for which accelerated relaxation causes strain softening, but for some of these models, *high stresses* not high strain rates, speed up the relaxation. In many of these models, the strain softening occurs gradually, not instantaneously, and the response to a step strain is not separable into time- and strain-dependent parts.

We consider two categories of constitutive models: the Giesekus and Leonov models and the network models. The former models have a term that is quadratic in stress. This term allows realistic prediction of stresses in shearing flows, but is less accurate for extensional flows. There is a wide variety of network models. They generalize the Green and Tobolsky model by allowing strand creation and destruction probabilities to be functions of strand extension, of stress, of strain rate, or of the histories of stress or strain rate. These functional dependencies are chosen empirically and can make the network models fit data in extensional as well as in shear flow.

6.2 GIESEKUS AND LEONOV MODELS

6.2.1 Giesekus Model

In 1966 Giesekus[1] proposed a class of constitutive equations based on *anisotropic drag*. The motivation for the model comes from considering how the relaxation of an elastic dumbbell is modified when it is surrounded by other oriented dumbbells. Thus Giesekus attempted to derive a theory for concentrated solutions or melts by starting from the simple dumbbell theory for dilute solutions.

For a dumbbell in dilute solution, the appropriate constitutive equation (Section 2.6.3) is the upper-convected Maxwell (UCM) equation,

$$\overset{\nabla}{\underset{\approx}{\sigma}} + \frac{8kT\beta^2}{\varsigma}(\underset{\approx}{\sigma} - G\underset{\approx}{\delta}) = \underset{\approx}{0} \tag{6-1a}$$

Here ς is the bead friction coefficient; the term $8kT\beta^2/\varsigma$ is the reciprocal relaxation time, $1/\lambda$. Giesekus speculated that the surrounding sea of oriented molecules creates an anisotropic drag; thus ς becomes *direction dependent*! To represent this, Giesekus replaced $8kT\beta^2/\varsigma$ by an *anisotropic* mobility tensor, $\underset{\approx}{B}/\lambda$:

$$\lambda\overset{\nabla}{\underset{\approx}{\sigma}} + \underset{\approx}{B}\cdot(\underset{\approx}{\sigma} - G\underset{\approx}{\delta}) = \underset{\approx}{0} \tag{6-1b}$$

$\underset{\approx}{B}$ depends on the anisotropy of the orientation of surrounding molecules and thus depends on the state of stress in the material. At equilibrium the stress is isotropic, so $\underset{\approx}{B}$ must be isotropic also; i.e. $\underset{\approx}{B} = \underset{\approx}{\delta}$ when $\underset{\approx}{\sigma} = G\underset{\approx}{\delta}$. When the stress is anisotropic, so is $\underset{\approx}{B}$. Giesekus assumed that the anisotropy in $\underset{\approx}{B}$ is proportional to the anisotropy in $\underset{\approx}{\sigma}$:

$$\underset{\approx}{B} - \underset{\approx}{\delta} = \alpha(\underset{\approx}{\sigma}/G - \underset{\approx}{\delta}) \tag{6-2}$$

Here α is an empirical constant of proportionality. The resulting constitutive equation is

$$\lambda\overset{\nabla}{\underset{\approx}{\tau}} + \underset{\approx}{\tau} + \frac{\alpha}{G}\underset{\approx}{\tau}^2 = 2G\lambda\underset{\approx}{D} \tag{6-3}$$

where $\underset{\approx}{\tau} \equiv \underset{\approx}{\sigma} - G\underset{\approx}{\delta}$ is a stress tensor that differs from $\underset{\approx}{\sigma}$ by the isotropic constant, $G\underset{\approx}{\delta}$. For incompressible fluids, the stress tensor is only defined to within an isotropic constant, so either $\underset{\approx}{\sigma}$ or $\underset{\approx}{\tau}$ can be used. The simple Giesekus equation, Equation 6-3, differs from the UCM equation by a term quadratic in stress. A more general form of the Giesekus equation can be found in Ref. 1.

The maximum and minimum anisotropies correspond to $\alpha = 1$ and $\alpha = 0$, respectively. α may not lie outside this range; if it does, stress magnitudes can grow rather than relax when deformation ceases. When $\alpha = 0$ (isotropic drag), the upper-convected Maxwell (UCM) equation is recovered. Amazingly, when α is set to unity, which corresponds to the most severely anisotropic drag, the Giesekus model predicts *steady-state* shear and extensional stresses that are identical to those of the corotational Maxwell (CRM) equation. (The transient properties of the Giesekus and CRM equations are different, however.) Thus as α is increased from 0 to 1, strain softening increases from that of the UCM equation, which we define to have no strain softening, to strain softening typical of the CRM equation.

The Giesekus equation and a related equation of Leonov discussed in the next section fit steady and transient shear data for commercial melts better than any other constitutive equation except that of Wagner.[2] The Giesekus and Leonov models do well in predicting steady-state shear viscosity, first and second normal stress differences, growth and relaxation of shear viscosity and first normal stress difference, and are reasonably accurate in single- and double-step shears.[3] Furthermore, good or excellent fits are often obtained with as few as two or three relaxation modes. Better constitutive equations[4a], including more general forms of the Giesekus equation[1,4b], can be found for predicting stresses in extensional flows, however.

Because of their similarity, we discuss the predictions of both the Giesekus and the Leonov equations after a discussion of the theory underlying the Leonov model.

6.2.2 Leonov model

Leonov's constitutive equation in its simplest form is similar to Giesekus', but is derived from a thermodynamic rather than a molecular approach.[5–7]

6.2.2.1 Basis of the Leonov Model Leonov's approach is to relate the stress tensor to the elastic strain stored in the material. Leonov was not the first to propose this idea; Weissenberg[8] suggested it as early as 1935. In Leonov's formulation, the stored elastic strain is allowed to dissipate or "leak away" by nonequilibrium processes (i.e., relaxation). We first present the simplest single-mode version of Leonov's model. The generalization to multiple modes is made in the usual way (Section 3.4.4.1).

In the single-mode version, the stored *elastic strain* is precisely the strain that would be recovered if *all* loads were suddenly removed from the sample. In shearing, the elastic strain is *not* the same as the recoverable shear, because in measuring shear recovery the normal forces are *not* removed, but instead the sample is *constrained* so that only shear strain can be recovered. Leonov's *elastic strain* is a tensorial quantity that describes the change in shape of the material element in *free recovery*.[9] In free recovery, all forces, shear and normal, are suddenly removed and the sample is free to recoil by both shear and extensional deformations. Free recovery in shear is akin to *extrudate swell*,[9] because when the viscoelastic liquid emerges after being sheared in the die, both shearing and normal stresses are suddenly removed, and *swelling*, which is an extensional deformation, occurs.

The change in shape of the material element in free recovery can be described by a Finger tensor, $\underset{\approx}{C}_e^{-1}$. $\underset{\approx}{C}_e^{-1}$ is the *elastic strain tensor* of Leonov.*

There are now two crucial steps to obtain the Leonov constitutive equation. The first is to relate the stress tensor to the elastic strain tensor. The second is to describe how the elastic strain tensor depends on the deformation history. In

* Leonov used $\underset{\approx}{C}_e$ to denote the elastic Finger tensor; $\underset{\approx}{C}_e^{-1}$ is used here because it is consistent with the notation for the Finger tensor used elsewhere in the book.

this way the dependence of the stress tensor on the deformation history is indirectly established.

6.2.2.2 Relating Stress and Elastic Strain The first crucial assumption of the Leonov model is that *the stress in a viscoelastic liquid is that of a rubber with the same elastic strain as the viscoelastic liquid.* Thus the *nonequilibrium* stress in a viscoelastic liquid at a particular instant is approximated by the *equilibrium* stress in a rubber with the same free recovery as the viscoelastic liquid has at that instant. Mathematically,

$$\underset{\approx}{\sigma} = 2\frac{\partial W}{\partial I_1}\underset{\approx}{C}_e^{-1} - 2\frac{\partial W}{\partial I_2}\underset{\approx}{C}_e + \text{isotropic term} \tag{6-4}$$

where W, the elastic potential, depends on the invariants of $\underset{\approx}{C}_e^{-1}$, here I_1 and I_2. In the simplest case, W is given by the classical theory of rubber elasticity; i.e., $W = ½G I_1$. Then

$$\underset{\approx}{\sigma} = G\underset{\approx}{C}_e^{-1} \tag{6-5}$$

to within an isotropic term. Hence *the stress tensor is proportional to the elastic strain tensor.*

This direct linkage between the elastic strain and the stress gives the Leonov equation an advantage over the Kaye-BKZ equation in describing recoverable strains. In the Kaye-BKZ superposition integral, there is no direct connection between the magnitude of the stress and the recoverable strain; thus the recoverable strain can continue to increase even though the stress reaches a plateau. As a result, the Kaye-BKZ equation often *overpredicts* recoverable strains, especially in elongational deformations; see Section 5.5.1. With the Leonov equation, if the stress reaches a plateau, so does the recoverable strain. Experimentally, plateaus in recoverable strain seem to occur when the stress reaches a plateau; thus more accurate predictions of recoverable strain are obtained with the Leonov model than with the Kaye-BKZ equation.[3]

6.2.2.3 Relating Elastic Strain and the Deformation History If there were no mechanisms for elastic strain to relax — that is, if the material were truly a rubber — then $\underset{\approx}{C}_e^{-1}$ would satisfy the identity (see Sections 1.5.6, 3.4.4),

$$\underset{\approx}{0} = \overset{\nabla}{\underset{\approx}{C}}{}_e^{-1} = \overset{0}{\underset{\approx}{C}}{}_e^{-1} - \underset{\approx}{D}\cdot\underset{\approx}{C}_e^{-1} - \underset{\approx}{C}_e^{-1}\cdot\underset{\approx}{D} \tag{6-6}$$

Leonov assumed that dissipative processes act to produce an internal or *irreversible rate of strain,* $\underset{\approx}{D}_p$, that spontaneously reduces the rate of accumulation of elastic strain. Thus the rate of change of the elastic strain is determined by $\underset{\approx}{D}_e \equiv \underset{\approx}{D} - \underset{\approx}{D}_p$, the *elastic* rate of deformation. For viscoelastic liquids, Equation 6-6 is therefore modified to

$$\underset{\approx}{0} = \overset{0}{\underset{\approx}{C}}{}_e^{-1} - \underset{\approx}{D}_e \cdot \underset{\approx}{C}_e^{-1} - \underset{\approx}{C}_e^{-1} \cdot \underset{\approx}{D}_e$$

$$= \overset{0}{\underset{\approx}{C}}{}_e^{-1} - (\underset{\approx}{D} - \underset{\approx}{D}_p) \cdot \underset{\approx}{C}_e^{-1} - \underset{\approx}{C}_e^{-1} \cdot (\underset{\approx}{D} - \underset{\approx}{D}_p)$$

Thus

$$\underset{\approx}{0} = \overset{\nabla}{\underset{\approx}{C}}{}_e^{-1} + \underset{\approx}{D}_p \cdot \underset{\approx}{C}_e^{-1} + \underset{\approx}{C}_e^{-1} \cdot \underset{\approx}{D}_p \tag{6-7}$$

Leonov assumed that $\underset{\approx}{D}_p$ could be related to the stress tensor by an analog of Equation 6-4,

$$\underset{\approx}{D}_p = 2\frac{\partial \psi}{\partial I_1}\underset{\approx}{C}_e^{-1} - 2\frac{\partial \psi}{\partial I_2}\underset{\approx}{C}_e + \text{isotropic term} \tag{6-8}$$

Leonov called $\psi(I_1, I_2)$ the "nonequilibrium potential." The isotropic term is determined by the requirement of incompressibility, namely that the irreversible strain rate be traceless. Hence

$$\underset{\approx}{D}_p = 2\frac{\partial \psi}{\partial I_1}(\underset{\approx}{C}_e^{-1} - \frac{1}{3}I_1 \underset{\approx}{\delta}) - 2\frac{\partial \psi}{\partial I_2}(\underset{\approx}{C}_e - \frac{1}{3}I_2 \underset{\approx}{\delta}) \tag{6-9}$$

$\underset{\approx}{D}_p$, as given above is traceless, since $I_1 = \mathrm{tr}\underset{\approx}{C}_e^{-1}$, $I_2 = \mathrm{tr}\underset{\approx}{C}_e$, and $\mathrm{tr}\underset{\approx}{\delta} = 3$.

The last step is to specify the nonequilibrium potential, $\psi(I_1, I_2)$. Seemingly arbitrarily, Leonov assumed that $\psi(I_1, I_2)$ is a function of the *symmetrized elastic potential*, W:

$$\psi(I_1, I_2) = \psi(W_s) \tag{6-10}$$

where

$$W_s = ½\,[W(I_1, I_2) + W(I_2, I_1)] \tag{6-11}$$

Leonov chose the simplest dependence of ψ on W_s:

$$\psi = \frac{1}{2G\lambda} W_s = \frac{1}{8\lambda}(I_1 + I_2 - 6) \tag{6-12}$$

where W was obtained from the classical theory of rubber elasticity, $W = ½ I_1$. The value of the constant $1/2G\lambda$ in Equation 6-12, was chosen so that in the constitutive equation that finally emerges, λ is the relaxation time.

With these assumptions, Equation 6-9 reduces to

$$\underset{\approx}{D}_p = \frac{1}{4\lambda}[\underset{\approx}{C}_e^{-1} - \underset{\approx}{C}_e + \frac{1}{3}(I_2 - I_1)\underset{\approx}{\delta}] \tag{6-13}$$

6.2.2.4 The Constitutive Equation For unsteady properties, the constitutive equation is obtained by inserting Equation 6-13 into Equation 6-7. When $\underset{\approx}{C}_e^{-1}$ is replaced by $\underset{\approx}{\sigma}/G$, the result is the simplest Leonov constitutive equation:

$$\overset{\nabla}{\underset{\approx}{\sigma}} + \frac{1}{2G\lambda}\left[\underset{\approx}{\sigma}\cdot\underset{\approx}{\sigma} - G^2\underset{\approx}{\delta} - \frac{1}{3}[\mathrm{tr}\underset{\approx}{\sigma} - G^2\mathrm{tr}(\underset{\approx}{\sigma}^{-1})]\,\underset{\approx}{\sigma}\right] = \underset{\approx}{0} \qquad (6\text{-}14)$$

6.2.2.5 Relationship to the Giesekus Equation In incompressible planar deformations, $\mathrm{tr}\underset{\approx}{C}_e^{-1} = \mathrm{tr}\underset{\approx}{C}_e$; hence $\mathrm{tr}\underset{\approx}{\sigma} = G^2\,\mathrm{tr}(\underset{\approx}{\sigma})^{-1}$ (see Equation 6-5), and the Leonov equation simplifies to

$$\lambda\overset{\nabla}{\underset{\approx}{\tau}} + \underset{\approx}{\tau} + \frac{1}{2G}\underset{\approx}{\tau}\cdot\underset{\approx}{\tau} = 2G\lambda\underset{\approx}{D} \qquad (6\text{-}15)$$

where $\underset{\approx}{\tau} \equiv \underset{\approx}{\sigma} - G\underset{\approx}{\delta}$. This result is identical to the Giesekus equation, Equation 6-3, with $\alpha = ½$. Thus the simple Leonov equation is equivalent to a special case of the Giesekus equation in all planar deformations, including shear.

The Leonov and Giesekus models can be extended in the usual way to include a spectrum of relaxation times. Apart from the linear viscoelastic constants, the Giesekus model has a single parameter, α, and the Leonov model has none at all.

6.2.2.6 Relationship to the Kaye-BKZ Equation Leonov's theory in its general form can be regarded as an alternative to the Kaye-BKZ approach. Both the Leonov and the Kaye-BKZ theories start from a general elastic potential for rubbery solids and then allow relaxation. The Kaye-BKZ equation is an ***integral superposition*** that linearly adds up the contributions to the stress from past strains. This approach is good when all the strains are in the same direction, but is often poor when there are both positive and negative contributions. Leonov's method of obtaining the stress from a relaxing elastic strain is better suited to predicting recoverable strains and possibly reversing double-step strains.[3] However, the general form of the Leonov model is much harder to work with than the Kaye-BKZ equation, because the Leonov potentials $W(I_1, I_2)$, and $\psi(I_1, I_2)$, unlike the Kaye-BKZ potential $u(I_1, I_2, t-t')$, are not directly obtainable from the stresses in simple deformations such as step strains.

6.2.2.7 Numerical Advantages of the Leonov Equation The simple form of the Leonov model with multiple modes has proved to be accurate and easy to compute with, not only in simple shear but also in shear-dominated flows, such as transient flows in channels of varying cross section.[10] Flows that are strongly elongational are, however, poorly described by the Leonov equation, at least in its simplest form.

Because the stress tensor for each mode, $\underset{\approx}{\sigma}_i$, is proportional to a strain tensor, $\underset{\approx}{C}_e^{-1}$, which is constrained to have unit determinant, there is a constraint on the components of $\underset{\approx}{\sigma}_i$ that reduces by one the number of differential equations that must be solved for that mode. Furthermore, only two or three modes are usually

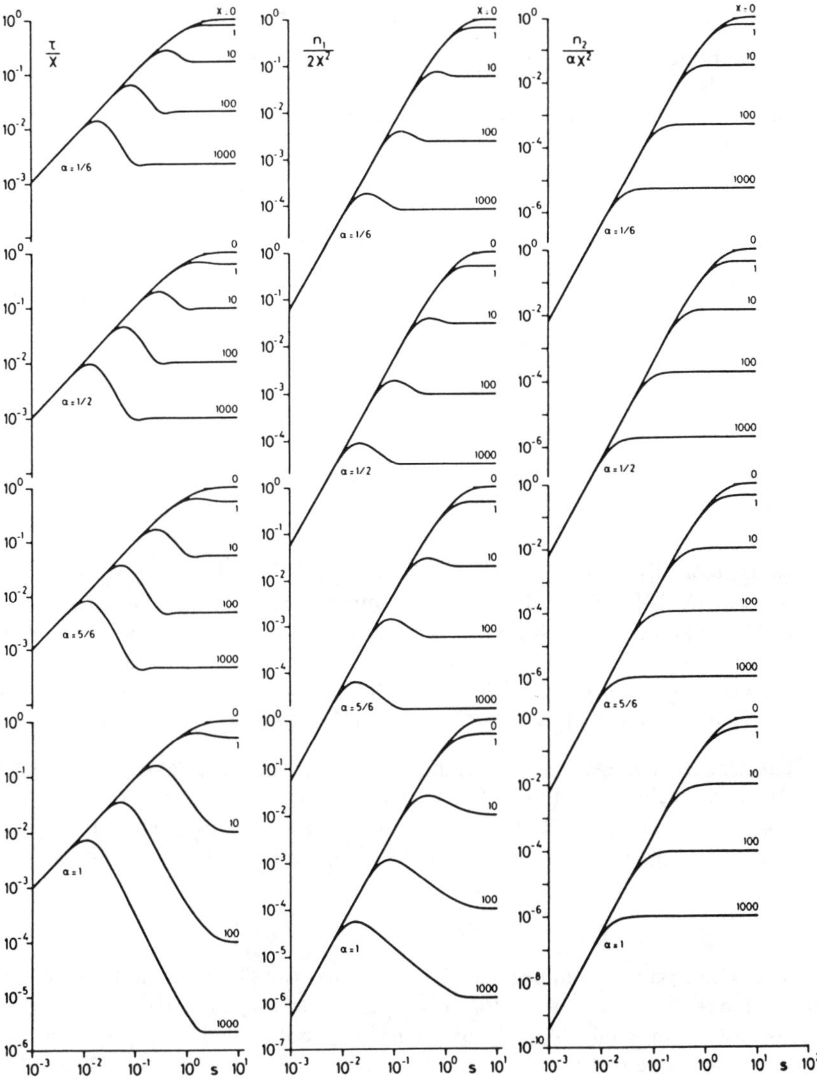

Fig. 6-1 Growth of dimensionless shear viscosity $\tau/\chi = \eta/\eta_0$, first normal stress coefficient $n_1/2\chi^2 = \psi_1/\psi_{10}$, and second normal stress coefficient $n_2/\alpha\chi^2 = \psi_2/\psi_{20}$, with dimensionless time $s = t/\lambda$ after start-up of steady shearing at dimensionless shear rate $\chi = \dot{\gamma}\lambda$ for the Giesekus model with α values of 1/6, 1/2, 5/6, and 1. (from Giesekus[11], reproduced with permission from the J. Non-Newtonian Fluid Mech.).

required to describe adequately the nonlinear response.[3] Since multidimensional viscoelastic flows usually tax computational resources to the limit, the reduced number of unknowns afforded by the Leonov equation is welcome.

6.2.3 Predictions of the Leonov and Giesekus Models

6.2.3.1 Shear The Giesekus model with $\alpha = 1/2$ is identical to the Leonov model in planar flows including shear. The predicted shear viscosity and normal stress coefficients for the Leonov or Giesekus model with $\alpha = 1/2$ are[3]

$$\begin{aligned} \eta &= \frac{2\eta_0}{1+X} \\ \Psi_1\dot{\gamma}^2 &= N_1 = \frac{\sqrt{2}\,G\,(X-1)}{\sqrt{1+X}} \\ \Psi_2\dot{\gamma}^2 &= N_2 = -G + \frac{\sqrt{2}\,G}{\sqrt{1+X}} \end{aligned} \tag{6-16}$$

where

$$X \equiv \sqrt{1+4\dot{\gamma}^2\lambda^2} \quad \text{and} \quad \eta_0 \equiv \lambda G \tag{6-17}$$

Asymptotically at large $\dot{\gamma}$, $\eta \equiv \eta_0(\lambda\dot{\gamma})^{-1}$, $\Psi_1 = 2\eta_0\lambda(\lambda\dot{\gamma})^{-3/2}$, and $\Psi_2 = -\eta_0\lambda\,(\lambda\dot{\gamma})^{-2}$. At small $\dot{\gamma}$, N_2/N_1 approaches –1/4. For the Giesekus model with other values of α, $N_2/N_1 \sim -\alpha/2$ at small $\dot{\gamma}$. Thus, by varying α, the Giesekus model can produce any value of N_2/N_1 between 0 and –1/2. Experimental values generally fall in the range –0.05 to –0.3. For $0<\alpha<1$, the Giesekus model predicts the scaling $\eta \propto \dot{\gamma}^{-1}$, $\Psi_1 \propto \dot{\gamma}^{-3/2}$, and $\Psi_2 \propto -\dot{\gamma}^{-2}$ at high $\dot{\gamma}$.

The transient shear viscosities and normal stress coefficients after start-up of steady shearing for $\alpha = 1/6$, 1/2, 5/6, and 1 appear in Figure 6-1. Overshoots in both η and Ψ_1 are predicted. The qualitative agreement of the predicted transients in η and Ψ_1 with the measured ones, Figure 3-1 and 3-2, is impressive.

6.2.3.2 Extension In nonplanar deformations, the Giesekus and Leonov models are no longer identical, even for $\alpha = 1/2$. However, in start-up of any steady extension — uniaxial, biaxial, or planar — both models show a monotonic increase with time of the corresponding extensional viscosity, up to a steady-state value. For the Giesekus model, the steady-state value at high $\dot{\epsilon}$ is $2\eta_0/\alpha$ for each of the extensional viscosities. For the Leonov model, the steady-state extensional viscosities at high extension rate, in contrast with those of the Giesekus model with $\alpha = 1/2$, are as follows

Extensional Viscosities at High $\dot{\epsilon}$

		Giesekus	**Leonov**
Uniaxial extensional viscosity,	$\bar{\eta}_u$	$4\eta_0$	$6\eta_0$
Planar extensional viscosity,	$\bar{\eta}_{p,1}$	$4\eta_0$	$4\eta_0$
Biaxial extensional viscosity,	$\bar{\eta}_b$	$4\eta_0$	$\sqrt{12}\,\eta_0(\dot{\epsilon}\lambda)^{-1/2}$

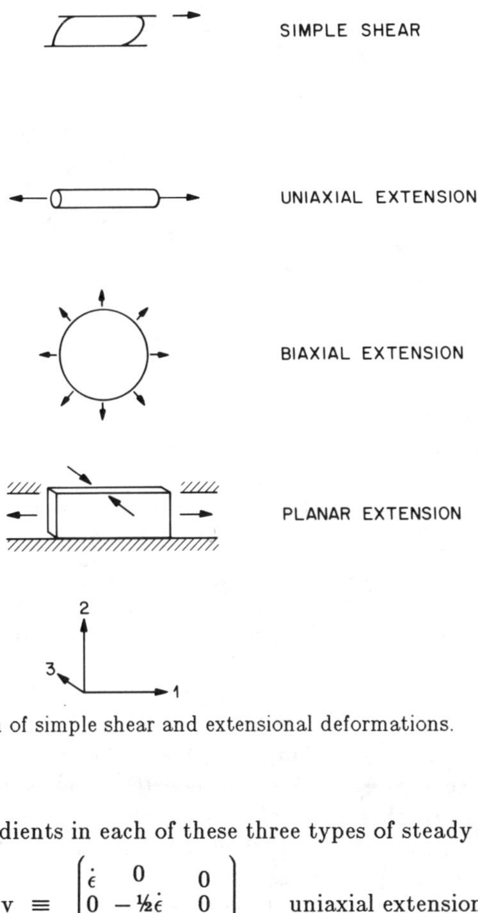

Fig. 6-2 Illustration of simple shear and extensional deformations.

The velocity gradients in each of these three types of steady extension are

$$\underset{\sim\sim}{\nabla \mathrm{v}} \equiv \begin{pmatrix} \dot{\epsilon} & 0 & 0 \\ 0 & -\tfrac{1}{2}\dot{\epsilon} & 0 \\ 0 & 0 & -\tfrac{1}{2}\dot{\epsilon} \end{pmatrix} \qquad \text{uniaxial extension}$$

$$\underset{\sim\sim}{\nabla \mathrm{v}} \equiv \begin{pmatrix} \dot{\epsilon} & 0 & 0 \\ 0 & 0 & 0 \\ 0 & 0 & -\dot{\epsilon} \end{pmatrix} \qquad \text{planar extension}$$

$$\underset{\sim\sim}{\nabla \mathrm{v}} \equiv \begin{pmatrix} \dot{\epsilon} & 0 & 0 \\ 0 & \dot{\epsilon} & 0 \\ 0 & 0 & -2\dot{\epsilon} \end{pmatrix} \qquad \text{biaxial extension}$$

These three types of extensional deformation, along with simple shear, are depicted in Figure 6-2. The uniaxial and biaxial extensional viscosities are defined as $(\sigma_{11} - \sigma_{33})/\dot{\epsilon}$. Planar extension has two extensional viscosities;

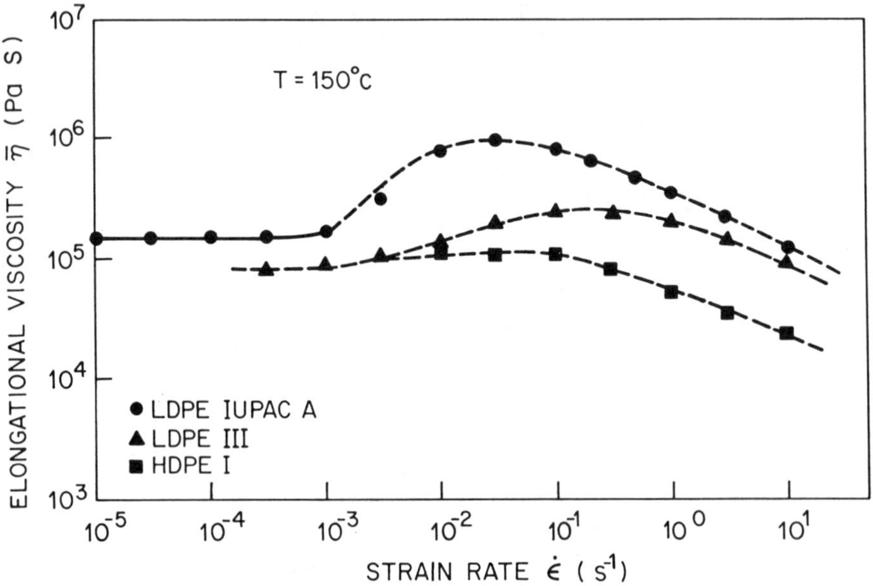

Fig. 6-3 Apparent steady-state elongational viscosities for three melts (from Laun[14], reproduced with permission from the Proc. Ninth Int. Congress Rheol.).

$\overline{\eta}_{p,1} \equiv (\sigma_{11} - \sigma_{33})/\dot{\epsilon}$, and $\overline{\eta}_{p,2} \equiv (\sigma_{22} - \sigma_{33})/\dot{\epsilon}$. The above table displays the dominant planar viscosity, $\overline{\eta}_{p,1}$. The limiting extensional viscosities of the Leonov model are more realistic than those of Giesekus in that experimental values of $\overline{\eta}_u$ tend to be greater than those of $\overline{\eta}_b$ at high extension rates.[12,13] The monotonic increase of the uniaxial extensional viscosity with extension rate is not characteristic of most experimental data, however. Experiments usually show a maximum in the apparent steady-state value of $\overline{\eta}_u$ as $\dot{\epsilon}$ increases;[14] see Figure 6-3. This maximum is large if the polymer molecules have long side branches or if the molecular weight distribution has a pronounced high-molecular-weight tail.[15]

Thus both the simplest Giesekus and Leonov models are accurate in their predictions of shear flows and inaccurate in their description of uniaxial extensional flows. A detailed comparison of the Giesekus model with experimental melt data appears in Chapter 7.

6.2.3.3 Step Strains The Giesekus and Leonov models both involve the upper-convected derivative. The immediate response to a step strain is therefore governed by

$$\overset{\nabla}{\underset{\approx}{\sigma}} = \underset{\approx}{0} \tag{6-18}$$

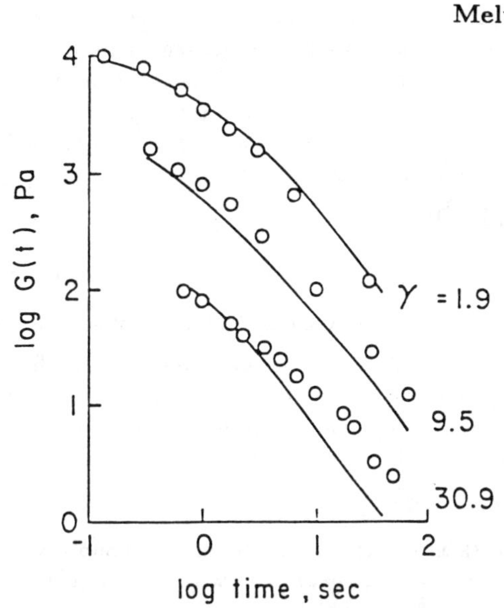

Fig. 6-4 Nonlinear shear modulus after step shear strains of magnitude γ for Melt I and the Leonov model. For Melt I, G(t) for each strain is parallel to G(t) for other strains; that is, the data show time-strain separability. The Leonov model is *approximately* time-strain separable up to a strain of 9.5 (from Upadhyay et al.[3], reproduced with permission from Rheologica Acta).

which implies that $\underset{\approx}{\sigma} = \mathrm{G}\underset{\approx}{\mathrm{C}}^{-1}$. The equation governing stress relaxation following the step is nonlinear, for example,

$$\lambda\dot{\underset{\approx}{\sigma}} + \underset{\approx}{\sigma} + \frac{\alpha}{\mathrm{G}}(\underset{\approx}{\sigma} - \mathrm{G}\underset{\approx}{\delta})^2 = \mathrm{G}\underset{\approx}{\delta} \tag{6-19}$$

for the Giesekus model. Thus the relaxation is *not* a simple exponential decay. Following a step extension, with stretch ratios λ_1, λ_2, and λ_3, the stress relaxation is given by[1]

$$\sigma_{\mathrm{ii}} = \frac{(\lambda_{\mathrm{i}}^2 - 1)\exp(-\mathrm{t}/\lambda)}{1 + \alpha(\lambda_{\mathrm{i}}^2 - 1)\exp(-\mathrm{t}/\lambda)} \tag{6-20}$$

for each normal stress component σ_{ii}. Equation 6-20 can*not* be factored into time- and strain-dependent parts. Thus $\underset{\approx}{\sigma} \neq \mathrm{G(t)}\underset{\approx}{\mathrm{E}}$ for the Giesekus model. Nonfactorability occurs whenever the relaxation terms in the constitutive equation are nonlinear in stress. This seems to contradict experimental data that show approximate factorability (Section 3.5.4), but when the broad spectrum of relaxation times typical of commercial melts is considered, the modes of the Giesekus and Leonov equations average together to produce a nearly factorable response. Figure 6-4 compares the stress relaxation after a step shear for Melt I, which exhibits a factorable response, with the response predicted by the Leonov model with a distribution of relaxation times appropriate for Melt I. The agreement is good for strains of about 10 or less.

6.3 NETWORK MODELS

We have seen that the upper-convected derivative $\overset{\nabla}{\underset{\approx}{\sigma}}$ when combined with the linear stress relaxation, $\lambda^{-1}(\underset{\approx}{\sigma} - G\underset{\approx}{\delta})$ predicts no shear thinning and singular extensional viscosities. In the Leonov and Giesekus models, this behavior is "softened" by the presence of nonseparable relaxation, specifically relaxation that is quadratic in the stress tensor. Nonseparable stress relaxation is also a source of strain softening in many of the modern *network models.*

6.3.1 Yamamoto's Model

The earliest network model is that of Green and Tobolsky;[16] see Chapter 2. It results in the *Lodge constitutive equation,*[17] which is equivalent to the UCM equation. In 1956, Yamamoto[18] presented a generalization of Green and Tobolosky's model, which allows *non-Gaussian chains,* and *a dependence of the breakage probability on chain extension.*

The generalization to non-Gaussian chains is necessary only at high chain extensions; by itself it leads to *positive second normal stress differences* and *shear thickening,*[19] which are not seen experimentally. We therefore drop consideration of non-Gaussian behavior and focus on nonconstant strand-breakage probability.

For Gaussian strands in affine motion, the stress tensor according to Section 2.3 is

$$\underset{\approx}{\sigma} = 2\nu kT\beta^2 \langle \underset{\sim}{R}\underset{\sim}{R} \rangle = 2\nu kT\beta^2 \int \underset{\sim}{R}\underset{\sim}{R}\psi \, dR^3 \qquad (6\text{-}21)$$

The stress can be evaluated once the probability distribution function, $\psi(\underset{\sim}{R})$, is known. For simplicity we assume that all strands are statistically the same. This leads to an equation with only one relaxation time, but the generalization to multiple relaxation times is straightforward.

According to Yamamoto, the distribution function obeys the time-evolution equation

$$\frac{\partial \psi}{\partial t} = -\frac{\partial}{\partial \underset{\sim}{R}} \cdot (\underset{\sim}{R} \cdot \nabla \underset{\sim}{v} \psi) + g - h\psi \qquad (6\text{-}22)$$

The first term on the right side is the usual one for affine motion. The second, g, the *strand creation probability,* is the probability per unit time that a strand with end-to-end vector, $\underset{\sim}{R}$, is created. The last term, $h\psi$, is the *strand destruction probability*; it is the probability per unit time that a strand with end-to-end vector, $\underset{\sim}{R}$, is destroyed.

Phan-Thien and Tanner[20] developed a network model in which they allowed ***nonaffine motion*** of the Gordon-Schowalter type. In their model,

$$\frac{\partial \psi}{\partial t} = -\frac{\partial}{\partial \underset{\sim}{R}} \cdot [\underset{\sim}{R} \cdot (\nabla \underset{\sim\sim}{v} - \xi \underset{\approx}{D}) \psi] + g - h\psi \tag{6-23}$$

The affine limit is recovered when $\xi = 0$.

All network models discussed here have Equation 6-23 in common. They differ in the assumptions made regarding the dependencies of g and h on chain extension and stress or strain rate. Multiplying Equation 6-23 by $\underset{\sim\sim}{RR}$ and integrating over all configurations, we obtain

$$\overset{\square}{\langle \underset{\sim\sim}{RR} \rangle}_a = \int g\underset{\sim\sim}{RR}\, dR^3 - \langle h\underset{\sim\sim}{RR} \rangle \tag{6-24}$$

where $\overset{\square}{\langle \underset{\sim\sim}{RR} \rangle}_a$ is the Gordon-Schowalter nonaffine convected derivative defined in Section 5.2.2.

In Yamamoto's original model, g and h depend on the ***magnitude of the end-to-end vector of the strand;*** thus $g = g(R)$ and $h = h(R)$. From Equation 6-22, one finds that ***at equilibrium in the absence of flow,***

$$g(R) - h(R)\psi_0 = 0 \tag{6-25}$$

where ψ_0 is the equilibrium distribution function. Hence

$$g(R) = h(R)\,\psi_0 \tag{6-26}$$

Thus the rate of creation of equilibrium strands must equal the rate of destruction of non-equilibrium strands. An alternative interpretation is that $g(R) = h(R)\psi_0$ is the probability per unit time that a strand suddenly escapes its entanglements and leaves the nonequilibrium population of strands to join the equilibrium population. Thus the total number of strands is conserved. Equation 6-23 then simplifies to

$$\frac{\partial \psi}{\partial t} + \frac{\partial}{\partial \underset{\sim}{R}} \cdot [\underset{\sim}{R} \cdot (\nabla \underset{\sim\sim}{v} - \xi \underset{\approx}{D}) \psi] = h(\psi_0 - \psi) \tag{6-27}$$

If h is a constant and $g = h\psi_0$, then the last two terms of Equation 6-24 reduce to

$$\frac{1}{2} h\, \beta^{-2}\, \underset{\approx}{\delta} - h \langle \underset{\sim\sim}{RR} \rangle$$

where we have taken ψ_0 to be Gaussian, $\psi_0 = (\beta/\sqrt{\pi})^3 \exp(-\beta^2 R^2)$. If we consider affine motion, $\xi = 0$, Equation 6-24 becomes

$$\overset{\nabla}{\underset{\sim\sim}{\langle RR \rangle}} = \frac{1}{2} h \beta^{-2} \underset{\approx}{\delta} - h \underset{\sim\sim}{\langle RR \rangle} \qquad (6\text{-}28)$$

Since the stress tensor $\underset{\approx}{\sigma}$ equals $2G\beta^2 \underset{\sim\sim}{\langle RR \rangle}$, we recognize Equation 6-28 as the UCM equation

$$\overset{\nabla}{\underset{\approx}{\sigma}} = \frac{1}{\lambda}(G\underset{\approx}{\delta} - \underset{\approx}{\sigma}) \qquad (6\text{-}29)$$

with $1/\lambda = h$. The destruction probability, h, is an *inverse relaxation time.* That the UCM equation should arise out of Yamamoto's theory when the strains are Gaussian and the destruction probability is constant is no surprise, since this limit of Yamamoto's theory is just Green and Tobolsky's original theory, which also yielded the UCM equation or its equivalent, the Lodge equation; see Chapter 2.

6.3.1.1 The Problem of Closure The much more interesting situation occurs when h is not constant. When h depends on R, however, the right side of Equation 6-24 cannot be replaced by terms involving the stress tensor. To obtain the stress tensor, we must solve Equation 6-27 for $\psi(\underset{\sim}{R})$, and then integrate $\underset{\sim\sim}{RR}\psi$ over configuration space. Computing the stresses therefore becomes a difficult numerical problem even in simple flows, to say nothing of computing them in complex multidimensional flows.

Thus the Yamamoto theory does not produce a closed constitutive equation — that is an equation that relates the stress tensor to kinematic variables alone — without requiring any averaging over microscopic variables. As a result of the closure problem, Yamamoto's theory has been largely overlooked as a model of polymer rheology. To date, only *closed* constitutive equations have been considered practical for complex polymer flow calculations. Even the Doi-Edwards equation is ordinarily used in numerical calculations only after it has been brought to closed form with the aid of the independent alignment approximation and the approximate Currie potential (Chapter 4).

6.3.1.2 Predictions of the Yamamoto Model It is nevertheless instructive to consider the behavior of the Yamamoto equation in simple flows for which computations, though difficult, are feasible. First, the function h(R) must be specified. A simple choice is that made by Fuller and Leal:[21]

$$h = h_0(1 + 2\epsilon\beta^2 R^2) \qquad (6\text{-}30)$$

in which the breakage probability *increases* as the strand extension, R^2, increases. ϵ is a positive parameter.

Fuller and Leal solved Equation 6-27 by the method of characteristics for shear flows and planar extensional flows and then carried out the integral in Equation 6-21 to obtain the stress tensor. They found that the steady-state values of $\eta(\dot{\gamma})$ and $\Psi_1(\dot{\gamma})$ decrease with increased $\dot{\gamma}$, and the transient values of both η and Ψ_1 have overshoots. The most interesting result is that the planar extensional viscosity $\overline{\eta}_{p,1}$ *shows a maximum as a function of $\dot{\epsilon}$ at steady state and has a maximum as a function of time in start-up of steady extension.*

6.3.2 Phan-Thien / Tanner Model

6.3.2.1 Method of Closure The closure problem of Yamamoto's theory was circumvented by Phan Thien and Tanner[20] by letting h depend on $<R^2>$ rather than on R^2. h can then be removed from the integrals on the right side of Equation 6-24 to give

$$\overset{\square}{<\underset{\sim\sim}{RR}>}_a = h[<R^2>]\,(\tfrac{1}{2}\beta^{-2}\underset{\approx}{\delta} - <\underset{\sim\sim}{RR}>) \tag{6-31}$$

The stress tensor is

$$\underset{\approx}{\sigma} = 2a\nu kT\beta^2<\underset{\sim\sim}{RR}> \tag{6-32}$$

(The factor $a = 1-\xi$ is included in the formula for the stress tensor for reasons cited in Section 5.2.1. Phan Thien and Tanner neglected it.) Since $<R^2> = \mathrm{tr}<\underset{\sim\sim}{RR}>$, Equation 6-31 is equivalent to

$$\overset{\square}{\underset{\approx}{\sigma}}_a + h[\mathrm{tr}\underset{\approx}{\sigma}](\underset{\approx}{\sigma} - \frac{G}{a}\underset{\approx}{\delta}) = \underset{\approx}{0} \tag{6-33}$$

with $G \equiv a^2\nu kT$. Defining $\underset{\approx}{\tau} = \underset{\approx}{\sigma} - (G/a)\underset{\approx}{\delta}$ transforms Equation 6-33 to the usual form of the *Phan-Thien/Tanner (PTT) constitutive equation*[20]:

$$\lambda\overset{\square}{\underset{\approx}{\tau}}_a + Y[\mathrm{tr}\underset{\approx}{\tau}]\,\underset{\approx}{\tau} = 2\eta_0\underset{\approx}{D} \tag{6-34}$$

where $Y \equiv \lambda h$ and $\eta_0 \equiv G\lambda$.

6.3.2.2 Predictions of the Phan-Thien / Tanner Model Specific predictions can be made for the PTT equation once the function $Y[\mathrm{tr}\underset{\approx}{\tau}]$ is specified. Phan-Thien and Tanner suggested two possibilities:

$$Y[\mathrm{tr}\underset{\approx}{\tau}] = 1 + \frac{\epsilon}{G}\mathrm{tr}\underset{\approx}{\tau} \tag{6-35a}$$

and

$$Y[\mathrm{tr}\underset{\approx}{\tau}] = \exp[\frac{\epsilon}{G}\mathrm{tr}\underset{\approx}{\tau}] \tag{6-35b}$$

With either choice, the PTT equation has two parameters, ξ and ϵ, apart from the usual linear viscoelastic constants. If ϵ is much smaller than ξ, then ϵ has little influence on shear properties and serves mainly to blunt the extensional singularity that would otherwise be present.[20] ξ then controls the level of shear thinning.

Section 5.3.2.3 discusses at length the disadvantages of using the Gordon-Schowalter convected derivative, $\overset{\square}{\underset{\approx}{\tau}}_a$, for melts. These disadvantages are inher-

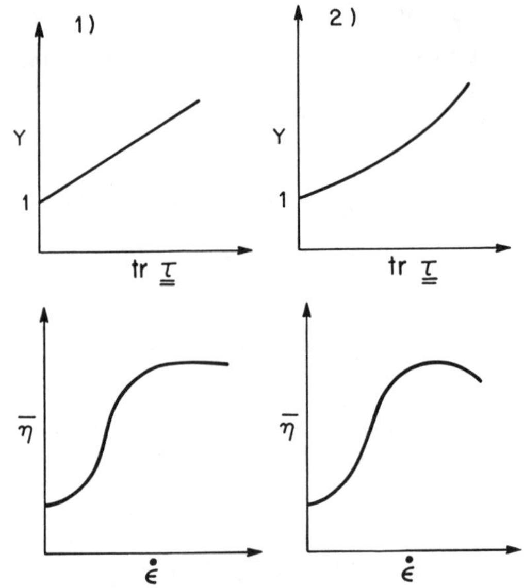

Fig. 6-5 Elongational viscosity of the Phan-Thien/Tanner model versus shear rate for the two proposed forms of the function $Y(\mathrm{tr}\,\underset{\approx}{\tau})$ (from Phan Thien,[20] reprinted from the Journal of Rheology ©1978 Soc. of Rheol. by permission of John Wiley & Sons, Inc.).

ited by the PTT model if $\xi \neq 0$. Primary among these disadvantages is that unphysical oscillations in η and Ψ_1 occur during start-up of steady shear at large rates of shear. These oscillations may cause confusion in the development of numerical routines for solving complex flows, because other spurious oscillations, caused by ***numerical instabilities***, are often seen in numerical studies. The presence of oscillations produced by the constitutive equation could be confused with those that are ***numerically induced.***

If ξ in the PTT model is set to zero –that is, the motion is assumed to be affine– the oscillations disappear. Shear thinning is still present because of the term involving ϵ. The PTT model with $\xi = 0$ and an arbitrary dependence of Y on the invariants of $\underset{\approx}{\tau}$ belongs to a family of constitutive equations introduced by Kaye[22] in 1966. The PTT equation with $\xi = 0$ and the linear form for $Y(\mathrm{tr}\,\underset{\approx}{\tau})$, Equation 35a, is similar in its predictions to the Giesekus equation,[23] except in its prediction that $N_2 = 0$ in shear and its prediction of no overshoot in growth of the first normal stress difference in start-up of steady shearing. As is true for the Giesekus equation, the extensional viscosities have no overshoot in their growth under steady extension and the steady-state values reach plateaus at high extension rates. When the exponential form, Equation 35b, is chosen for $Y[\mathrm{tr}\,\underset{\approx}{\tau}]$, the steady-state extensional viscosities as functions of $\dot{\epsilon}$ have maxima; see Figure 6-5. The exponential form for $Y[\mathrm{tr}\,\underset{\approx}{\tau}]$ thus leads to the most favorable comparisons with uniaxial extensional data for melts.

A detailed comparison of this latter form of the Phan-Thien / Tanner model with polymer melt data is presented in the next chapter.

6.3.3 Criticisms of the Phan-Thien / Tanner Model

The crucial step leading to the PTT constitutive equation is the assumption that h depends on $\langle R^2 \rangle$ rather than on R^2. There are two ways that this assumption can be justified.

6.3.3.1 Preaveraging Approximation The first is by a preaveraging approximation applied to Yamamoto's theory. If h is given by Equation 6-30, for example, then in the preaveraging approximation $2\epsilon\beta^2 R^2$ is replaced by its average value $2\epsilon\beta^2 \langle R^2 \rangle = \epsilon \, \mathrm{tr}\underset{\approx}{\sigma}/a\nu kT$. This method of closure has been criticized by Fuller and Leal.[21] They found that Yamamoto's theory *without preaveraging* produces a large overshoot in time-dependent planar viscosity; this overshoot does not appear in the PTT model because it is eliminated by the preaveraging approximation. Preaveraging is only a reliable approximation when the distribution function, ψ, is strongly peaked about a nonzero value of $\underset{\sim}{R}$. This does *not* occur in strongly stretching flows if the strands are Gaussian (that is infinitely extensible); instead the distribution function broadens with the peak remaining at $R = 0$.*

6.3.3.2 Mean-Field Approximation A second possible justification for letting h depend on $\langle R^2 \rangle$ instead of R^2 is that the stress in surrounding strands could have as much or more to do with the breakage of a given strand than does the stress in the given strand. Thus the breakage rate should depend on R averaged in some way over surrounding strands. $\langle R^2 \rangle$ is an average over many surrounding strands. This justification has been criticized by Mewis and Denn.[24] The gist of their criticism is as follows. The identity $g = h\psi_0$ must hold at equilibrium, but if h and g are not functions of R alone, then this identity need not hold away from equilibrium.

Mewis and Denn proposed that the rate of formation minus the rate of destruction be given by

$$g - h\psi = h(\psi_0 - \psi) - p\psi \tag{6-36}$$

where p must go to zero at equilibrium but need not be zero away from equilibrium.

If this *net* creation rate is incorporated into Equation 6-27, the result is

* For *finitely* extensible strands in strongly stretching flows, the distribution becomes sharply peaked near the maximum end-to-end extension of the strand, and pre-averaging becomes a good approximation.

$$\dot{\psi} = \frac{\partial}{\partial \underset{\sim}{R}} \cdot [\underset{\sim}{R} \cdot \nabla \underset{\sim}{v}\, \psi] + h(\psi_0 - \psi) - p\psi \tag{6-37}$$

Because of the term involving p, the configuration distribution function no longer normalizes to unity; there is a net loss of strands or equivalently of entanglements. The strands do not leave the nonequilibrium distribution as fast as they join the equilibrium distribution. Integrating Equation 6-37 over configuration space, dR^3,

$$\dot{x} = h(1 - x) - px \tag{6-38}$$

where

$$x \equiv \int \psi \, dR^3 \tag{6-39}$$

is less than or equal to unity. At equilibrium, $x = 1$. The product xG is a *non-equilibrium modulus.*

6.3.4 Model of Acierno, La Mantia, Marrucci, and Titomanlio

6.3.4.1 Basis of the Model Equation 6-38 is a *structural evolution equation* that describes how the nonequilibrium modulus varies during deformation or relaxation. Acierno and co-workers[25] developed an ad hoc constitutive equation based on a structural evolution equation of this kind. They proposed a simple Maxwell equation

$$\lambda \overset{\nabla}{\langle \underset{\sim\sim}{RR} \rangle} + \langle \underset{\sim\sim}{RR} \rangle = \tfrac{1}{2}\beta^{-2} \underset{\approx}{\delta} \tag{6-40}$$

where $\langle \underset{\sim\sim}{RR} \rangle$ is an average over a distribution function that is normalized to unity. The stress tensor, however, is

$$\underset{\approx}{\tau} = G[2\beta^2 \langle \underset{\sim\sim}{RR} \rangle - \underset{\approx}{\delta}] \tag{6-41}$$

with

$$G = G_0 x; \qquad \lambda = \lambda_0 x^{1.4} \tag{6-42}$$

This choice of the dependence of λ on x was made so that the zero-shear viscosity would be proportional to the polymer concentration to the 3.4 power. The constitutive equation is therefore

$$\frac{\lambda_0 x^{1.4}}{G_0} \overset{\nabla}{\left(\frac{1}{x} \underset{\approx}{\tau}\right)} + \frac{1}{G_0 x} \underset{\approx}{\tau} = 2\lambda_0 x^{1.4} \underset{\approx}{D} \tag{6-43}$$

The structural evolution equation is

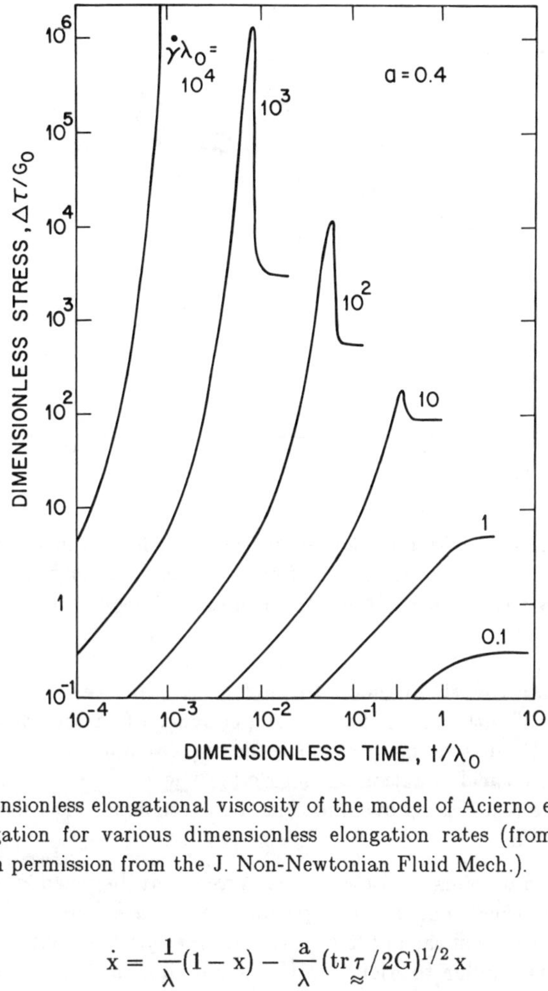

Fig. 6-6 Dimensionless elongational viscosity of the model of Acierno et al. after start-up of steady elongation for various dimensionless elongation rates (from Acierno et al.[25], reproduced with permission from the J. Non-Newtonian Fluid Mech.).

$$\dot{x} = \frac{1}{\lambda}(1 - x) - \frac{a}{\lambda}(\mathrm{tr}\,\underset{\approx}{\tau}/2G)^{1/2}\,x \tag{6-44}$$

where a is an empirical constant. Equation 6-44 is obtained from Equation 6-38 by making a specific choice for the dependence of p on $\mathrm{tr}\,\underset{\approx}{\tau}$, namely $p = a(\mathrm{tr}\,\underset{\approx}{\tau}/2G)^{1/2}/\lambda$.

This model, often called "the Marrucci model," has some inconsistencies explored by Mewis and Denn.[24] But the level of empiricism involved in the model, and indeed in all the network models, is so high that one should not take their molecular or structural derivations too seriously. These models ought to stand or fall by their predictions of stress, not by the strength of their molecular-theoretical underpinnings.

6.3.4.2 Predictions of the Model of Acierno, et al. The most distinctive prediction of the Marrucci model is that the elongational stress, $\Delta\tau = \tau_{11} - \tau_{22}$, overshoots before reaching a steady-state value, Figure 6-6. The overshoot is

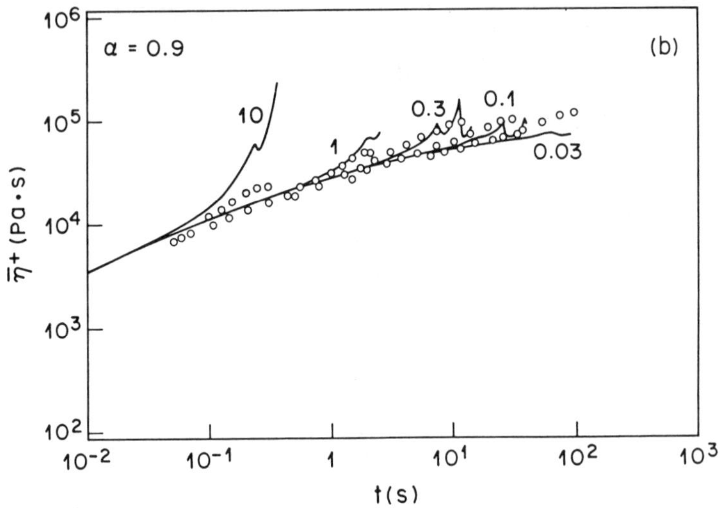

Fig. 6-7 Oscillations in growth of elongational viscosity for the model of Acierno et al. with multiple modes. Data are for Melt I (from Khan and Larson,[27] reprinted from the Journal of Rheology ©1987 Soc. of Rheol. by permission of John Wiley & Sons, Inc.).

similar to that found in the Yamamoto model. It is interesting that if the PTT model is "improved" either by allowing a dependence of the creation and destruction functions on R^2 instead of $\langle R^2 \rangle$ (as in the Yamamoto model), or by allowing a separate structural equation to determine the modulus (as in the model of Acierno et al.), the result is an overshoot in the extensional viscosity as a function of time.

Acierno and co-workers reported good agreement between a multiple-mode version of the Marrucci model and experimental data in shear and elongational stress growth as well as shear and elongational creep and recovery for low-density polyethylene melts similar to Melt I.[26] Khan and Larson,[27] however, found that the model of Acierno et al. shows pronounced *oscillations* in elongational stress growth and stress relaxation after step shearing when results from multiple relaxation modes are added together in the usual way. The oscillations in elongational stress growth occur because the maxima in viscosity do not occur at exactly the same strain for the different modes; when superposed, the sharp maxima associated with the various modes are spaced apart from each other and when taken together appear as an oscillation. Spacing of relaxation times as close as a factor of 5 apart does not eliminate the oscillation. This oscillation, depicted in Figure 6-7, could severely limit the usefulness of the Marrucci equation.

6.3.5 Other Structural Models

6.3.5.1 Giesekus' Model with Mobility Relaxation In the model of Acierno et al., the *modulus* is given by a structural evolution equation. Giesekus[28] generalized his model by suggesting an evolution equation for the anisotropic *mobility tensor*,

$$\underset{\approx}{B} - \underset{\approx}{\delta} + \lambda_m (\underset{\approx}{B} - \underset{\approx}{\delta})^{\nabla} = \alpha(\underset{\approx}{\sigma}/G - \underset{\approx}{\delta}) \tag{6-45}$$

Here the upper-convected derivative ∇ operates on $\underset{\approx}{B} - \underset{\approx}{\delta}$. λ_m is a *mobility retardation time* distinct from the stress relaxation time λ. Equation 6-45 reverts to the earlier form, Equation 6-2, when $\lambda_m = 0$.

When $\lambda_m > 0$, stress overshoots appear in start-up of extensional flow and in start-up of shearing the overshoots already present in the simple Giesekus model intensify.

6.3.5.2 Doi-Edwards Model with Incomplete Retraction As discussed Section 4.6.3.4, large overshoots observed in the first normal stress difference seen after start-up of steady shearing are *not* predicted by the Doi-Edwards model, probably because the model assumes complete and instantaneous *retraction* of polymer strands during deformation. This assumption is not expected to be valid when the shear rate $\dot{\gamma}$ approaches or exceeds $1/\lambda_k$, the reciprocal of the *retraction time*.

Marrucci and Grizzuti[29] calculated the effects of incomplete retraction in steady flows. More recently Pearson and Kiss[30] proposed a simple modification of the Doi-Edwards consititutive model that accounts in an approximate way for incomplete retraction in *any* flow. The Pearson-Kiss equation is

$$\underset{\approx}{\sigma} = G_0 L^2 \underset{\approx}{S} \tag{6-46}$$

where

$$G_0 \underset{\approx}{S} \equiv G_0 \int_{-\infty}^{t} m(t-t') \underset{\approx}{Q}(t',t)\, dt' \tag{6-47}$$

is the stress tensor in the Doi-Edwards theory. G_0 is the plateau modulus. In the above L is the length of a polymer strand under flow relative to its length at equilibrium; its evolution equation is

$$\dot{L} = \underbrace{\underset{\approx}{D} : \underset{\approx}{S}\, L}_{\text{(stretching)}} + \underbrace{\frac{1}{\lambda_k}(1 - L)}_{\text{(retraction)}} \tag{6-48}$$

From equations 6-46 to 6-48, overshoots in the first normal stress difference can be predicted.

6.3.5.3 Structural Model of Liu, Soong, and Williams Liu, Soong, and Williams[31] proposed a kinetic network model based on a *structural viscosity*. In their model, the structural parameter obeys

$$\dot{x} = \frac{k_1}{b}(1 - x) - k_1 \dot{\gamma}^m x \tag{6-49}$$

where $\dot{\gamma}$, the shear rate in shearing deformations, is defined as $\dot{\gamma} \equiv \sqrt{II_D}$, and $II_D \equiv 2\underset{\approx}{D}:\underset{\approx}{D}$ is the second invariant of the strain rate tensor. In shear, $II_D = \dot{\gamma}^2$. The stress tensor satisfies the UCM equation, Equation 6-29, with

$$\lambda = x\lambda_0 \; ; \qquad G = G_0 \tag{6-50}$$

The *modulus* is constant, and the *viscosity* $\eta = \lambda G$ is proportional to the structural constant, x. This dependence, $\eta \propto x$, is weaker than that of Acierno et al., namely $\eta \propto x^{2.4}$. As a result, the model of Liu et al. doesn't have the severe overshoots in viscosity present in the model of Acierno et al.

The model was designed for predicting steady shearing flows and does surprisingly well, *with only one relaxation time*, not only at steady-state, but also in start-up of steady shear and in stress relaxation after cessation of steady shear. It can describe stress relaxation with only one *stress* relaxation time, because that relaxation time is structure dependent and *becomes larger as the relaxation progresses*. Thus the relaxation occurs over a much broader range of time than a simple exponential decay. The relaxation time *itself* relaxes according to Equation 6-49; the time required for the relaxation time to relax is the *structural* relaxation time b/k_1. Thus this model with a single stress relaxation time mimics the effect of a broad distribution of relaxation times.

The model of Liu et al. is not adequate for extensional flow predictions, since it shows singular viscosities at finite extension rates.

6.3.6 General Network Model: Differential and Integral Form

The functions g and h in the general network equation, Equations 6-23 and 6-24, need not be functions of $\mathrm{tr}\,\underset{\approx}{\tau}$ only. Closed-form constitutive equations can also be derived if these functions depend on any invariants of $\underset{\approx}{\sigma}$ or of $\underset{\approx}{D}$. They can even depend on the history of the invariants, as, for example, in the structural evolution model of Acierno et al. If $g = G\psi_0$, with G and h independent of R^2, Equation 6-24 can be reduced to

$$\overset{\square}{\underset{\approx}{\sigma}}_a + \frac{1}{\lambda}\underset{\approx}{\sigma} = G\underset{\approx}{\delta} \tag{6-51}$$

where $\lambda = h^{-1}$ and G can depend on the history of the invariants of $\underset{\approx}{\sigma}$ and $\underset{\approx}{D}$. This equation includes the PTT, the White-Metzner, and the Marrucci equations. Phan-Thien and Tanner[20] noticed that it can be cast into an integral form

$$\underset{\approx}{\sigma} = \int_{-\infty}^{t} \frac{G(t')}{\lambda(t')} \exp\left[-\int_{t'}^{t} \frac{dt''}{\lambda(t'')}\right] \underset{\approx}{E}(t,t')\, dt' \tag{6-52}$$

where $G(t')$ and $\lambda(t'')$ are the modulus and relaxation time at times t' and t'', respectively. $\underset{\approx}{E}$ is the elastic strain measure; see Section 5.3.1. The equation can be generalized to allow a distribution of relaxation times;

$$\underset{\approx}{\sigma} = \int_{-\infty}^{t} \sum_i \left\{ \frac{G_i(t')}{\lambda_i(t')} \exp\left[-\int_{t'}^{t} \frac{dt''}{\lambda_i(t'')}\right] \right\} \underset{\approx}{E}(t,t')\,dt' \tag{6-53}$$

6.3.7 The Equations of Bird and Carreau

An earlier rate-dependent constitutive equation is that of Bird and Carreau,[32] which is obtained from the general equation, Equation 6-53, by setting

$$\underset{\approx}{E} = \left(1+\frac{\epsilon}{2}\right)\underset{\approx}{C}^{-1} + \frac{\epsilon}{2}\underset{\approx}{C} \tag{6-54}$$

Here ϵ is an empirical constant that introduces a second normal stress difference in steady shearing: $N_2/N_1 = \epsilon/2$. Because of the second term, this equation has no simple differential equivalent. Bird and Carreau chose each relaxation time λ_i to be constant and assigned a dependence of each modulus G_i on $\dot{\gamma}$ equivalent to

$$G_i = \frac{G_i^0}{1+(a_i\dot{\gamma})^2} \tag{6-55}$$

where each a_i is an empirical constant. Bird and Carreau designed the equation to fit *steady-state* shear data; it describes the steady-state very accurately, but performs poorly in transient flows, such as start-up of steady shearing.

If the shear rate is suddenly increased, the modulus, Equation 6-55, of the Bird-Carreau equation *instantaneously* decreases. Thus the structure instantly breaks down as soon as a high strain rate is imposed. Since the moduli of the Bird-Carreau equation depend on strain *rate* and not on strain, the Bird-Carreau equation reduces to the equation of linear viscoelasticity, Equation 3-4, at small strains only if the strain rate is small.[33,34] It has been observed experimentally, however, that linear viscoelasticity holds for melts and solutions at ***any strain rate***, large or small, when the *strains* are small.[34–36] Thus in oscillatory shearing, if the strain amplitude is small (less than 30-70%), then the equation of linear viscoelasticity holds no matter how high the frequency of oscillation is; see Figure 6-8. This result accords with the theory of viscoelastic simple fluids, which predicts that the transition from linear to nonlinear viscoelasticity occurs at a finite strain no matter how high the strain rate is. Contrary to this theory and to experiment, in the Bird-Carreau equation the strain amplitude at which the transition from linear to nonlinear viscoelasticity occurs decreases towards zero as the frequency increases. Therefore, the Bird-Carreau equation has not enjoyed much favor lately. Nevertheless, the equation may be useful in describing thixotropic (or structured) fluids for which the structure depends on the instantaneous strain rate.

The Bird-Carreau equation contrasts with the White-Metzner[37] equation, in which each modulus is constant but each relaxation time depends on the strain rate. The WM equation does *not* violate the equation of linear viscoelasticity at small strains and large strain rates, if λ_i decreases with increased $\dot{\gamma}$ no more steeply than $\dot{\gamma}^{-1}$. The dependence assumed by Ide and White,[38] Equation 5-59, falls within this limitation. In the *Carreau-B equation*,[39] both G_i and λ_i depend on the strain-rate invariant, $\dot{\gamma}$. The Carreau-B equation describes stresses in start-

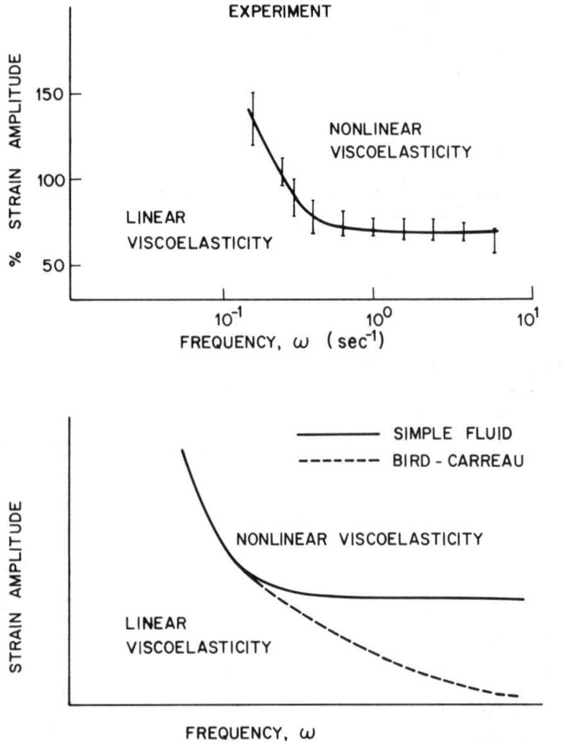

Fig. 6-8 Strain amplitude as a function of frequency at which there is a transition from linear to nonlinear viscoelasticity. (top) experiments of Jongschaap et al.[39] (bottom) prediction of the Coleman and Noll theory of viscoelastic simple fluids and the Bird-Carreau equation.

up of shear better than does the Bird-Carreau equation, but it remains subject to the criticism that it does not reduce to the equation of linear viscoelasticity at small strains.

In the Kaye-BKZ or Wagner equation, the moduli depend on *strain*, and the relaxation times are constant. As a result, deviations from linear viscoelasticity occur only when at a critical value of the strain is exceeded.

Other possible dependencies of G_i and λ_i on stress, strain, or strain rate invariants are possible. Some of these are explored by Takahashi et al.[40]

6.4 CONFIGURATION DISTRIBUTION FUNCTIONS

We have seen how stress constitutive equations for network models are derived from evolution equations for ψ, the probability distribution function. One seeks closed-form equations for the stress tensor $\underset{\approx}{\sigma} \propto \int \psi \underset{\sim}{R}\underset{\sim}{R}\, dR^3$ which is the second moment of ψ. Once an exact or approximate closed-form equation is obtained, one can solve the equation for $\underset{\approx}{\sigma}$ without ever calculating ψ.

There are situations, however, in which one might desire to know ψ. One might, for example, label portions of the macromolecules with infrared absorbers and use infrared dichroism to determine the orientation of portions of the macromolecules. Such an experiment would allow one to subject a molecular model to tests that are more discriminating than is possible with stress or birefringence measurements. Moreover, there are physical phenomena, such as flow-induced phase separation and welding of polymer-polymer contact lines, that depend on averages over ψ other than the second moment. Thus it may be important to evaluate ψ.

Fortunately, it turns out that *whenever a closed-form equation for $\underset{\approx}{\sigma}$ can be obtained*, either exactly or by preaveraging, *an equation for ψ can also be obtained.*[41]

6.4.1 Green-Tobolsky Network Model

Consider for example the simplest network model, that of Green and Tobolsky.[16] The evolution of ψ satisfies a special case of Equation 6-27:

$$\frac{\partial \psi}{\partial t} + \frac{\partial}{\partial \underset{\sim}{R}} \cdot (\underset{\sim}{R} \cdot \nabla \underset{\sim}{v}\, \psi) = h(\psi_0 - \psi) \qquad (6\text{-}56)$$

with h constant. The constitutive equation for $\underset{\approx}{\sigma}$ in this limit is the UCM equation or the Lodge equation with one relaxation time. The probability distribution function, ψ, for the Green-Tobolsky (GT) model is

$$\psi(\underset{\sim}{R},t) = \pi^{-3/2} \int_{-\infty}^{t} \frac{1}{\lambda} e^{-(t-t')/\lambda} \exp[-\underset{\sim}{R} \cdot \beta^2 \underset{\approx}{C}(t,t') \cdot \underset{\sim}{R}]\, dt' \qquad (6\text{-}57)$$

Equation 6-57 is an integral over past times, i.e., a *history integral.*

6.4.2 Rouse-Zimm Dumbbell Model

The distribution function for the elastic dumbbell model is governed by Equation 2-45,

$$\frac{\partial \psi}{\partial t} + \frac{\partial}{\partial \underset{\sim}{R}} \cdot [\underset{\sim}{R} \cdot \nabla \underset{\sim}{v}\, \psi - \frac{4kT\beta^2}{\varsigma} \underset{\sim}{R} \psi - \frac{2kT}{\varsigma} \frac{\partial \psi}{\partial \underset{\sim}{R}}] = 0 \qquad (6\text{-}58)$$

The constitutive equation for the elastic dumbbell model is identical to that for the GT model; namely the UCM or Lodge equation. *The configuration distribution function for the dumbbell is different, however.* It is an *anisotropic Gaussian:*[42,43]

$$\psi = \pi^{-3/2} (\det \underset{\approx}{S})^{-1/2} \exp[-\underset{\sim}{R} \cdot \underset{\approx}{S}^{-1} \cdot \underset{\sim}{R}] \qquad (6\text{-}59)$$

where $\underset{\approx}{S}$, the covariance tensor, is

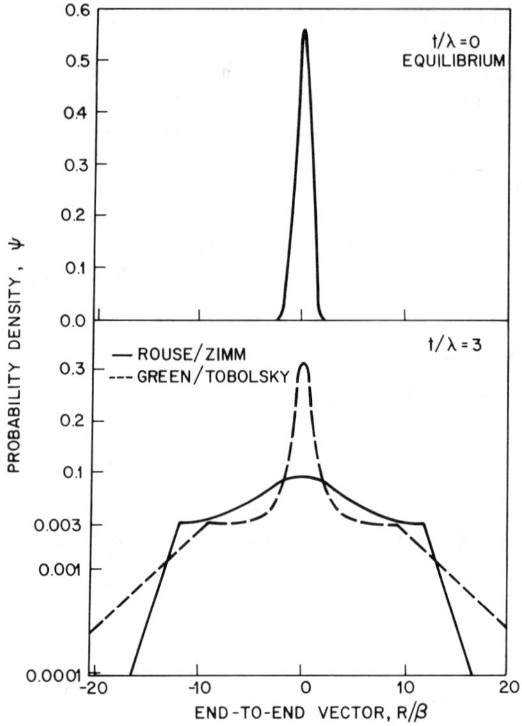

Fig. 6-9 Configuration distribution function for the Rouse-Zimm and Green-Tobolsky models. (top) At equilibrium both are the same. (bottom) After a time, $t/\lambda = 3$, in an elongational flow, the Rouse-Zimm distribution is a stretched Gaussian; the Green-Tobolsky distribution has a non-Gaussian tail.

$$\underset{\approx}{S} \equiv 2 < \underset{\sim\sim}{RR} > \tag{6-60}$$

Surfaces of equal probability are spheres at equilibrium; after any deformation history they are *ellipsoids*. The lengths of the axes of the ellipsoids are proportional to the eigenvalues of the covariance tensor $\underset{\approx}{S}$. $\underset{\approx}{S}$ is proportional to the stress tensor, $\underset{\approx}{\sigma} = 2\nu kT\beta^2 < \underset{\sim\sim}{RR} > = \nu kT\beta^2 \underset{\approx}{S}$. ψ for the dumbbell model can therefore be obtained once the constitutive equation is solved for $\underset{\approx}{\sigma}$.

Figure 6-9 illustrates the difference between the two distribution functions in a strong elongational flow, with $\dot{\epsilon}\lambda = 1$. At this strain rate, the stress grows without bound in time after start-up of steady extension. The two distribution functions are equal at equilibrium (top) –that is at t=0, just before the deformation starts. At dimensionless time $t/\lambda = 3$ (bottom), the Rouse-Zimm distribution appears much broader than that of Green and Tobolsky, although the two distributions have the same second moment.[41] The apparent narrowness of the GT distribution is offset by its *long non-Gaussian tail,* which is evident in the logarithmic blowup of the low-probability region at the bottom of Figure 6-9. Thus ψ for the Green-Tobolsky model exceeds ψ for the Rouse-Zimm model at large strand extensions, R. These few highly extended strands contribute most of the stress in the Green-Tobolsky model.

Thus the dilute solution of dumbbells and the elastic network have the same constitutive equation but different configuration distribution functions. Though ψ is different for the two models, the second moment of ψ, $<\underset{\sim}{R}\underset{\sim}{R}>$, is the same.

6.4.3 Other Models

The configuration distribution function can be derived for other models also. For the general network model, Equation 6-23, the solution is[41]

$$\psi(\underset{\sim}{R},t) = \int_{-\infty}^{t} g(t')\exp\{-\int_{t'}^{t} h(t'')\,dt''\}\,\psi_c(\underset{\sim}{R},t,t')\,dt' \qquad (6\text{-}61)$$

where

$$\psi_c(\underset{\sim}{R},t,t') \equiv \pi^{-3/2}\,[\det\underset{\approx}{E}(t,t')]^{-1/2}\exp[-\underset{\sim}{R}\cdot\beta^2\underset{\approx}{E}^{-1}(t,t')\cdot\underset{\sim}{R}] \qquad (6\text{-}62)$$

$\underset{\approx}{E}$ is the elastic strain measure.

If ψ satisfies a Smoluchowski equation that is linear in $\underset{\sim}{R}$ (e.g., Equation 6-58), then the solution for ψ is *not* a history integral but an *anisotropic Gaussian.* Consider, for example, the Giesekus model, which has an *anisotropic mobility* $\underset{\approx}{B}$. The Smoluchowski equation in this case is

$$\frac{\partial\psi}{\partial t} + \frac{\partial}{\partial\underset{\sim}{R}}\cdot[\underset{\sim}{R}\cdot\underset{\sim}{\nabla}\underset{\sim}{v}\psi - \frac{1}{2}\underset{\approx}{B}\cdot(\underset{\sim}{R}\psi - \frac{1}{2}\frac{\partial\psi}{\partial\underset{\sim}{R}})] = 0 \qquad (6\text{-}63)$$

Multiplying Equation 6-63 by $\underset{\sim}{R}\underset{\sim}{R}$ and integrating over dR^3 gives the Giesekus constitutive equation

$$\lambda\overset{\nabla}{\underset{\approx}{\sigma}} + \underset{\approx}{B}\cdot(\underset{\approx}{\sigma} - G\underset{\approx}{\delta}) = \underset{\approx}{0} \qquad (6\text{-}64)$$

ψ is again given by an anisotropic Gaussian, Equation 6-59, where the covariance tensor $\underset{\approx}{S}$ is again proportional to $\underset{\approx}{\sigma}$. For a given deformation history, the configuration distribution function for the Giesekus model is different from that for the isotropic dumbbell model because $\underset{\approx}{S}$ for the two models is different.

Thus for each closed-form constitutive equation, the configuration distribution function is derivable. It takes either of two forms: a history integral if ψ is given by a creation-destruction equation, or an anisotropic Gaussian if ψ satisfies a Smoluchowski equation that is linear in R. The difference between the two types

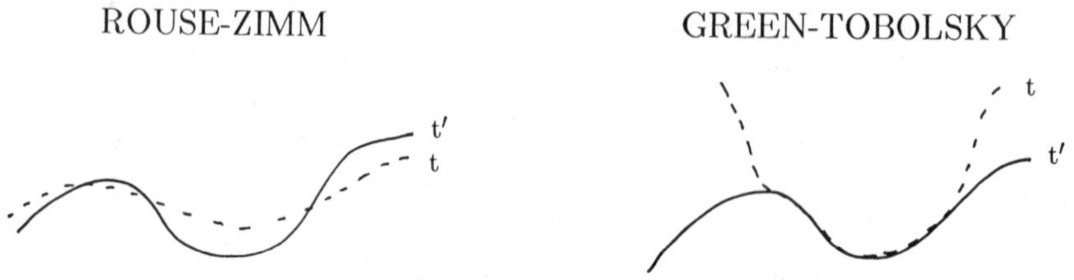

Fig. 6-10 In Rouse-Zimm diffusion, a molecule gradually changes its configuration. In Green-Tobolsky diffusion (such as that of the Doi-Edwards model), parts of the molecule suddenly change conformation.

of distribution functions comes from the difference in the postulated mechanisms of diffusion. For the anisotropic Gaussian, which occurs in dilute dumbbell or Rouse-Zimm theories, a strand of polymer ***continuously*** diffuses away from its initial conformation, Figure 6-10. For the history integral, which occurs in network models of concentrated solutions or melts such as that of Green and Tobolsky, a strand ***suddenly*** escapes its entanglements, leaves its conformation, and takes on another completely different conformation. The Rouse-Zimm type of diffusion best describes isolated coils in diluted solution. The Green-Tobolsky type of diffusion describes entangled systems, such as melts. The distribution function for the Doi-Edwards model for entangled systems is given by a history integral of the Green-Tobolsky type.

Conditional configuration distribution functions, $\psi(\tilde{R},t;\tilde{R}_0,t_0)$, can also be derived.[41] These describe the distribution function at time t of strands that at time t_0 had end-to-end vector $\tilde{R}_0$. If strands with end-to-end vector $\tilde{R}_0$ could be labeled at time t_0, then $\psi(\tilde{R},t;\tilde{R}_0,t_0)$ would describe the distribution of end-to-end vectors of the labeled strands at a later time, t, after an arbitrary deformation history.

6.5 SUMMARY

The constitutive models considered in this chapter involve nonlinear relaxation. They fall into two categories: the Giesekus-Leonov models, and the network models.

The Giesekus-Leonov equations differ from the upper-convected Maxwell equation in that they have a term that is ***quadratic in the stress tensor***. This term represents stress-induced acceleration of the relaxation process; it produces strain softening and shear thinning. The Giesekus model is derived from an ***elastic dumbbell*** theory in which high stresses ***anisotropically*** increase the rate of relaxation along the direction of molecular orientation. The Leonov model is a theory of ***rubber elasticity*** with an ***irreversible rate of strain***. The Leonov equation is identical to a special case of the Giesekus equation in planar deformations. For melts or concentrated solutions, the Giesekus and Leonov models give excellent predictions in shearing deformations but are inaccurate in extensional flows.

The network models are generalizations of Green and Tobolsky's model, in which strand creation and destruction probabilities can depend on strand extension, stress history, or strain-rate history. In Yamamoto's theory, the destruction probability depends on strand extension. This theory does not yield a closed-form constitutive equation, and so is not practical for analysis of complex viscoelastic flows. In the Kaye model and the Phan-Thien Tanner model, the destruction probability depends on the instantaneous *stress*. If this dependence is chosen carefully, the PTT model gives realistic predictions of extensional viscosities.

In the model of Acierno et al. ("the Marrucci model") the modulus and relaxation time depend on a structural parameter governed by a differential equation separate from the constitutive equation. This model can be thought of as a network model in which the creation and destruction probabilities depend on the history of the stress. The model of Acierno et al., like that of Yamamoto, shows large overshoots in the time-dependent extensional viscosities.

In the structural network model of Liu et al., the relaxation time depends on the history of the strain rate. This model, designed for describing shearing deformations, can with a single relaxation time mimic the effect of multiple relation times. In the Bird-Carreau models, the modulus depends on the instantaneous strain rate; in the Carreau-B model, both the modulus and the relaxation time depend on the strain rate. Constitutive equations in which the *modulus* depends on the instantaneous strain rate show departures from linear viscoelasticity at arbitrarily small strain amplitudes if the strain rate is high enough. This is contrary to the theory of viscoelastic simple fluids and to experimental evidence that shows departure from linear viscoelasticity only after a critical strain amplitude is exceeded.

All models that allow derivation of a closed-form constitutive equation also have formulas for the strand configuration distribution function. These take two forms. In models with diffusion of the Rouse-Zimm variety, such as dilute solution theories, the configuration distribution function is an anisotropic Gaussian. In models with diffusion of the Green and Tobolsky type, such as the Doi-Edwards or network models for concentrated polymeric systems, the configuration distribution function is given by a history integral. In Rouse-Zimm diffusion, each strand continuously drifts from one configuration to another; in Green-Tobolsky diffusion, the strand suddenly leaves one conformation and takes on another.

References

1. H. Giesekus, Rheol. Acta, *5*, 29 (1966); J. Non-Newtonian Fluid Mech., *11*, 69 (1982).

2. H.M. Laun, Rheol. Acta, *17*, 1 (1978).

3. R.K. Upadhyay, A.I. Isayev, and S.F. Shen, Rheol. Acta, *20*, 443 (1981); J. Rheol., *27*, 155 (1983).

4. (a) R.G. Larson, Rheol. Acta, *22*, 435 (1983); (b) M. Grmela and P.J. Carreau, J. Non-Newt. Fluid Mech., *23*, 271 (1987).

5. A.I. Leonov, Rheol. Acta, *15*, 85 (1976).

6. A.I. Leonov and A.N. Prokunin, Rheol. Acta, *19*, 393 (1980).

7. P.A. Dashner and W.E. Van Arsdale, J. Non-Newtonian Fluid Mech., *8*, 59 (1981).

8. K. Weissenberg, Arch. Sci. Phys. Natur, *140*[5], 130 (1935).

9. A.S. Lodge, *Elastic Liquids*, Academic Press, New York (1964).

10. R.K. Upadhyay and A.I. Isayev, Rheol. Acta, *25*, 80 (1986).

11. H. Giesekus, J. Non-Newtonian Fluid Mech., *12*, 367 (1983).

12. J. Meissner, S.E. Stephenson, A. Demarmels, and P. Portmann, J. Non-Newtonian Fluid Mech., *11*, 221 (1982).

13. S. Chatrei C.W. Macosko, and H.H. Winter, J. Rheol., *25*, 433 (1981).

14. H.M. Laun, Proc. Ninth Int. Congress Rheol., Acapulco, Mexico, 1984.

15. H.M. Münstedt, J. Rheol., *24*, 847 (1980).

16. M.S. Green and A.V. Tobolsky, J. Chem. Phys., *14*, 80 (1946).

17. A.S. Lodge, Trans. Faraday Soc., *52*, 120 (1956).

18. M. Yamamoto, J. Phys. Soc. Japan, *11*, 413 (1956); *12*, 1148 (1957); *13*, 1200 (1958).

19. A.S. Lodge, Rheol. Acta, *7*, 379 (1968).

20. N. Phan Thien and R.I. Tanner, J. Non-Newtonian Fluid Mech., *2*, 353 (1977); N. Phan Thien, J. Rheol., *22*, 259 (1978).

21. G.G. Fuller and L.G. Leal, J. Polym. Sci., Polym. Phys. Ed., *19*, 531 (1981).

22. A. Kaye, Brit. J. Appl. Phys., *17*, 803 (1966).

23. R.G. Larson, J. Non-Newtonian Fluid Mech., *23*, 249 (1987).

24. J. Mewis and M.M. Denn, J. Non-Newtonian Fluid Mech., *12*, 69 (1983).

25. D. Acierno, F.P. La Mantia, G. Marrucci, and G. Titomanlio, J. Non-Newtonian Fluid Mech., *1*, 125 (1976).

26. D. Acierno, F.P. La Mantia, G. Marrucci, G. Rizzo, and G. Titomanlio, J. Non-Newtonian Fluid Mech., *1*, 147 (1976); D. Acierno, F.P. La Mantia, G. Marrucci, ibid., *2*, 271 (1977).

27. S.A. Khan and R.G. Larson, J. Rheol., *31*, 207 (1987).

28. H. Giesekus, Proc. Ninth Intern. Congress on Rheology, Vol. 1, p.39 (1985); J. Non-Newtonian Fluid Mech., *17*, 349 (1985).

29. G. Marrucci and N. Grizzuti, submitted to Gazzeta Chimica Italiana (1987).

30. D.S. Pearson and A.D. Kiss, presentation at the March Meeting of the Amer. Phys. Soc., New York, 1987.

31. T.Y. Liu, D.S. Soong, and M.C. Williams, Polym. Engng. Sci., *21*, 675 (1981).

32. R.B. Bird and P.J. Carreau, Chem. Eng. Sci., *23*, 427 (1968); P.J. Carreau, I.F. Macdonald, and R.B. Bird, ibid., *23*, 901 (1968).

33. L.H. Gross and B. Maxwell, Trans. Soc. Rheol., *16*, 577 (1972).

34. G. Astarita and G. Marrucci, *Principles of Non-Newtonian Fluid Mechanics*, McGraw-Hill, London (1974).

35. W. Philippoff, Trans. Soc. Rheol., *10*, 317 (1966).

36. R.J.J. Jongschaap, K.H. Knapper, and J.S. Lopulissa, Polym. Engng. Sci., *18*, 788 (1978).

37. J.L. White and A.B. Metzner, J. Appl. Polym. Sci., *8*, 1367 (1963).

38. Y. Ide and J.L. White, J. Non-Newtonian Fluid Mech., *2*, 281 (1977).

39. P.J. Carreau, Trans. Soc. Rheol., *16*, 99 (1972).

40. M. Takahashi, T. Masuda, and S. Onogi, J. Rheol., *22*, 285 (1978).

41. R.G. Larson, Proc. Symp. on Recent Developments in Structured Continua, Univ. of Windsor, Ontario May 29–31 (1985).

42. P.H. van Wiechen and H.C. Booij, J. Eng. Math., *5*, 89 (1971).

43. A.S. Lodge and Y. Wu, Rheol. Acta, *10*, 539 (1971).

CHAPTER 7

Comparison of Constitutive Equations For Melts

"He that is first in his own cause seemeth just; but his neighbour cometh and searcheth him."

Proverbs 18:17

7.1 INTRODUCTION

No single choice of constitutive equation is best for all purposes. This chapter compares and contrasts different types of constitutive equations, hopefully guiding readers in their selection of an equation that best suits their purposes.

7.1.1 Considerations Affecting the Choice of Constitutive Equation

A constitutive equation should be chosen with several factors in mind. One is the *type of flow* to be considered. Is the flow predominantly shear, predominantly extensional, or are both shear and extension equally important? Is the flow nearly steady or highly transient? Does strain reversal occur? Is the flow planar?

A second consideration is the polymeric *material* to be described. Is the material a melt, a concentrated solution, or a dilute solution? Is the molecular-weight distribution nearly monodisperse or polydisperse? Do the molecules have long side branches?

A third factor is the *numerical or analytic scheme* to be used to solve the flow problem of interest. Does the scheme place constraints on the mathematical form of the constitutive equation, requiring, for example, a differential equation as opposed to an integral equation? Are the nonlinearities required to be of a certain form? Does the prospective constitutive equation have singularities or double-valued stresses or strain rates that might limit the conditions under which the numerical technique can converge to a solution? How many relaxation modes can the scheme accommodate?*

* Multiple modes act to reduce the level of shear thinning. Therefore, if only a single relaxation mode can be afforded in the numerical scheme, it is better *not* to choose a constitutive equation that is designed to be realistic when many modes are allowed, for it would display too much shear thinning when only one mode is permitted. It is better instead to choose a constitutive equation that with one mode displays a level of shear thinning comparable to that displayed by realistic constitutive equations when several modes are allowed.

Finally, one must decide what *phenomena* one hopes to predict or describe and with what accuracy. It is not obvious a priori which constitutive equations will be most successful in predicting complex phenomena such as extrudate swell, vortex formation, or transitions to time-dependent flow. But one can make judgments about *simple* flows. What are the simple flows that one wants the constitutive equation to describe? Should the equation predict stress overshoots? Is it important that the equation have a nonzero second normal stress difference? Should the equation predict that the steady-state elongational viscosity pass through a maximum as a function of extension rate? These considerations can be influenced strongly by personal preference or bias.

7.1.2 Approach Taken in This Chapter

With the above considerations in mind, we now examine the differences between some popular constitutive equations. We take the following approach. First we compare and contrast the properties of general *categories* of constitutive equations. The categories contrasted are integral versus differential and separable versus nonseparable equations. We consider whether the simple integral equations have any distinctive predictions not shared by the simple differential equations. Can a differential equation that is similar or identical to any given integral equation be found, and vice versa? We compare several of the simplest differential constitutive equations with experimental melt data in a variety of deformations, assess the successes and failures of the models, and suggest possibilities for improvement.

7.2 THE RELATIONSHIP BETWEEN INTEGRAL AND DIFFERENTIAL CONSTITUTIVE EQUATIONS

We here consider the possibility of finding integral equivalents to common differential constitutive equations, and vice versa.

7.2.1 Network Integral Equations

In Section 6.3.6, we saw that there is an integral equivalent for the general differential equation

$$\overset{\square}{\underset{\approx}{\sigma}}_{a} + \frac{1}{\lambda}\underset{\approx}{\sigma} = \frac{G}{a\lambda}\underset{\approx}{\delta} \tag{7-1}$$

where λ and G can depend on the invariants or the history of the invariants of $\underset{\approx}{\sigma}$ and $\underset{\approx}{D}$. This class of equations, which includes the models of Phan-Thien and Tanner[1] (PTT), of Acierno et al.[2] (Marrucci model), and of White and Metzner[3] (WM) as special cases, can be integrated:

$$\underset{\approx}{\sigma} = \int_{-\infty}^{t} \frac{G(t')}{\lambda(t')} \exp\left[-\int_{t'}^{t} \frac{dt''}{\lambda(t'')}\right] \underset{\approx}{E}(t, t')\, dt' \tag{7-2}$$

Equation 7-2 can easily be generalized to include a spectrum of relaxation times; see Sec. 6.3.6. In contrast, the differential versions accommodate only a discrete relaxation spectrum and then only if a separate differential equation is written for each relaxation time in the spectrum, Section 3.4.4.1. Thus the integral form removes a major disadvantage of the differential models for use in complex numerical flow simulations.

In the above integral, Unless $a = 1$ or -1 in the above integral, the elastic strain tensor, $\underset{\approx}{E}$, must be found by solving a separate differential equation

$$\underset{\approx}{0} = \overset{\square}{\underset{\approx}{E}}_a \equiv \overset{0}{\underset{\approx}{E}} + a(\underset{\approx}{D} \cdot \underset{\approx}{E} + \underset{\approx}{E} \cdot \underset{\approx}{D}) \qquad (7\text{-}3)$$

If the moduli or relaxation times depend on stress or the history of the stress, as in the PTT and Marrucci models, then the stress tensor appears on *both* sides of Equation 7-2, and cannot be calculated directly from the deformation history alone. Only when $\underset{\approx}{E} = \underset{\approx}{C}^{-1}$ and G and λ depend on $\underset{\approx}{D}$ only, as is true for the WM equation, can the right hand side be evaluated directly from the deformation history without solving any other equation.

It is also interesting and useful to seek differential equivalents to *Kaye-BKZ* integral equations.[4,5] Exact differential equivalents exist when the Kaye-BKZ equation is separable, the memory function is a sum of exponentials, and the elastic strain measure is $\underset{\approx}{C}^{-1}$ or $\underset{\approx}{C}$. The differential equivalents are the upper- and lower-convected Maxwell equations, respectively. There are no other known exact equivalents.

7.2.2 Differential Analogs for Separable Kaye-BKZ Equations

There are Kaye-BKZ equations, however, for which *differential analogs*[6] can be found; these differential analogs behave the same as the Kaye-BKZ equations in stress relaxation after step strains. Kaye-BKZ equations of the form

$$\underset{\approx}{\sigma} = \int_{-\infty}^{t} \frac{G}{\lambda} e^{(t'-t)/\lambda}\, h(I_1)\, \underset{\approx}{C}^{-1}(t,t')\, dt' \qquad (7\text{-}4)$$

have differential analogs. Equations of this type include an approximation to the Doi-Edwards constitutive equation; see Section 4.5.2. After a *step strain*, Equation 7-4 predicts that the stress relaxation is

$$\underset{\approx}{\sigma} = G e^{-t/\lambda} h(I_1)\, \underset{\approx}{C}^{-1} + G(1 - e^{-t/\lambda})\, \underset{\approx}{\delta} \qquad (7\text{-}5)$$

where the Finger tensor, $\underset{\approx}{C}^{-1}$, describes the change in shape that occurs in the step strain. A differential equation that also predicts this response to a step strain is

$$\lambda \overset{\blacksquare}{\underset{\approx}{\sigma}} + \underset{\approx}{\sigma} = G \underset{\approx}{\delta} \qquad (7\text{-}6)$$

where $\overset{\blacksquare}{\underset{\approx}{\sigma}}$ is a convected derivative that must be chosen so that $\overset{\blacksquare}{\underset{\approx}{E}} = \underset{\approx}{0}$, with $\underset{\approx}{E} \equiv h(I_1)\, \underset{\approx}{C}^{-1}$. The appropriate convected derivative has the form[6]

$$\overset{\blacksquare}{\underset{\approx}{\sigma}} = \overset{\nabla}{\underset{\approx}{\sigma}} + g\underset{\approx}{D}:\underset{\approx}{\sigma}\underset{\approx}{\sigma} \tag{7-7a}$$

where g is a function of an invariant of $\underset{\approx}{\sigma}$ and can be obtained from the shear damping function, $h(\gamma)$. Thus

$$g = \frac{-dh/d\gamma}{Gh^2\gamma} \tag{7-7b}$$

Equation 7-7b is derived in the appendix to this chapter. Particular examples are

$$h(I_1) = [1 + \frac{\alpha}{3}(I_1 - 3)]^{-1} \rightarrow g = \frac{2\alpha}{3G} \tag{7-8a}$$

$$h(I_1) = (\frac{I_1}{3})^{-\xi} \rightarrow g = \frac{2\xi}{tr\underset{\approx}{\sigma}} \tag{7-8b}$$

$$h(I_1) = \exp[-n(\sqrt{I_1} - \sqrt{3})] \rightarrow g = \frac{3n^2(\det\underset{\approx}{\sigma})^{-1/3}}{3\sqrt{3}n - \log\det(\underset{\approx}{\sigma}/G)} \tag{7-8c}$$

This latter damping function in Equation 7-8c closely approximates the damping function of Wagner, namely $h = \exp(-n\gamma)$, except at small strains, γ. The difference at small γ is required to keep g bounded as $\gamma \rightarrow 0$. Unlike the first two damping functions in Equation 7-8, the third leads to an overshoot in the primary normal stress difference during start-up of steady shearing. Such an overshoot is often seen experimentally.[7]

The existence of these differential analogs provides an alternative to the procedure of fitting step shear data to single integral equations.[7] In this procedure, the shear stress, σ_{12}, is measured after a step shearing strain. If this stress is factorable into time- and strain-dependent functions,

$$\sigma_{12}(t,\gamma) = G(t)\,h(\gamma)\,\gamma \tag{7-9}$$

then a factorable integral equation

$$\sigma_{12} = \int_{-\infty}^{t} -\frac{dG}{dt}(t-t')\,h[\gamma(t,t')]\,\gamma(t,t')\,dt' \tag{7-10}$$

is proposed as a constitutive equation for the material. Equation 7-10 is then used to predict, with varying degrees of accuracy, the shear stress in other shear histories, such as stress growth in start-up of steady shearing, double-step shear, etc.

An alternative suggested here is to use Equation 7-7b to find the function $g(\underset{\approx}{\sigma})$ that corresponds to $h(I_1)$, with $I_1 = 3 + \gamma^2$ in shear. With this function, a *differential* constitutive equation is defined by Equations 7-6 and 7-7a that can then be tested in other deformation histories and compared to the predictions of the Kaye-BKZ equation. In strain histories with strain reversal, such as reversing double-step strains and stress growth followed by elastic recovery, this differential equation is in better accord with experimental data than is the corresponding integral equation.[8,9]

For integral equations with a damping function, h, that depends on I_2, corresponding convected derivatives are not generally known. There is a convected derivative, however, that corresponds to $h(\mathrm{tr}\underset{\approx}{C}^{-1/2})\underset{\approx}{C}^{-1/2}$; see Section 5.3.3 for the simplest example.

7.2.3 Differential Analogs for Nonseparable Kaye-BKZ Equations

For nonseparable Kaye-BKZ equations –that is, equations in which the stress after a step strain is not factorable into time- and strain-dependent parts– there is no known procedure for finding differential equations that behave the same in stress relaxation after step strains. One can, however, find Kaye-BKZ analogs for some nonseparable differential equations. To do this, one must find the response of the differential equation to an arbitrary step strain; from this a unique Kaye-BKZ equation is determined.

Consider, for instance, the following version of the PTT equation,

$$\lambda \overset{\nabla}{\underset{\approx}{\tau}} + \frac{\alpha}{G}\mathrm{tr}(\underset{\approx}{\tau})\underset{\approx}{\tau} + \underset{\approx}{\tau} = 2G\lambda\underset{\approx}{D} \tag{7-11}$$

The stress tensor $\underset{\approx}{\tau}$ differs from the stress tensor $\underset{\approx}{\sigma}$ used heretofore in the chapter in that $\underset{\approx}{\tau}$ is defined to be the zero tensor at equilibrium, $\underset{\approx}{\tau} = 0$. The stress tensor $\underset{\approx}{\sigma}$ equals $G\underset{\approx}{\delta}$ at equilibrium, where $\underset{\approx}{\delta}$ is the unit tensor. Either stress tensor is acceptable, since for incompressible fluids the stress is only determined to within a term proportional to $\underset{\approx}{\delta}$. Constitutive equations involving $\underset{\approx}{\sigma}$ can be converted to equations involving $\underset{\approx}{\tau}$ by replacing each occurrence of $\underset{\approx}{\sigma}$ with $\underset{\approx}{\tau} + G\underset{\approx}{\delta}$.

In stress relaxation after a step strain, Equation 7-11 has the solution

$$\underset{\approx}{\tau} = \frac{G(\underset{\approx}{C}^{-1} - \underset{\approx}{\delta})\exp(-t/\lambda)}{1 + \alpha(I_1 - 3)[1 - \exp(-t/\lambda)]} \tag{7-12}$$

A Kaye-BKZ equation with the same response to a step strain is

$$\underset{\approx}{\tau} = \int_{-\infty}^{t} \frac{(1 + \alpha(I_1 - 3))(\underset{\approx}{C}^{-1} - \underset{\approx}{\delta})\exp[(t' - t)/\lambda]\,dt'}{\{1 + \alpha(I_1 - 3)(1 - \exp[(t' - t)/\lambda])\}^2} \tag{7-13}$$

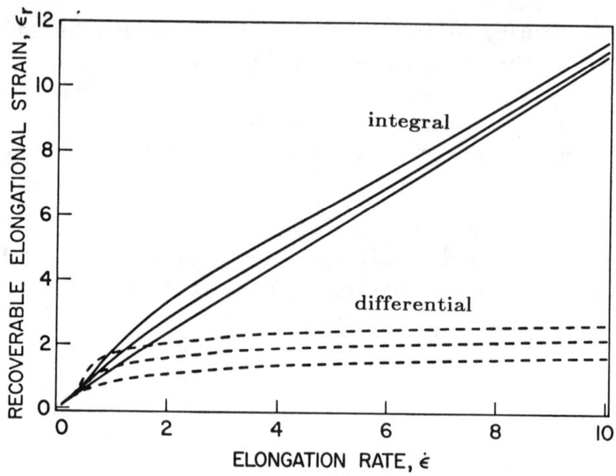

Fig. 7-1 Comparison of separable integral (-) and differential (- -) equations in uniaxial extensional strain recovery following a prior steady-state extension. Integral and differential equations are given by Equations 7-4, 7-6 and 7-7a with step-strain response defined by Equation 7-8a; α = 0.03, 0.12, 0.60, from top curves to bottom (from Larson,[9] reproduced with permission from J. Non-Newtonian Fluid Mech.).

7.2.4 Comparison of Kaye-BKZ Equations with Their Differential Analogs

Although identical in their response to a step strain, the analog differential and integral equations presented above are not identical in other deformation histories.

The *separable* integral and differential equations are similar to each other in stress growth after inception of steady straining,[9] but differ greatly in their prediction of elastic recovery. The separable differential equations predict a tendency of the recoverable strain to reach a plateau, especially in elongation, at large prior steady-state strain rates. The separable integral equations don't show a plateau; see Figure 7-1. Experimental data[10] favor the separable differential equations in this respect, although Wagner has shown how an *irreversible damping function* in separable integral equations can bring the integral equation into agreement with the data; see Sec. 5.5.1.

The nonseparable differential and integral equations are very different in their predictions of both recoverable strain and of stress growth, even though their response to a step strain is identical. Thus a preference between nonseparable integral and nonseparable differential constitutive equations could be determined by choosing a nonseparable differential equation and a nonseparable integral equation that give an identical fit to experimental step strain data, and then comparing the predictions of the two equations to experimental stress growth data.

7.2.5 Comparison of Separable and Nonseparable Differential Constitutive Equations

We focus now on differential constitutive equations and seek to discover which of the two categories, separable or nonseparable, is qualitatively more realistic.

A key difference between separable and nonseparable differential constitutive equations is that in the separable equations strain softening comes about because of the nonlinear *convected derivative*, whereas in the nonseparable equation it is the nonlinearity in the *relaxation* that produces softening. We say that a constitutive equation shows strain softening if any of its stress predictions fall below the corresponding predictions of the upper-convected Maxwell (UCM) equation. At high strain rates, the nonlinear convected derivative produces strain softening after a certain level of strain has been applied, *no matter how high the strain rate* becomes. Strain softening occurs in nonseparable equations, however, when the stress is large enough that the nonlinear relaxation overwhelms the convected derivative. As the rate of increase in the stress grows, the convected derivative is overwhelmed at progressively higher stresses. Thus in a strong flow, which initially has an exponentially increasing rate of stress growth, a nonseparable constitutive equation often produces higher stresses than in a weak flow. A separable equation is not influenced by flow strength to the same degree. Table 7-1 illustrates the trends.[9]

Table 7-1 Asymptotics of Differential Models ($\dot{\gamma} = \dot{\epsilon} \rightarrow \infty$)

	Separable Larson ($\alpha > 0$)	Nonseparable Phan-Thien/Tanner ($\xi = 0$)	Nonseparable Giesekus ($0 < \alpha < 1$)	Nonseparable Leonov
$\Delta\tau_u / N_1$	1	$\dot{\epsilon}^{1/3}$	$\dot{\epsilon}^{1/2}$	$\dot{\epsilon}^{1/2}$
$\Delta\tau_u / \Delta\tau_b$	1	2	1	$\dot{\epsilon}^{1/2}$
$\Delta\tau_{p,1} / N_1$	1	$\dot{\epsilon}^{1/3}$	$\dot{\epsilon}^{1/2}$	$\dot{\epsilon}^{1/2}$

Tabulated in the first row are values of the ratio $\Delta\tau_u / N_1$ at asymptotically high strain rates for separable and nonseparable differential equations. $\Delta\tau_u$ is the normal stress difference in *uniaxial* extension; N_1 is the primary normal stress difference in shear. The strain rates for extension and shear are taken to be equal and large; $\dot{\epsilon} = \dot{\gamma} \rightarrow \infty$. Only the scalings with $\dot{\epsilon}$ are shown in Table 7-1; constant coefficients are dropped. The version of the PTT model considered in Table 7-1 has a linear dependence of Y on $\mathrm{tr}\,\underset{\approx}{\tau}$; see Equation 6-35a. At high strain rates, the nonseparable models predict that the ratio $\Delta\tau_u / N_1$ becomes large. Thus the nonseparable equations produce higher stresses in uniaxial extension than in shear. The separable equation predicts stresses of equal magnitude at high rates of strain. This result, by itself, would tend to favor the nonseparable equations, because experimentally $\Delta\tau_u$ at a high elongation rate is usually found to be larger

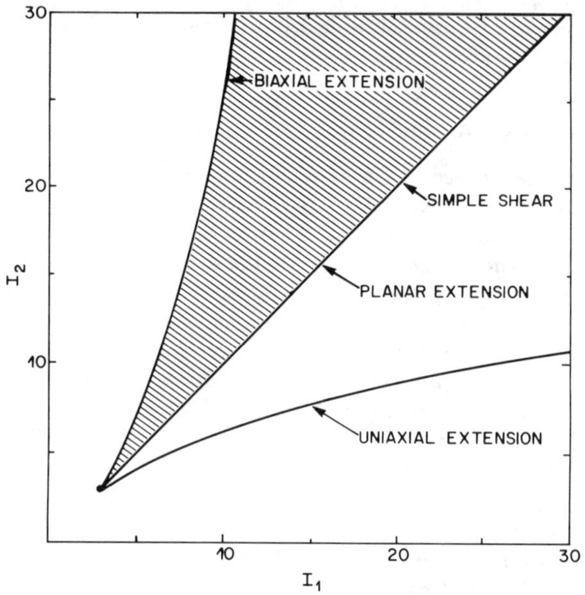

Fig. 7-2 Relation between I_1 and I_2 for particular types of strain histories (from Treloar,[13] reproduced with permission from *The Physics of Rubber Elasticity*).

than N_1 at a corresponding shear rate for polymer melts. For IUPAC A (or Melt I*), for example, $\Delta\tau_u$ at $\dot{\epsilon} = 1\ \text{sec}^{-1}$ reaches an apparent steady-state value of 5 x 10^6 dyn/cm^2, which is 10 times the value of N_1 at the corresponding shear rate[11] $\dot{\gamma} = 1\ \text{sec}^{-1}$.

7.2.6 Alignment Strength versus Flow Strength

Nonseparable equations may distinguish between uniaxial extension and shear for the wrong reason, however. Steady uniaxial extension differs from steady shear in two important ways. First, steady uniaxial extension is a *strong* flow; some material linear grow exponentially in time. Steady shear is *weak*; material lines grow at most linearly in time.[12] Second, uniaxial extension deforms material elements in a way that is *geometrically different* than does shear. A material element in uniaxial extension stretches in one direction and shrinks equally in the other two. In shear, the material element stretches in one direction, shrinks in another, and remains constant in the third direction.

---

* Melt I and IUPAC A are two different batches of low density polyethylene that have the same viscoelastic properties, including the same linear viscoelastic spectrum[10].

These differences are quantified by the strain invariants I_1 and I_2. Figure 7-2 shows that for volume-conserving deformations, uniaxial extension lies at one geometric extreme, biaxial extension at the other, and shear or planar extensions are intermediate. Planar extension and simple shear are *geometrically equivalent*, because both have a neutral axis, with stretching and compression along the other two axes. Uniaxial extension most strongly aligns molecules along a single axis, because stretching occurs in one direction only; we say that it is *strongly aligning*. Biaxial extension stretches equally in two directions; it is *weakly aligning*. These distinctions are made quantitative by the definitions

$$\begin{aligned} I_1 - I_2 &> 0 \quad \text{strongly aligning} \\ I_1 - I_2 &= 0 \quad \text{neutrally aligning} \\ I_1 - I_2 &< 0 \quad \text{weakly aligning} \end{aligned} \tag{7-14}$$

We call the parameter $I_1 - I_2$ the *alignment strength*.

A general Kaye-BKZ equation is capable of exhibiting sensitivity to flow strength, because it has an arbitrary time dependence in the memory function and is therefore capable of showing sensitivity to differences between exponential and nonexponential stretching. It is also capable of showing sensitivity to alignment strength, since it has in its memory function an arbitrary dependence on I_1 *and* I_2.

Some experimental evidence suggests that for *melts*, the high stress levels often seen in uniaxial extension may be related as much or more to *alignment strength* as to flow strength. (The converse is probably true for dilute solutions; see Chapter 8.) For a polyisobutylene melt studied by Meissner et al.,[14] the stresses in planar and biaxial extension do not attain the high levels often found in uniaxial extension,[14] Figure 7-3, though all three of these are strong flows.

The high stresses predicted for steady-state uniaxial extension by the Giesekus[15] and PTT models are caused by flow strength, not alignment strength, however.[8] This is shown in the second row of Table 7-1, which compares stresses for two strong flows, uniaxial and biaxial extension, that have *opposite extremes* of alignment strength. The Giesekus and PTT models show almost no contrast between the stress difference in biaxial extension and that in uniaxial extension. Thus these models are not sensitive to alignment strength. The third row of Table 7-1 compares the stresses for flows of the *same alignment strength* but different flow strength, namely planar extension and shear. The entries show that the PTT and Giesekus equations are *sensitive to flow strength*. Thus the high stresses in uniaxial extension predicted by these models are caused by a sensitivity to flow strength, not alignment strength. Experiments with polyisobutylene, however, suggest that high stress levels correlate with alignment strength.

The Leonov[16] equation shows sensitivity to both flow strength and alignment strength and, in this regard, is more realistic than the PTT and Giesekus equations. The reason for this is discussed later.

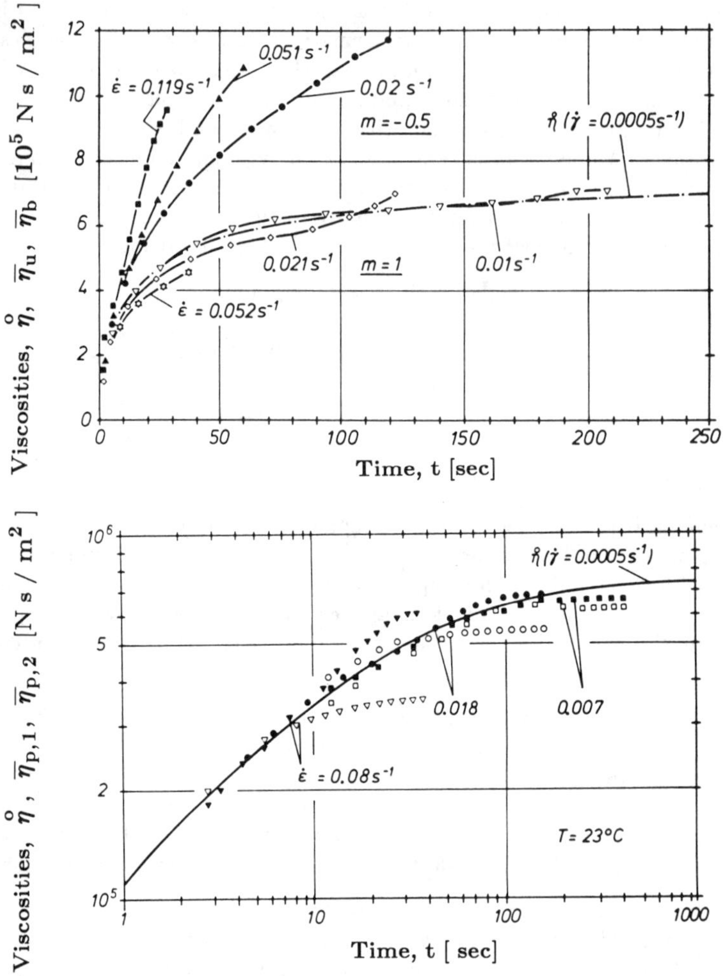

Fig. 7-3 Extensional viscosities as functions of time after start-up of steady extension, at particular rates, $\dot{\epsilon}$. (top) Uniaxial viscosity, $\bar{\eta}_u$ (solid symbols) and biaxial viscosity, $\bar{\eta}_b$ (open symbols); (bottom) planar viscosities, $\bar{\eta}_{p,1}$ (solid symbols) and $\bar{\eta}_{p,2}$ (open symbols). Viscosities are defined in Section 6.2.3.2. $\overset{o}{\eta}$ is the linear viscoelastic response (from Meissner et al.[14] reproduced with permission from J. Non-Newtonian Fluid Mech.).

7.3 COMPARING CONSTITUTIVE EQUATIONS TO MELT DATA

The proof of a constitutive equation is its ability to fit rheological data. In this section the strengths and weaknesses of some of the simple differential equations are evaluated by comparing their predictions to rheological data for three commercial melts. The first melt considered is a high-density polyethylene (HDPE) that has few, if any, long side branches. Later some comparisons to a low-density polyethylene (LDPE) with long side branches, namely IUPAC A, and to a polystyrene (PS) will be made. The molecular weight distributions of these melts are tabulated in Table 7-2.[17]

Table 7-2 Melt Properties

	M_w	M_w/M_n
HDPE	152,000	13
LDPE	472,000	24.9
PS	398,000	2.9

First we consider some of the simplest differential constitutive equations that have only a single nonlinear parameter, namely the models of Johnson and Segalman[18] (JS), White and Metzner[3] (WM), Acierno et al.,[2] Giesekus,[15] Larson,[19] and Phan-Thien and Tanner[1] (PTT).

Khan and Larson[20] compared these models to time-dependent data in relaxation after *step shear*,[17] stress growth in *start-up of steady elongation*,[21] and relaxation after a *step biaxial extension*.* The motivation for choosing these particular deformation histories is as follows. First, these three deformation types include the two extremes and the mean among the possible deformation geometries depicted in Figure 7-2; they include a strongly aligning, a weakly aligning, and a neutrally aligning deformation. Two of the three deformations are *step straining*, which are particularly easy to interpret because of time-strain separability. The other is a *strong flow*. No attempt is made here to fit shear stresses in steady shearing, because shearing viscosities are much easier to fit than step shear data. For example, even the corotational Maxwell equation, which fits *step* shear data extremely poorly, does reasonably well in its prediction of *steady-state* shear viscosity; see Sec. 3.4.4.1. If a good fit to the step shear data is obtained, then transient or steady-state shear viscosities are likely to be predicted reasonably well.[7] Hence, the step shear experiment is a more sensitive test of the constitutive equation than is the measurement of steady-state or transient shear viscosity.

Each constitutive equation considered has only a single nonlinear parameter. All linear parameters were determined by linear viscoelastic measurements. The linear spectra determined for the HDPE and the LDPE are given in Table 7-3.[24] To attempt to fit simultaneously all data for all three types of deformation history with a single value of the adjustable parameter is a severe test of a

* The biaxial data were obtained using a lubricated squeezing technique developed by Chatrei and Macosko.[22] Soskey and Winter were the first to do step biaxial extension using this technique.[23]

constitutive equation.

It is crucial to note that conclusions drawn regarding the relative strengths of the constitutive equations reflect their predictive capacity *for the deformation histories chosen.* The conclusions might be different if a different set of deformation histories were chosen. The set of deformation histories considered here was chosen to test the ability of the constitutive equations to predict stresses in *a broad range of deformation types.* This choice may or may not suit a given application. If one is concerned with planar flows or nearly steady flows, for example, a different set of tests would be more relevant, and the conclusions regarding the relative merits of the various models might be different.

Table 7-3 Linear Spectra for HDPE and LDPE

λ_i (sec)	HDPE G_i (Pa)	IUPAC A G_i (Pa)
0.001	2.058×10^5	1.52×10^5
0.005	7.82×10^4	4.005×10^4
0.028	4.26×10^4	3.326×10^4
0.14	1.59×10^4	1.659×10^4
0.7	6.32×10^3	8.69×10^3
3.8	1.835×10^3	3.151×10^3
20	3.4×10^2	8.596×10^2
100	4.85×10^1	1.283×10^2
500	1.02×10^0	1.849×10^0
1000	1.5×10^0	0×10^0

7.3.1 Differential Constitutive Equations

7.3.1.1 Johnson-Segalman and White-Metzner Equations The models are presented in order from the least to the most successful in fitting the experimental data for HDPE. A multiple-mode version of each model, with the relaxation spectrum given in Table 7-3, is considered. Thus the total stress tensor is given by the sum of contributions from all the modes, $\underset{\approx}{\tau} = \sum \underset{\approx}{\tau}_i$.

Figures 7-4 and 7-5 compare the experimental step shear data for the HDPE to predictions of the equations of Johnson and Segalman,

$$\overset{\nabla}{\underset{\approx}{\tau}_i} + \alpha(\underset{\approx}{D} \cdot \underset{\approx}{\tau}_i + \underset{\approx}{\tau}_i \cdot \underset{\approx}{D}) + \frac{1}{\lambda_i}\underset{\approx}{\tau}_i = 2G_i\underset{\approx}{D} \tag{7-15}$$

and of White and Metzner,

$$\overset{\nabla}{\underset{\approx}{\tau}_i} + \frac{1}{\Lambda_i}\underset{\approx}{\tau}_i = 2G_i\underset{\approx}{D} \tag{7-16a}$$

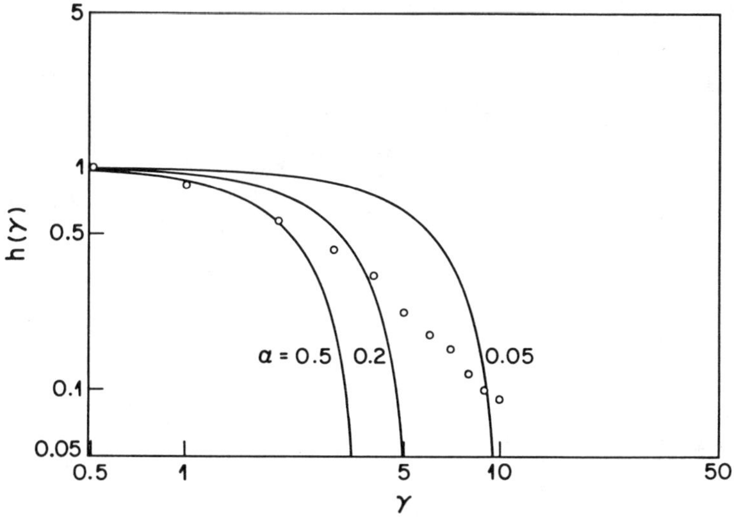

Fig. 7-4 Comparison of the shear damping function of HPDE (o) with the predictions of the Johnson-Segalman equation, Equation 7-15 (from Khan and Larson,[20] reprinted from the Journal of Rheology ©1978 Soc. of Rheol. by permission of John Wiley & Sons, Inc.).

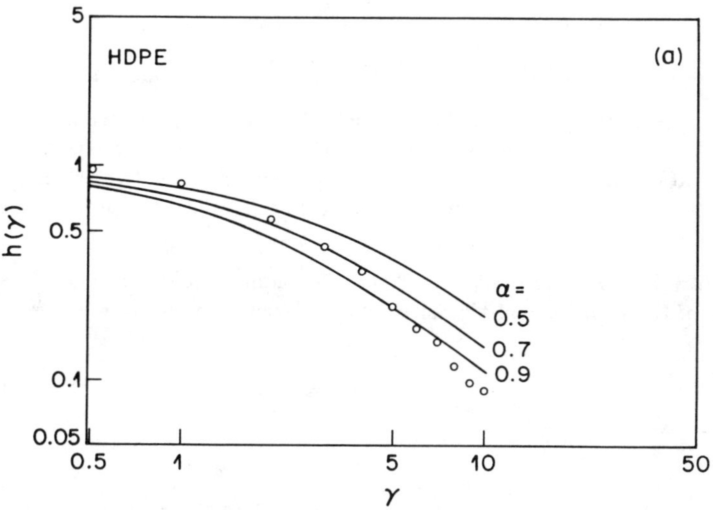

Fig. 7-5 Comparison of the shear damping function of HPDE (o) with the predictions of the White-Metzner equation, Equation 7-16 (from Khan and Larson,[20] reprinted from the Journal of Rheology ©1978 Soc. of Rheol. by permission of John Wiley & Sons, Inc.).

with[26]

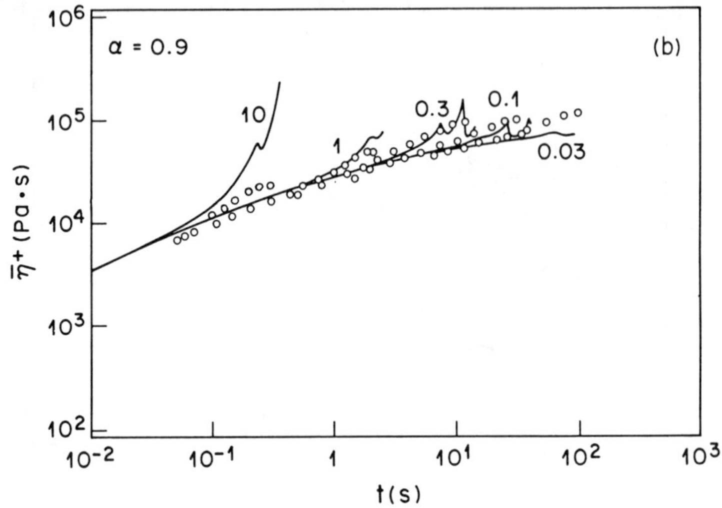

Fig. 7-6 Comparison of elongational stress growth at various strain rates, $\dot{\epsilon}$, for HDPE (o) with the predictions of the model of Acierno et al, Equation 7-17 (from Khan and Larson,[20] reprinted from the Journal of Rheology ©1978 Soc. of Rheol. by permission of John Wiley & Sons, Inc.).

$$\Lambda_i = \frac{\lambda_i}{1 + \alpha\lambda_i(2\mathrm{tr}\underset{\approx}{D} : \underset{\approx}{D})^{1/2}} \tag{7-16b}$$

In neither case is the fit particularly good. The damping function for the JS model reaches negative values at large strains. Neither model obeys the Lodge-Meissner relationship $N_1 = \gamma\tau_{12}$. Both models have singularities in the steady-state extensional viscosities for low and moderate values of the nonlinear parameter.

7.3.1.2 Model of Acierno et al. Figure 7-6 compares experimental data for the growth of the elongational viscosity to the Marrucci model,

$$x_i^{1.4}\left(\frac{1}{x}\overset{\nabla}{\underset{\approx}{\tau}}\right) + \frac{1}{x_i\lambda_i}\underset{\approx}{\tau_i} = 2x_i^{1.4}G_i\underset{\approx}{D} \tag{7-17a}$$

where x_i satisfies

$$\dot{x}_i = \frac{1}{\lambda_i}x_i^{-1.4}(1 - x_i) - \frac{\alpha}{\lambda_i}x_i^{-0.4}\left(\frac{\mathrm{tr}\,\underset{\approx}{\tau_i}}{2G_ix_i}\right)^{1/2} \tag{7-17b}$$

The Marrucci model shows oscillatory stresses, as discussed in Section 6.3.4.2. The oscillations are not smoothed out even though 10 relaxation modes spaced a half-decade apart in time are used; this spacing gives a near perfect fit to the

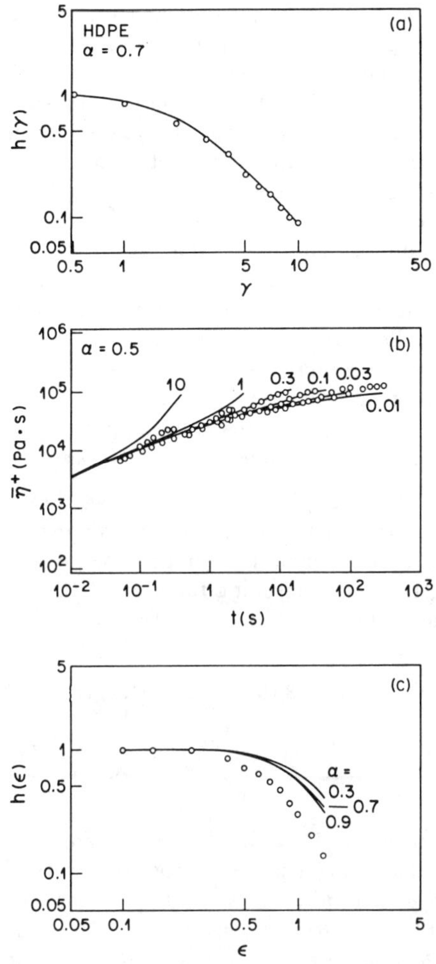

Fig. 7-7 Comparison of the predictions of the Giesekus model, Equation 7-18, with HDPE data (o) in step shear (top), steady elongation (middle), and step biaxial extension (bottom) (from Khan and Larson,[20] reprinted from the Journal of Rheology © 1978 Soc. of Rheol. by permission of John Wiley & Sons, Inc.).

linear data. Oscillations also appear in the step-shear predictions.

7.3.1.3 Giesekus Model The Giesekus equation is

$$\overset{\nabla}{\underset{\approx}{\tau}}_{i} + \frac{\alpha}{\lambda_i G_i}\underset{\approx}{\tau}_{i}^{2} + \frac{1}{\lambda_i} = 2G_i \underset{\approx}{D} \tag{7-18}$$

Though each mode of the Giesekus equation is nonseparable (the stress after a step strain cannot be factored into time- and strain-dependent functions), when all modes are added together the result is nearly factorable; see Section 6.2.3.3.

Thus for a given relaxation spectrum, a damping function can be obtained from the Giesekus model by a procedure analogous to that used to extract the damping function from the experimental data. That is, stress relaxation after a step strain is computed and the vertical shift factor required to superpose the nonlinear response onto the linear response is found. The damping function is the reciprocal of the shift factor. The shear damping function obtained in this way from the Giesekus equation describes the experimental damping function well, Figure 7-7 (top). The Giesekus model is less accurate in its description of elongational stress growth, Figure 7-7 (middle). The model does not show the experimental maximum in the apparent steady-state or plateau value of the elongational viscosity as $\dot{\epsilon}$ increases. Also, the value of the nonlinear parameter α that best fits the elongational data is smaller than the value that best fits the shear data. Since strain softening increases as α increases, we can use the value of α that best fits a given type of deformation as a measure of the amount of strain softening that occurs in that deformation type. Thus according to the Giesekus model, *the material is more strain softening in shear than it is in uniaxial extension.*

The biaxial damping function, h_ϵ, is defined as $\Delta\tau_b \,/\, (e^{2\epsilon} - e^{-4\epsilon})G(t)$, where ϵ is the biaxial Hencky strain. By this definition, h_ϵ is unity for the UCM equation. Thus strain softening occurs when h decreases below unity. In biaxial step strains, Figure 7-7 (bottom), the damping function of the Giesekus model fails to show enough strain softening to fit the experimental data for any value of α in the acceptable range, $0 \leq \alpha \leq 1$.

7.3.1.4 Larson Model Figure 7-8 (left) shows that the model of Larson,

$$\overset{\nabla}{\underset{\approx}{\tau}}_i + \frac{2}{3}\frac{\alpha}{G_i}\underset{\approx}{D} : \underset{\approx}{\tau}_i(\underset{\approx}{\tau}_i + G_i\underset{\approx}{\delta}) + \frac{1}{\lambda_i} = 2G_i\underset{\approx}{D} \tag{7-19}$$

gives a reasonable description of all the data for a single value of the nonlinear parameter, $\alpha = 0.35$. This model is more consistent with the elongational data than is the Giesekus model in that it predicts a maximum in the apparent steady-state value of the elongational viscosity as a function of elongation rate $\dot{\epsilon}$. As with the Giesekus model, however, *the uniaxial data show less strain softening* than do the shear data; that is, the value of α that would best fit the uniaxial data is less than that which would give the best fit to the shear data.

7.3.1.5 Phan-Thien /Tanner Model The version of the PTT model with an *exponential* relaxation coefficient,

$$\overset{\nabla}{\underset{\approx}{\tau}}_i + \frac{1}{\lambda_i}\exp\left(\frac{\alpha}{G_i}\,\mathrm{tr}\,\underset{\approx}{\tau}_i\right)\underset{\approx}{\tau}_i = 2G_i\underset{\approx}{D} \tag{7-20}$$

was chosen because the exponential term results in a maximum in the steady-state elongational viscosity and thus is in better accord with the experimental data than is the version with a linear coefficient. This version of the PTT equation gives agreement with the data that is more or less equivalent to that of Larson's model; see Figure 7-8 (right).

As is true of the models of Giesekus and Larson, when the parameter is chosen to obtain a best fit in shear and biaxial strains, the prediction of the elongational data is too "soft"; i.e., it underpredicts the apparent steady-state elongational

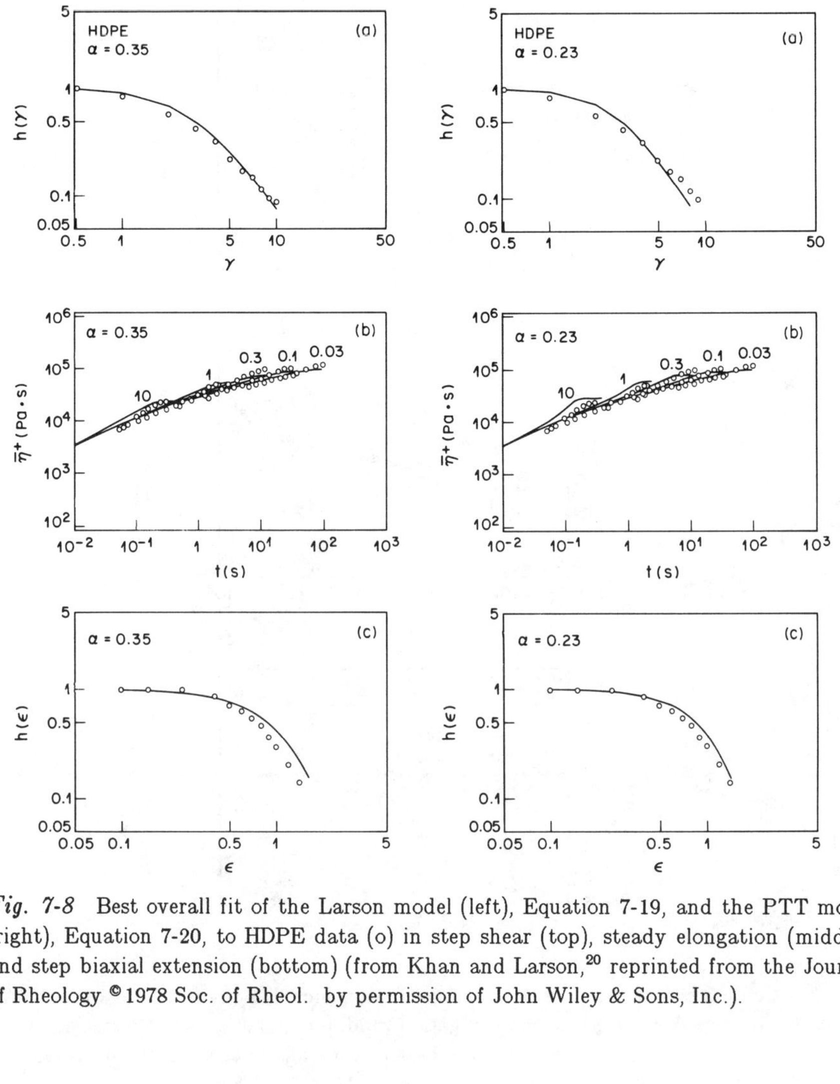

Fig. 7-8 Best overall fit of the Larson model (left), Equation 7-19, and the PTT model (right), Equation 7-20, to HDPE data (o) in step shear (top), steady elongation (middle), and step biaxial extension (bottom) (from Khan and Larson,[20] reprinted from the Journal of Rheology © 1978 Soc. of Rheol. by permission of John Wiley & Sons, Inc.).

stress.

Thus the models of Phan-Thien and Tanner and Larson make relatively good predictions for all three sets of data for the HDPE. The weakest aspect of both models is that they cannot predict enough strain softening in step shear and step biaxial extension without predicting too much in steady uniaxial extension.

7.3.1.6 Comparison to IUPAC A The under-prediction of uniaxial extensional data for the HDPE also occurs, and is much more severe, for LDPE IUPAC A, a polymer with long side branches. Figure 7-9 shows that the value of α for the model of Larson that best fits the steady uniaxial extensional data is a factor

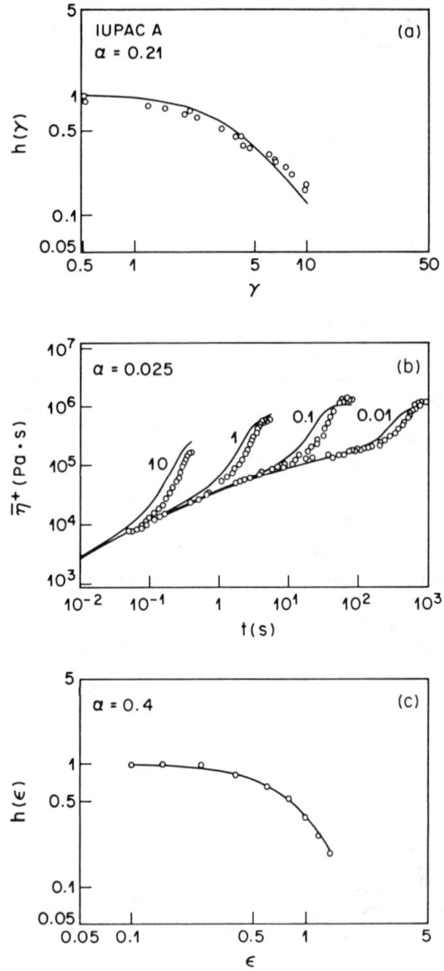

Fig. 7-9 Comparison of the predictions of the Larson model, Equation 7-19, with IUPAC A data (o) in step shear (top), steady elongation (middle), and step biaxial extension (bottom). Each value of α given represents the best fit for each deformation (from Khan and Larson,[20] reprinted from the Journal of Rheology ©1978 Soc. of Rheol. by permission of John Wiley & Sons, Inc.).

of 10 or more lower than the values required to fit step shear or step biaxial relaxation data. A similar result is found with the PTT model. For IUPAC A, unlike the HDPE, no intermediate value of α can be chosen that would give reasonable agreement with all three types of deformations. For IUPAC A, step biaxial extension is fit by the largest value of α, step shear by an intermediate value, and uniaxial extension by the smallest value. Using α as an index of strain softening, we therefore find for these deformation histories that ***strain softening decreases as the alignment strength increases.***

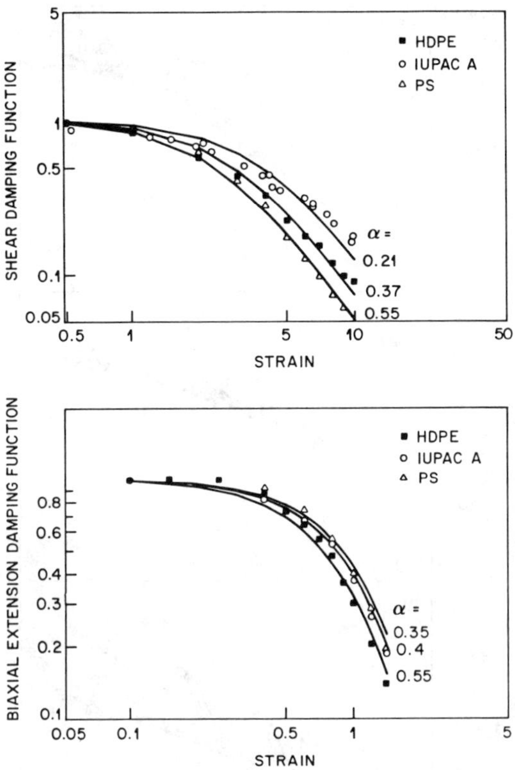

Fig. 7-10 Shear and biaxial extension damping functions as functions of strain for IUPAC A, HDPE, and PS compared to fits with the Larson model (from Khan et al.,[17] reproduced with permission from Rheol. Acta).

Figure 7-10 shows the measured damping functions in step shear and step biaxial extension for the HDPE, the LDPE, and the third material, an unbranched commercial polystyrene (PS). Figure 7-11 shows the uniaxial stress growth for the same materials. The greatest difference among the materials is seen in steady uniaxial extension — the strongly aligning deformation, and the least difference in step biaxial extension — the weakly aligning deformation. Unfortunately, the deformation that most strongly distinguishes materials, namely steady uniaxial extension, is the one for which reliable measurements are the hardest to obtain.

The materials without long side branches, PS and HDPE, act similarly in each deformation, and are most different from the branched IUPAC A in uniaxial extension. The small level of strain softening in uniaxial extension is not peculiar to IUPAC A; it occurs in other branched LDPE's[21,27]. All the results on differential models presented in this section are taken from Khan and Larson[20] and are summarized in Table 7-4 in a format used by Tanner.[25]

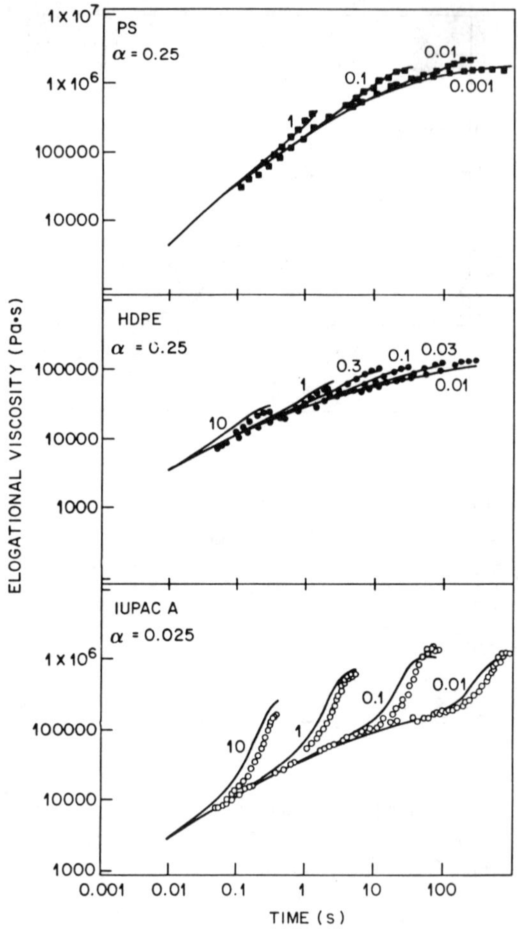

Fig. 7-11 Growth of elongational viscosity after start-up of steady elongation for IUPAC A, HDPE, and PS compared to fits with the Larson model (from Khan et al.,[17] reproduced with permission from Rheol. Acta).

7.3.2 Alignment Strength and the Damping Function

The failure of the models of Phan-Thien/Tanner and Larson to predict the difference in strain softening between uniaxial extension and shear or biaxial extension might be due in part to the insensitivity of these models to *alignment strength*.

Table 7-4

Summary of Simple Differential Constitutive Equations (from Khan and Larson[20])

Constitutive Models	Step Shear	Elongation Start up	Step Biaxial Extension	Remarks
Johnson-Segalman (Equation 7-15)	P-U	P-U	—	Predicts unrealistic (negative) damping function in shear. Singularity in uniaxial extension.
White-Metzner (Equation 7-16)	P	M-P	—	Poor damping function prediction. Singularity in elongation at large strains.
Marrucci (Equation 7-17)	U-P	P-U	—	Oscillations observed in stress relaxation and elongational viscosity
Giesekus (Equation 7-18)	E	M-G	P	Cannot predict strain softening in biaxial extension
Larson (Equation 7-19)	G	G	G	Fails to predict all deformations of branched melt (IUPAC A) with a single value of the adjustable parameter
Phan-Thien Tanner (Equation 7-20)	M-G	G	E	

E = excellent; G = good; M = moderate;
P = possible result but not in good agreement;
— = result not compared;
U = physically unrealistic result.

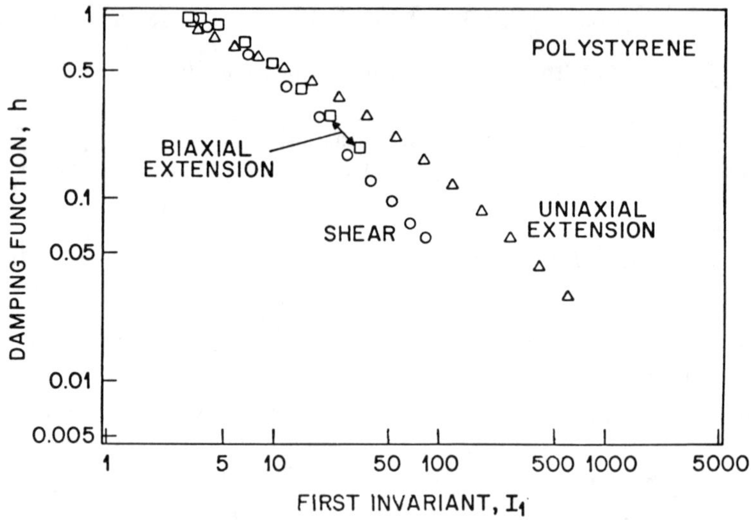

Fig. 7-12 Damping function, h, versus the first invariant, I_1, in shear, biaxial extension, and uniaxial extension for the polystyrene melt (from Larson,[9] reproduced with permission from J. Non-Newtonian Fluid Mech.).

For both of these models, each relaxation mode's contribution to the stress after a step deformation of any kind obeys[9]

$$\underset{\approx}{\tau}_i = G_i H(t/\lambda_i, I_1)\,(\underset{\approx}{C}^{-1} - \underset{\approx}{\delta}) \tag{7-21}$$

to within an isotropic constant. Here $H(t/\lambda_i, I_1)$ is a function that for the separable model of Larson is factorable; $H(t/\lambda_i, I_1) = G(t/\lambda_i)h(I_1)$. When all modes for a commercial material are added together, the nonseparable PTT equation also gives an approximately factorable response. The crucial point is that H *does not depend on* I_2. Thus, it has no dependence on alignment strength $I_1 - I_2$. Hence, according to these models, *the damping function should be a universal function of* I_1 *only.*

Figure 7-12 shows the damping functions in all three types of deformation plotted against I_1 for PS. The damping function in uniaxial extension was inferred by Wagner from the stress growth data by finding the damping function that, when inserted into a separable integral equation, predicts the stress growth data. From Figure 7-12, one sees that the damping function is *not* a universal function of I_1 (although the difference between h_{uniax} and h_{shear} or h_{biax} might be partly due to the indirect way in which h_{uniax} was obtained). Thus there appears to be a dependence on alignment strength. The dependence is not a severe one; a single function lying between the uniaxial data and the shear and biaxial data could roughly describe all three sets of data. A similar result holds for the HDPE, which explains why the models of Larson and Phan-Thien/Tanner can describe this material reasonably well with a single value of the nonlinear parameter.

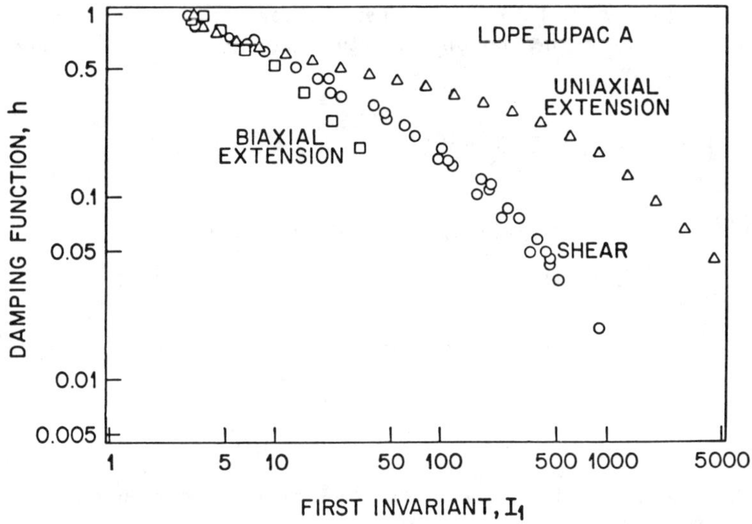

Fig. 7-13 Damping function, h, versus the first invariant, I_1, in shear, biaxial extension, and uniaxial extension for IUPAC A (from Larson,[9] reproduced with permission from J. Non-Newtonian Fluid Mech.).

For the branched LDPE, however, the deviation among the three damping functions for the three deformation types is great, Figure 7-13. The damping function apparently depends strongly on the alignment strength.

7.3.3 Constitutive Equations with a Dependence on Alignment Strength

It is evident, therefore, that a constitutive equation that will fit data in all types of deformation, particularly for materials with long side branches, should have a dependence on I_2. The PTT and Larson models lack this dependence. For the Giesekus model relaxation after a step strain shows a slight dependence on I_2, but one too weak to be significant. The Giesekus model's lack of dependence on alignment strength is also evident in steady flows; see Table 7-1.

7.3.3.1 The Leonov Model A dependence on alignment strength is present if the invariant $\mathrm{tr}(\underset{\approx}{\sigma}^{-1})$ appears in the relaxation term of a differential constitutive equation. If only $\mathrm{tr}\underset{\approx}{\sigma}$ appears, as in the PTT equation, then the relaxation after a step strain depends on I_1 alone (if the convected derivative is the upper-convected derivative or that of Larson). The Leonov model,[16] however, has a dependence on $\mathrm{tr}(\underset{\approx}{\sigma}^{-1})$;

$$\overset{\nabla}{\underset{\approx}{\sigma}} + \frac{1}{2G\lambda}\left[\underset{\approx}{\sigma}\cdot\underset{\approx}{\sigma} - G^2\underset{\approx}{\delta} - \frac{1}{3}[\mathrm{tr}\underset{\approx}{\sigma} - G^2\mathrm{tr}(\underset{\approx}{\sigma}^{-1})]\,\underset{\approx}{\sigma}\right] = \underset{\approx}{0} \qquad (7\text{-}22)$$

As a result, at high strain rates it shows a dependence on both flow strength and alignment strength, Table 7-1. The effect is too weak, however, to obtain agreement with IUPAC A data.

7.3.3.2 Wagner Equation In Wagner's integral constitutive equation (the Wagner I equation), the damping function has a direct dependence on I_2. Wagner's equation[28] is

$$\underset{\approx}{\sigma} = \int_{-\infty}^{t} m(t-t')\, h(I_1,I_2)\, \underset{\approx}{C}^{-1}(t,t')\, dt' \tag{7-23}$$

The damping function $h(I_1,I_2)$ proposed by Wagner is

$$h(I) = f\exp(-n_1\sqrt{I-3}) + (1-f)\exp(-n_2\sqrt{I-3}) \tag{7-24}$$

where I is a "generalized strain invariant," which is a linear combination of I_1 and I_2:

$$I \equiv \beta I_1 + (1-\beta)I_2 \tag{7-25}$$

In shear, $I = I_1 = I_2 = 1+\gamma^2$. Values of the three parameters, namely

$$f = 0.57, \quad n_1 = 0.310, \quad \text{and} \quad n_2 = 0.106,$$

in Equation 7-24 were obtained by fits to the measured shear damping function of IUPAC A. Since the fit to the shear damping function is independent of the value of β in Equation 7-25, the uniaxial extensional damping function inferred from steady uniaxial extension for IUPAC A was used to obtain a best fit for the value of β in Equation 7-25, $\beta = 0.032$. The function $h(I_1,I_2)$ with these parameter values fits shear and uniaxial data well. Its fit of the recently acquired step biaxial data, however, is not as good.[17,23] Figure 7-14 shows that it is too strain softening.

7.3.3.3 Equation of Papanastasiou, Scriven, and Macosko A simpler damping function was proposed by Papanastasiou et al.,[29]

$$h = \frac{1}{1+\alpha(I-3)} \tag{7-26}$$

where I is the generalized invariant given in Equation 7-25. This damping function gives a good fit to step shear and steady uniaxial data for IUPAC A, but like Wagner's damping function it is too strain-softening for the step biaxial data.

7.3.3.4 Kaye-BKZ Potential Function for IUPAC A The equations of Wagner or Papanastasiou et al. would have some drawbacks even if they did fit all three types of deformation. These equations can be obtained from the separable Kaye-BKZ equation

$$\underset{\approx}{\sigma} = \int_{-\infty}^{t} m(t-t')\,[\phi_1(I_1,I_2)\,\underset{\approx}{C}^{-1}(t,t') - \phi_2(I_1,I_2)\underset{\approx}{C}(t,t')]\, dt' \tag{7-27}$$

by dropping the term involving the Cauchy tensor $\underset{\approx}{C}$. But since the functions ϕ_1 and ϕ_2 are derived from a potential,

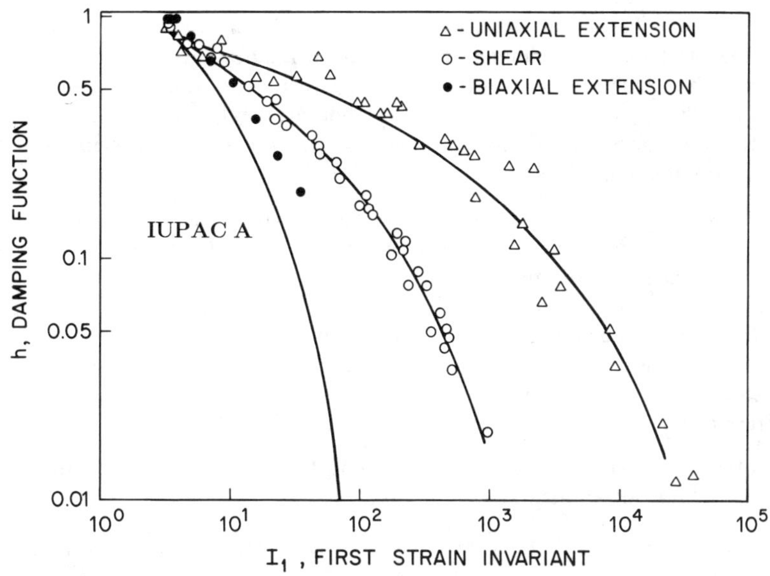

Fig. 7-14 Damping function, h, versus the first invariant, I_1, in shear, biaxial extension, and uniaxial extension for IUPAC A compared to the equation of Wagner et al., Equations 7-24 and 7-25 (from Larson and Monroe,[31b] reproduced with permission from Rheol. Acta).

$$\phi_1 = 2\frac{\partial U}{\partial I_1}\ ; \qquad \phi_2 = 2\frac{\partial U}{\partial I_2} \tag{7-28}$$

the Cauchy term in the Kaye-BKZ equation can really only be dropped if U depends on I_1 alone, so $\psi_2 = 2\partial U/\partial I_2 = 0$. However, to fit both shear and elongational data, Wagner and Papanastasiou et al. retain a strong dependence of the first coefficient, $h = 2\phi_1$, on I_2. (In these models, the dependence of h on I_2 is actually much *stronger* than the dependence on I_1; see Equation 7-25, for example.) Thus by dropping the second term, the stress constitutive equations of Wagner or of Papanastasiou can no longer be derived from an elastic potential and therefore cease to be Kaye-BKZ equations.[30]

There are two significant consequences of this. First consider a *hypothetical* deformation history that is *infinitely fast* so that relaxation becomes negligible. The stress during such a deformation history according to the constitutive equations of Wagner (or Papanastasiou et al.) is

$$\underset{\approx}{\sigma} = G(0)\,h(I_1,I_2)\,\underset{\approx}{C}^{-1} \tag{7-29}$$

where G(0) is the zero-time relaxation modulus. If one computes the work required for the deformation, it will be *path dependent*; that is it will depend not only on the total strain $\underset{\approx}{C}^{-1}$ but also on the sequence of shapes through which the material element passes. This is so because there is no elastic potential function.

Then if one carries out an infinitely fast *cyclic strain* –that is, a strain history that brings the material element back to its original shape– the required work will not be zero; it will be positive in one direction and *negative* in the other direction. Thus the absence of an elastic potential implies that in the limit of fast cyclic deformations, the material can do work on its environment, which is a thermodynamic impossibility.[30]

This thermodynamic consequence of dropping the Cauchy term in the Kaye-BKZ equation can be avoided if the damping function is made *irreversible*, as proposed by Wagner and discussed in Section 5.5.1. A second consequence of dropping the Cauchy term is *not* thereby overcome, however. Without the Cauchy term, *the second normal stress difference must be zero*. Experimentally it is almost always found to be nonzero and negative, though it is often small compared to the first normal stress difference. Thus to obtain a nonzero second normal stress difference, we should retain both tensors in Equation 7-25.

A potential function, U, that yields a good fit, not only to shear and uniaxial extensional data but to biaxial* as well, is[31b]

$$U = \frac{3}{2\alpha} \log(1 + \frac{\alpha}{3}(I - 3)) \tag{7-30a}$$

where

$$\alpha \equiv c_0 + c_2 \arctan[\frac{c_1 (I_2 - I_1)^3}{1 + (I_2 - I_1)^2}] \tag{7-30b}$$

and

$$I \equiv (1 - \beta) I_1 + (1 + 2\beta I_2)^{1/2} - 1 \tag{7-30c}$$

with

$$c_0 = 0.20, \quad c_1 = 0.05, \quad c_2 = 0.121, \quad \text{and} \quad \beta = 0.1.$$

A potential function similar to Equation 7-30a emerged earlier in a theory of Larson[19] (Section 5.4.3). In that theory, α appears as a constant (see Equation 7-19) that allows fits to IUPAC A data in uniaxial extension, shear, and biaxial extension for values of α of 0.025, 0.21, and 0.4, respectively; see Figure 7-9. Equation 7-30 generalizes this potential function; in this generalization, α is no longer a constant but is sensitive to the alignment strength, $I_1 - I_2$, in such a way that for large uniaxial, shear, and biaxial deformations, α takes the values 0.01, 0.20, and 0.39, respectively. The damping functions calculated from this potential function are compared to the experimental damping functions for IUPAC A in Figure 7-15.

---

* In ref. 31a, a potential function was given from which good agreement with both shear and uniaxial data for IUPAC A was claimed. The figure showing the fit, however, was erroneously plotted, and the fit to uniaxial extensional data was actually poor.

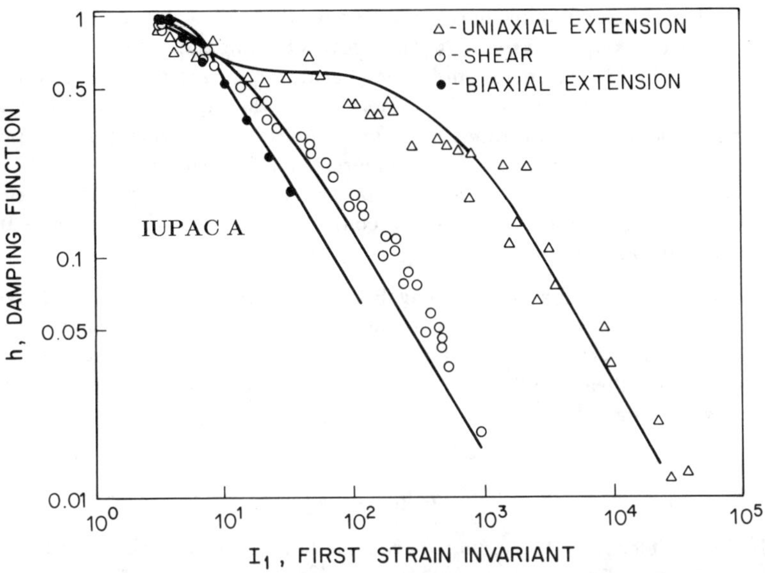

Fig. 7-15 Damping function, h, versus the first invariant, I_1, in shear, biaxial extension, and uniaxial extension for IUPAC A compared to the equation of Kaye-BKZ equation of Larson, Equation 7-30 (from Larson and Monroe,[31b] reproduced with permission from Rheol. Acta).

The correct amount of strain softening is predicted for all three types of deformation.

The Kaye-BKZ equation, with the potential function defined in Equation 7-30 gives a negative second normal stress difference; the predicted ratio of second to first normal stress differences decreases in magnitude from –0.11 at low shear rates to near zero at high shear rates. The magnitude of the second normal stress difference, N_2, is roughly proportional to the value of the parameter β. For moderate and large I_1, $h_{biax} < h_{uniax}$. There is, however, a predicted region at small I_1 in Figure 7-15 where $h_{biax} > h_{uniax}$. Such a region *must* occur at small strains if N_2 is negative. The size of this region grows with β and hence with the magnitude of N_2. The experimental data in Figure 7-16 show that the size of this region is small, which is consistent with a small magnitude for N_2.

7.4 SUMMARY

In this chapter a variety of integral and differential constitutive equations have been surveyed. First we noted that, apart from a few special cases, the simple integral equations, such as the Kaye-BKZ equations, do not have simple exact differential equivalents. However, the *separable* Kaye-BKZ integral equations in which the potential function depends only on the first strain invariant, I_1, have differential *analogs* that behave *identically* to the integral equations in relaxation after a step strain and behave similarly in *nonreversing* strain histories.

Most separable and nonseparable differential constitutive equations are not able to predict ***simultaneously*** high stress levels in uniaxial extension and modest stress levels in biaxial extension and shear because they are not sensitive to *alignment strength*, that is, to $I_1 - I_2$. Therefore these equations are not able to predict stresses over the full range of deformation types, especially for commercial melts with long side branches. For commercial melts without long side branches, the experimental dependence on alignment strength does not appear to be severe, and the Larson and PTT differential equations give credible agreement with experimental data.

A dependence on alignment strength can be introduced into the Kaye-BKZ integral equations by choosing a dependence of the potential function on *both* of the strain invariants, I_1 and I_2. If the choice is made carefully, agreement with step biaxial, step shear, and steady uniaxial data can be achieved even for IUPAC A, a melt with long side branches.

References

1. N. Phan Thien and R.I. Tanner, J. Non-Newtonian Fluid Mech., *2*, 353 (1977); N. Phan Thien, J. Rheol., *22*, 259 (1978).

2. D. Acierno, F.P. La Mantia, G. Marrucci, and G. Titomanlio, J. Non-Newtonian Fluid Mech., *1*, 125 (1976).

3. J.L. White and A.B. Metzner, J. Appl. Polym. Sci., *8*, 1367 (1963).

4. A. Kaye, College of Aeronautics, Cranford, U.K., Note No. 134 (1962).

5. B. Bernstein, E.A. Kearsley, and L.J. Zapas, Trans. Soc. Rheol., *7*, 391 (1963).

6. R.G. Larson, Proc. Ninth Int. Congress on Rheology, Acapulco, Mexico, Oct. 8-13, 1984.

7. H.M. Laun, Rheol. Acta, *17*, 1 (1978).

8. R.G. Larson and V.A. Valesano, J. Rheol., *30*, 1093 (1986).

9. R.G. Larson, J. Non-Newtonian Fluid Mech., *23*, 249 (1987).

10. M.H. Wagner, Rheol. Acta, *18*, 681 (1979).

11. M.H. Wagner, Rheol. Acta, *15*,136 (1976).

12. R.I. Tanner and R.R. Huilgol, Rheol. Acta, *14*, 959 (1975).

13. L.R.G. Treloar, *The Physics of Rubber Elasticity*, 3rd ed., Oxford Univ. Press, London, 1975.

14. J. Meissner, S.E. Stephenson, A. Demarmels, and P. Portmann, J. Non-Newtonian Fluid Mech., *11*, 221 (1982).

15. H. Giesekus, Rheol. Acta, *5*, 29 (1966); J. Non-Newtonian Fluid Mech., *11*, 69 (1982).

16. A.I. Leonov, Rheol. Acta, *15*, 85 (1976).

17. S.A. Khan, R.K. Prud'homme, and R.G. Larson, Rheol. Acta, *26*, 144 (1987).

18. M.W. Johnson, Jr. and D. Segalman, J. Non-Newtonian Fluid Mech., *2*, 225 (1977).
19. R.G. Larson, J. Rheol., *28*, 545 (1984).
20. S.A. Khan and R.G. Larson, J. Rheol., *31*, 207 (1987).
21. H.M. Laun, Proc. Ninth Int. Congress Rheol., Acapulco, Mexico, 1984.
22. S. Chatrei and C.W. Macosko, *25*, 433 (1981).
23. P.R. Soskey and H.H. Winter, J. Rheol., *29*, 493 (1985).
24. H.M. Laun, J. Rheol., *30*, 459 (1986).
25. R.I. Tanner, *Engineering Rheology,* Oxford Press, 1985.
26. Y. Ide and J.L. White, J. Non-Newtonian Fluid Mech., *2*, 281 (1977).
27. L.A. Utracki, A. Catani, J. Lara, and M.R. Kamal, Proceed. Europ. Meeting Polymer Processing Properties, Capri, Italy, June 13-16, 1983.
28. M.H. Wagner, T. Raible, and J. Meissner, Rheol. Acta, *18*, 427 (1979).
29. A.C. Papanastasiou, L.E. Scriven, and C.W. Macosko, J. Rheol., *27*, 387 (1983).
30. R.G. Larson, Rheol. Acta, *22*, 435 (1983).
31. (a) R.G. Larson and K. Monroe, Rheol. Acta, *23*, 10 (1984). (b) corrigendum, *26*, 208 (1987).

APPENDIX

Suppose the stress tensor in step strains with no relaxation is given by

$$\underset{\approx}{\sigma} = G\,h(I_1)\,\underset{\approx}{C}^{-1} \tag{7-A1}$$

Then a convected derivative, $\overset{\blacksquare}{\underset{\approx}{\sigma}}$, exists that, when set to zero,

$$\overset{\blacksquare}{\underset{\approx}{\sigma}} = \underset{\approx}{0} \tag{7-A2}$$

has the solution in Equation 7-A1. To derive the convected derivative, we note that h can be regarded as a function of U, the Kaye-BKZ potential from which $h(I_1)$ is obtained,

$$\underset{\approx}{\sigma} = 2G\frac{\partial U}{\partial I_1}\underset{\approx}{C}^{-1} = G\,h\,\underset{\approx}{C}^{-1} \tag{7-A3}$$

We take the time derivative of Equation 7-A1:

$$\dot{\underset{\approx}{\sigma}} = G\,h\,\dot{\underset{\approx}{C}}^{-1} + G\,\dot{h}\,\underset{\approx}{C}^{-1}$$

$$= Gh\,[\underset{\sim}{\nabla}\,\underset{\sim}{v}^{T}\cdot\underset{\approx}{C}^{-1} + \underset{\approx}{C}^{-1}\cdot\underset{\sim}{\nabla}\,\underset{\sim}{v}] + G\frac{dh}{dU}\dot{U}\,\underset{\approx}{C}^{-1}$$

$$= \underset{\sim}{\nabla}\,\underset{\sim}{v}^{T}\cdot\underset{\approx}{\sigma} + \underset{\approx}{\sigma}\cdot\underset{\sim}{\nabla}\,\underset{\sim}{v} - \frac{1}{h}\frac{dh}{dU}\dot{U}\,\underset{\approx}{\sigma} \tag{7-A4}$$

Since $G\dot{U} = \underset{\approx}{\sigma}:\underset{\approx}{D}$, Equation 7-A4 can be rearranged to

$$\overset{\blacksquare}{\underset{\approx}{\sigma}} = \overset{\nabla}{\underset{\approx}{\sigma}} + g\underset{\approx}{D}:\underset{\approx}{\sigma}\underset{\approx}{\sigma} \tag{7-A5}$$

where

$$g = \frac{-dh/dU}{Gh} \tag{7-A6}$$

Finally g must be expressed as a function of an invariant of $\underset{\approx}{\sigma}$. This is most easily done by considering a shearing deformation. Then since $dU/d\gamma = \sigma_{12}/G = h(\gamma)\,\gamma$,

$$g = \frac{-(dh/d\gamma)(dU/d\gamma)^{-1}}{Gh} = \frac{-dh/d\gamma}{G\,h^2\,\gamma} \tag{7-A7}$$

g can be expressed in terms of an invariant of $\underset{\approx}{\sigma}$, for example $\mathrm{tr}\underset{\approx}{\sigma}$, if $\mathrm{tr}\underset{\approx}{\sigma} = G\,h(\gamma)\,(3+\gamma^2)$ is a monotonic function of γ. If $\mathrm{tr}\underset{\approx}{\sigma}$ is not a monotonic function of γ, as is true for the damping function in Equation 7-8c, then g must be expressed as a function of $\det\underset{\approx}{\sigma}$.

CHAPTER 8

Viscoelasticity of Dilute Polymer Solutions

"Doth not wisdom cry?
and understanding put forth her voice?"

Proverbs 8:1

8.1 INTRODUCTION

Chapter 2 introduced the simplest models for dilute solutions of flexible polymer molecules, namely the elastic dumbbell and the Rouse chain of beads and springs. The stress tensors for both of these models exactly satisfy the upper-convected Maxwell (UCM) constitutive equation with a single relaxation time for the one-spring dumbbell model and a spectrum of relaxation times for the multiple-spring model.

Several of the assumptions on which these models are based are at best questionable. Among the questionable assumptions are the neglect of the following: hydrodynamic interaction, excluded volume, internal viscosity, and self-entanglements. Also, the solvent is treated as a simple Newtonian continuum, thus neglecting specific steric and energetic interactions between solvent molecules and polymeric moieties. The assumption of *Gaussian* submolecules (or linear springs) is reasonable unless the chain becomes highly extended.

Unfortunately, it has not been possible to relax any of the above assumptions without thereby precluding the derivation of an exact closed-form constitutive equation for the stress tensor. With the aid of approximations, it is sometimes possible to derive closed-form *approximate* constitutive equations, although even this has not yet been done for the model of multiple beads and springs with an accurate treatment of hydrodynamic interaction.

For these reasons the results presented in this chapter are mostly piecemeal. Small-amplitude oscillatory flows are considered first. For these flows, reasonable approximations allow the derivation of results that are in close agreement with data. For steady-state shear or extensional flows at high strain rates, experimental studies have uncovered important phenomena for which there are now plausible, though oversimplified, theories, but the development of a unified constitutive equation for dilute solutions is still in the speculative phase.

As discussed in Section 4.2.1, the radius, R, of a coil in dilute solution scales with molecular weight, M, according to a power law, $R \propto M^{\nu}$. If $\nu = 3/5$, the solvent is considered *good*; $\nu = 1/2$ in a *theta solvent*. In a theta solvent, the polymer molecule experiences a slight repulsive interaction towards the solvent that is just large enough to cancel the *excluded volume* interaction of any one part of the chain with the rest of the chain.[1] The configuration distribution of a chain in a theta solvent is therefore that of a random flight; it is *Gaussian*. Theta conditions are normally only achieved at a single temperature; good solvents are the norm.

A polymer solution is considered dilute if polymer coils rarely overlap. The viscoelastic properties of the solution are then governed by the behavior of a *single* coil in a bath of solvent. The threshold between dilute and concentrated solutions is reached when the volume of a coil, R^3, times the number of coils per unit volume, n, is roughly unity. Now $n = cN_A/M$, where c is the mass per unit volume of polymer in solution and N_A is Avogadro's number. Therefore c^*, the *overlap concentration*, scales as[2]

$$c^* = \frac{n^*M}{N_A} = \frac{R^{-3}M}{N_A} \sim M^{1-3\nu} = \begin{cases} M^{-1/2} & \text{(theta solvent)} \\ M^{-4/5} & \text{(good solvent)} \end{cases} \tag{8-1}$$

For polymers of high molecular weight, c^* is 10^{-2} gm/cm^3 or less.[3] Mechanical or optical measurements at such low concentrations require sensitive instrumentation. It is usually not possible to make measurements at concentrations that are sufficiently far below c^* that effects of coil overlap are completely negligible. The usual experimental procedure is to make measurements at a series of low concentrations, subtract any solvent contribution, and then *extrapolate* the measurements to zero concentration. The *intrinsic* steady-state shear viscosity, for example, is one such extrapolated quantity; it is defined as

$$[\eta] = \lim_{c \to 0} (\eta - \eta_s)/\eta_s c \tag{8-2}$$

The intrinsic viscosity is independent of solvent viscosity and polymer concentration. It depends only on the coil dimensions, and it is often measured as a means of determining the molecular weight of the polymer.

8.2 LINEAR VISCOELASTICITY

8.2.1 Rouse Model

8.2.1.1 Relaxation Spectrum The Rouse beads-and-springs model, presented in Chapter 2, yields a specific prediction for the distribution of relaxation times,

$$\lambda_p = \frac{\zeta}{16kT\beta^2 \sin^2(\frac{p\pi}{2(N+1)})} \approx \frac{\zeta N^2}{4\pi^2 kT\beta^2 p^2}, \quad G_p = G = nkT \; ; \; p = 1,..., N \tag{8-3}$$

where λ_p is the pth relaxation time. Here N is the number of springs, and $3\beta^{-2}/2$ is the mean square length of a spring at equilibrium; the spring force constant is $H = 2\beta^2 kT$. The longest relaxation time, λ_1, is the time required for relaxation of the entire chain; the shortest, λ_N, is approximately the time required for relaxation of a single submolecule. From the spectrum of relaxation times, G' and G'' can be predicted:

$$G' = G \sum_{p=1}^{N} \frac{\omega^2 \lambda_p^2}{1 + \omega^2 \lambda_p^2}$$

$$G'' = G \sum_{p=1}^{N} \frac{\omega \lambda_p}{1 + \omega^2 \lambda_p^2} + \eta_s \omega \tag{8-4}$$

where the modulus G equals nkT and n is the number of molecules per unit volume. G'' contains an additive contribution from the Newtonian solvent. This contribution is present even in the absence of polymer, when $n = 0$.

These predictions can readily be tested against experiment. The most distinctive prediction of the Rouse model is that for frequencies in the intermediate range, $1/\lambda_1 << \omega << 1/\lambda_N$, G' and G'' are given by

$$G' = G'' - \eta_s \omega = \left(\frac{\pi}{2\sqrt{2}}\right) G\,(\omega \lambda_1)^{1/2} \tag{8-5}$$

Thus in the intermediate-frequency regime, G' and $G'' - \eta_s \omega$ are predicted to be equal to each other and to be proportional to the square root of the frequency. This behavior is *not* seen experimentally in dilute solutions, however. Instead, the behavior of dilute high-molecular-weight polymers in good or theta solvents is typified by Figure 8-1. In this figure, the intermediate frequency regime over which the Rouse predictions of Equation 8-5 should apply extends from $\log \omega\lambda_1 \sim 1$, to $\log \omega\lambda_1 = 2$ and beyond. The high frequency regime, where $\omega \sim 1/\lambda_N$, is off the scale of this figure. Experimentally, G' and G'' in the intermediate-frequency regime of Figure 8-1 scale as $\omega^{2/3}$ rather than the predicted Rouse scaling, $\omega^{1/2}$.

8.2.1.2 Intrinsic Viscosity The Rouse model also fails in its prediction of the scaling of the low-shear viscosity, η_0, with molecular weight. Note that

$$\eta_0 = G \sum_{p=1}^{N} \lambda_p + \eta_s \approx nkT\lambda_1 \sum_{p=1}^{N} \frac{1}{p^2} + \eta_s = \frac{\pi^2}{6} nkT\lambda_1 + \eta_s$$

$$= \frac{\pi^2}{6} \frac{cN_A kT}{M} \lambda_1 + \eta_s \tag{8-6}$$

where $c = nM/N_A$ is the mass concentration per unit volume, N_A being Avagadro's number, and $\lambda_1 \doteq \varsigma N^2 / 4\pi^2 kT\beta^2$. Since $N \propto M$, the *zero-shear rate* intrinsic viscosity is proportional to the molecular weight,

$$[\eta]_0 \propto M \qquad \text{(Rouse)}$$

Experimental results can be fit to a power law

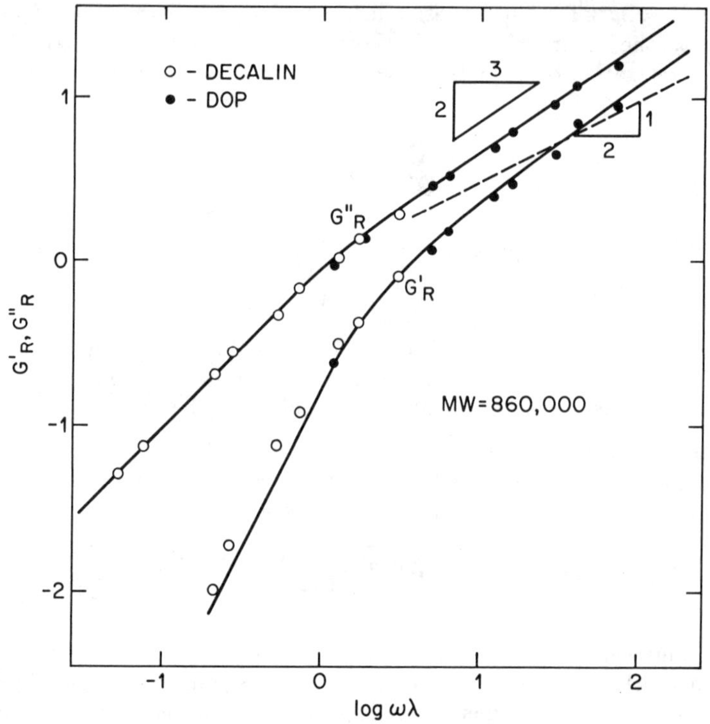

Fig. 8-1 Linear viscoelastic data for polystyrene in two theta solvents. For frequencies $\lambda\omega$ greater than 10, G' and G'' scale as $\omega^{2/3}$, in disagreement with the Rouse theory, which predicts a proportionality to $\omega^{1/2}$ in this regime. Here $G'_R \equiv G'/nkT$ and $G''_R \equiv (G'' - \eta_s\omega)/nkT$; $\lambda \equiv [\eta]_0\eta_s M/N_A kT$ (from Johnson et al.,[4] reproduced with permission from Polym. J.).

$$[\eta]_0 = KM^a \qquad \text{(experiment)}$$

but the exponent a is always less than unity and thus disagrees with the Rouse prediction. For theta solvents a is 0.5, and for good solvents[5] it is found to be no higher than 0.8.

It is convenient to define a *characteristic relaxation time*, λ, in terms of macroscopic quantities. We define

$$\lambda \equiv (\eta_0 - \eta_s)/G = \frac{[\eta]_0\eta_s c}{nkT} = \frac{[\eta]_0\eta_s M}{N_A kT}$$

λ is roughly equal to the longest relaxation time; for the Rouse model, $\lambda = (\pi^2/6)\lambda_1 = 1.645\lambda_1$. For the Zimm model with dominant hydrodynamic interaction discussed below, $\lambda = 2.369\lambda_1$.

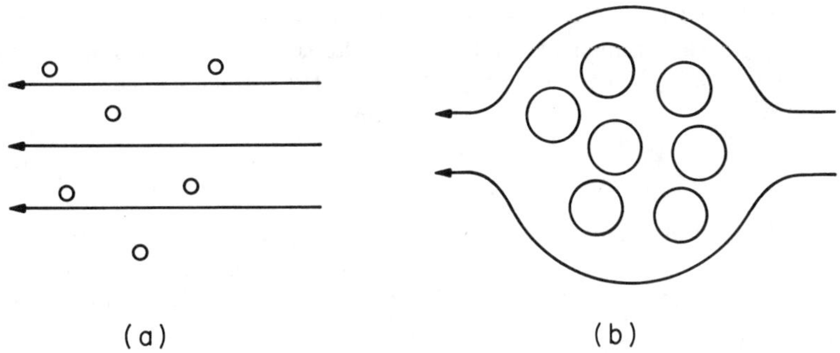

Fig. 8-2 A swarm of particles moving through a solvent. In the free-draining limit (a) the velocity field in the solvent penetrates the swarm; in the limit of dominant hydrodynamic interaction (b) it does not.

8.2.2 Hydrodynamic Interaction

The failure of the Rouse model to predict the correct frequency dependence of the linear viscoelastic functions, G' and G'', and the correct molecular-weight dependence of $[\eta]_0$ can be corrected by accounting for ***hydrodynamic interaction.***

In the Rouse model, it is assumed that the solvent velocity near one bead is unaffected by the motion of the other beads. Thus the drag on any one bead is independent of the drag on the others. In reality, the motion of one bead distorts the solvent velocity, and therefore the drag, on any other bead in its vicinity.

Consider, for example, a swarm of spheres moving en masse as depicted in Figure 8-2. Suppose that each sphere has radius a and that the entire swarm of N beads occupies a sphere of radius R. If the spheres are far apart, Figure 8-2a, the drag, f, on any one sphere is independent of the others and is given by the Stokes-Einstein formula, $f/v = 6\pi\eta a$, where v is the velocity of the swarm. The swarm is then said to be ***freely draining***, and the total drag on it is

$$f = 6\pi\eta_s Nav \qquad \text{(freely draining)} \tag{8-7}$$

If, however, the spheres are closely spaced, Figure 8-2b, then the occluded solvent is dragged along with the spheres, and the swarm with the occluded solvent acts like a single hydrodynamic entity with radius R, and therefore $f/v = 6\pi\eta_s R$. This is the Kirkwood-Riseman picture.[6] If the spheres are beads on a bead-spring chain, then the root mean square radius of the swarm is $R = N^\nu \ell$, where ℓ is proportional to the spacing between neighboring beads on the chain. $\nu = 1/2$ in a theta solvent and 3/5 in a good solvent. The drag coefficient, f/v, is then

$$f = 6\pi\eta_s N^\nu \ell v \qquad \text{(dominant hydrodynamic interaction)} \tag{8-8}$$

In this limit ***hydrodynamic interaction*** between the beads dominates.

8.2.2.1 Scaling Laws Now the molecular weight, M, of a polymer chain is proportional to N, the number of beads. From the theory of Brownian motion,[7] the drag coefficient, f/v, is inversely proportional to the diffusivity, D:

$$\frac{f}{v} = \frac{kT}{D} \tag{8-9}$$

Fundamental scaling laws relate the diffusivity to the characteristic relaxation time, λ and the zero-shear viscosity, η_0. By a dimensional argument, λ is proportional to R^2/D. Thus from Equations 8-7 and 8-9, for a freely draining chain

$$\lambda \sim \frac{R^2}{D} = \frac{6\pi\eta_s a\ell^2}{kT} N^2 \sim M^2 \quad \text{(freely draining, theta solvent)}$$

For a chain with dominant hydrodynamic interaction,

$$\lambda \sim \frac{R^2}{D} = \frac{6\pi\eta_s}{kT} R^3 \sim M^{3\nu} \quad \text{(dominant hydrodynamic interaction)}$$

The polymer contribution to the zero-shear viscosity, $\eta_{0,p}$, scales as $G\lambda$, where the modulus G is proportional to n, the number of chains per unit volume. But since $n \propto c/M$, $\eta_{0,p} \sim c\lambda/M$. Therefore

$$[\eta]_0 = \frac{\eta_0 - \eta_s}{c\eta_s} = \frac{\eta_{0,p}}{c\eta_s} \sim \frac{\lambda}{M}$$

Since $\lambda \propto M^2$ for a Rouse chain,

$$[\eta]_0 \sim M \quad \text{(freely draining)}$$

For dominant hydrodynamic interaction, $\lambda \propto M^{3\nu}$; thus

$$[\eta]_0 \sim M^{3\nu-1} = \begin{cases} M^{1/2} & \text{theta solvent} \\ M^{4/5} & \text{good solvent} \end{cases}$$

In theta solvents, experiments do show the $M^{1/2}$ scaling predicted for $[\eta]_0$ in the limit of dominant hydrodynamic interaction.* In good solvents, the measured exponents are generally slightly less than the value, 4/5, predicted for dominant hydrodynamic interaction. Table 8-1 summarizes the predicted scaling behavior

---

* The scaling law, $[\eta]_0 = KM^{1/2}$, for a flexible polymer in a theta solvent is often used to obtain molecular weights from viscosity measurements. Here the coefficient K depends on the chemical composition of the polymer and, perhaps to a minor extent, on the solvent. A tabulation of values of K for various polymers can be found in Flory's *Statistical Mechanics of Chain Molecules*.[8] For polystyrene, $K \approx 8 \times 10^{-4}$ deciliters g^{-1} $(g/mole)^{-1/2}$

in the freely draining and dominant hydrodynamic interaction limits. A more detailed discussion of these scaling laws can be found in de Gennes' book.[2] Experimental tests of the scaling laws in dilute solution generally conform to the limit of dominant hydrodynamic interaction, although in good solvents the agreement is not as close as it is in theta solvents. Theory suggests that hydrodynamic interaction should become ever more important as the molecular weight increases.[6] Even at low molecular weight, however, the limit of dominant hydrodynamic interaction has been found to be a better description of experimental data than is the Rouse freely draining limit.[9] A transition to Rouse behavior occurs as the concentration increases and the polymer coils begin to overlap.

Table 8-1 Predicted Scaling Laws in Dilute Solution

	Freely Draining Gaussian Chain	Dominant HI Theta Solvent	Dominant HI Good Solvent
Friction coefficient, f/v	M	$M^{1/2}$	$M^{3/5}$
Diffusivity, D	M^{-1}	$M^{-1/2}$	$M^{-3/5}$
Relaxation time, λ_1	M^2	$M^{3/2}$	$M^{9/5}$
Intrinsic viscosity, $[\eta]_0$	M	$M^{1/2}$	$M^{4/5}$

8.2.2.2 Intrinsic Viscosity and c* The intrinsic viscosity can be related to the entanglement threshold concentration, c^*, in the limit of dominant hydrodynamic interaction where the contribution of each polymer molecule to the solution viscosity is equivalent to that of an impenetrable sphere with a radius comparable to that of the polymer coil. The Einstein formula for the contribution of suspended spheres to the suspension viscosity is

$$\eta_p = 2.5\,\eta_s \phi$$

where ϕ is the volume fraction of spheres. From the definition of intrinsic viscosity,

$$[\eta]_0 = \frac{\eta_p}{\eta_s c} = \frac{2.5\phi}{c} \tag{8-10}$$

Now at c^* the spheres begin to overlap. As a crude estimate we take $\phi \sim 0.40$ at c^*. Then from Equation 8-10

$$c^* \sim \frac{1}{[\eta]_0}$$

and

$$\eta_p \sim \eta_s \qquad \text{at } c = c^*$$

8.2.2.3 Zimm model of linear viscoelasticity Not only does a consideration of hydrodynamic interaction account for the scaling laws for D, λ_1, and $[\eta]_0$, but it also allows an accurate prediction of the relaxation spectrum. To study the effect of hydrodynamic interaction on the relaxation spectrum, we must describe hydrodynamic interaction more precisely. The following discussion is largely taken from Zimm,[10] as presented in Bird et al.[11] Suppose a force, $\underset{\sim}{f}$, is somehow exerted on the Newtonian solvent at the origin. This force sets the surrounding solvent in motion; away from the origin at a point $\underset{\sim}{r}$ the solvent velocity, calculated from the Stokes equation, reaches the steady-state value

$$\underset{\sim}{v}' = \underset{\approx}{\Omega} \cdot \underset{\sim}{f} \tag{8-11}$$

where $\underset{\approx}{\Omega}$ is the *Oseen tensor*:[12]

$$\underset{\approx}{\Omega}(\underset{\sim}{r}) = \frac{1}{8\pi\eta_s r}\left[\underset{\approx}{\delta} + \frac{\underset{\sim}{r}\,\underset{\sim}{r}}{r^2}\right] \tag{8-12}$$

$\underset{\approx}{\delta}$ is the unit tensor. The influence of a force applied at a point in the solvent therefore decays as 1/r as the distance r from the point increases. Because this decay is so gradual, the drag force that a bead exerts on the solvent is effectively transmitted to a volume large enough to contain many other beads. The Oseen tensor is an exact expression for the influence of a force on the surrounding velocity field only if the force is concentrated at a single point. If the force is distributed on the surface of a sphere, then the Oseen tensor *approximates* the influence that the total drag force on that sphere has on the solvent velocity at distances, R, *that are large compared to the bead diameter.* Close to a bead, the Oseen tensor is inaccurate; a more accurate expression is the *Rotne-Prager tensor.*[13] For most purposes, the Oseen tensor is adequate, and only it will be considered here.

Suppose bead i is located at position $\underset{\sim}{r}_i$. The drag, $\underset{\sim}{f}_i^d$, that bead i exerts on the solvent is

$$\underset{\sim}{f}_i^d = \varsigma[\dot{\underset{\sim}{r}}_i - \underset{\sim}{v}_i]$$

where $\dot{\underset{\sim}{r}}_i$ is the velocity of bead i and $\underset{\sim}{v}_i$ is the velocity the solvent would have at position $\underset{\sim}{r}_i$ if bead i were not there. ς is the drag coefficient. In the Rouse model, $\underset{\sim}{v}_i$ is set equal to $\underset{\sim}{v}_{i,0}$, the velocity of the solvent at $\underset{\sim}{r}_i$ in the *absence of any beads,* either bead i or the other beads on the chain. To include the effects of hydrodynamic interaction, we set $\underset{\sim}{v}_i = \underset{\sim}{v}_{i0} + \underset{\sim}{v}'_i$, where $\underset{\sim}{v}'_i$ is the contribution to the velocity of the solvent at $\underset{\sim}{x}_i$ from all beads other than bead i. Thus

$$\underset{\sim}{v}'_i = \sum_j \underset{\approx}{\Omega}_{ij} \cdot \underset{\sim}{f}_j^d \tag{8-13}$$

where $\underset{\sim}{f}_j^d$ is the drag exerted on the solvent by some other bead j. Here we have defined

$$\underset{\approx}{\Omega}_{ij} \equiv \underset{\approx}{\Omega}(\underset{\sim}{r}_i - \underset{\sim}{r}_j) \qquad \text{for } i \neq j$$

$$\underset{\approx}{\Omega}_{ij} \equiv \underset{\approx}{0} \qquad \text{for } i = j \tag{8-14}$$

Now the drag exerted on bead i by the solvent, $-\underset{\sim}{f}_i^d$, is balanced by the spring forces exerted by the two springs that are attached to bead i and by the Brownian force. Thus

$$\underset{\sim}{f}_i^d = \varsigma[\dot{\underset{\sim}{r}}_i - (\underset{\sim}{v}_{i,0} + \underset{\sim}{v}'_i)] = 2kT\beta^2(\underset{\sim}{R}_i - \underset{\sim}{R}_{i-1}) - kT\frac{\partial}{\partial \underset{\sim}{r}_i}\log\psi \tag{8-15}$$

where $\underset{\sim}{R}_i \equiv \underset{\sim}{r}_{i+1} - \underset{\sim}{r}_i$ is the end-to-end vector from bead i to bead i+1. A result analogous to Equation 8-15 holds for the drag exerted on bead j. Thus Equation 8-13 becomes

$$\underset{\sim}{v}'_i = \sum_j \underset{\approx}{\Omega}_{ij} \cdot \underset{\sim}{f}_j^d = \sum_j \underset{\approx}{\Omega}_{ij} \cdot \left\{2kT\beta^2(\underset{\sim}{R}_j - \underset{\sim}{R}_{j-1}) - kT\frac{\partial}{\partial \underset{\sim}{r}_j}\log\psi\right\} \tag{8-16}$$

When this formula for $\underset{\sim}{v}'_i$ is substituted into Equation 8-15, the resulting equation contains only the distribution function, ψ, the known solvent velocity field in the absence of polymer, $\underset{\sim}{v}_{i,0}$, and the bead coordinates, $\{\underset{\sim}{r}_k\}$. Thus

$$\dot{\underset{\sim}{r}}_i = \underset{\sim}{v}_{i,0} + \sum_j \frac{kT}{\varsigma}(\delta_{ji}\underset{\approx}{\delta} + \varsigma\underset{\approx}{\Omega}_{ij}) \cdot [2kT\beta^2(\underset{\sim}{R}_j - \underset{\sim}{R}_{j-1}) - kT\frac{\partial}{\partial \underset{\sim}{r}_j}\log\psi] \tag{8-17}$$

where δ_{ji} is unity when i=j; otherwise it is zero. When this result is substituted into the probability continuity equation,

$$\frac{\partial\psi}{\partial t} + \sum_i \frac{\partial}{\partial \underset{\sim}{r}_i}(\dot{\underset{\sim}{r}}_i\psi) \tag{8-18}$$

the result is the *Smoluchowski equation*. In the Rouse treatment, the next step is to *diagonalize* the Smoluchowski equation, forming independent equations for the *normal modes*. When hydrodynamic interaction is present, this step is blocked, because of the dependence of $\underset{\approx}{\Omega}_{ij}$ on the bead coordinates. Zimm[10,11] circumvented the problem by *equilibrium averaging* the Oseen tensor; that is, he replaced $\underset{\approx}{\Omega}_{ij}$ by $<\underset{\approx}{\Omega}_{ij}>_0$, its average over the *equilibrium* distribution function, ψ_0:

$$\underset{\approx}{\Omega}_{ij} \approx \langle\underset{\approx}{\Omega}_{ij}\rangle_0 = \frac{1}{6\pi\eta_s}\left\langle\frac{1}{|\underset{\sim}{r}_i - \underset{\sim}{r}_j|}\right\rangle_0 \underset{\approx}{\delta}, \quad i \neq j \tag{8-19}$$

This step is clearly a poor approximation if the polymer coil is severely distorted. For linear viscoelastic measurements, however, the coil is always near equilibrium, and Equation 8-19 is a good approximation. In a theta solvent, the equilibrium distribution function is Gaussian and

$$\langle\underset{\approx}{\Omega}_{ij}\rangle_0 = \frac{1}{6\pi\eta_s}\sqrt{\frac{4\beta^2}{\pi|i-j|}}\,\underset{\approx}{\delta} = \frac{h^*}{\zeta}\sqrt{\frac{2}{|i-j|}}\,\underset{\approx}{\delta}, \quad i \neq j \tag{8-20}$$

where

$$h^* \equiv \frac{\zeta}{\eta_s}\sqrt{\frac{\beta^2}{18\pi^3}} \tag{8-21}$$

is the ***hydrodynamic interaction parameter.*** For spherical beads of radius a, $\zeta = 6\pi\eta_s a$. Also $\beta^2 = 3/2\ell^2$, where ℓ is the root mean square length of a submolecule. Therefore

$$h^* = 0.98a/\ell \tag{8-22}$$

Thus in the language of beads and springs h^* is approximately the ratio of the bead radius to the length of a submolecule. According to this picture, h^* must be less than 0.49, or else the beads overlap. A value of $h^* = 0.25$–0.30, however, is generally considered to be the maximum.[14]

After equilibrium averaging of the Oseen tensor, Zimm found a set of normal coordinates, $\{\underset{\sim}{R}'_i\}$ that reduces the Smoluchowski equation to a set of independent equations, one for each relaxation mode,

$$\frac{\partial\psi_i}{\partial t} + \frac{\partial}{\partial \underset{\sim}{R}'_i}\cdot\left\{\underset{\sim}{R}'_i\cdot\nabla\underset{\sim}{v}\psi_i - \frac{1}{4\hat{\lambda}_i}\left[2\beta^2\underset{\sim}{R}'_i + \frac{\partial \log\psi_i}{\partial\underset{\sim}{R}'_i}\right]\right\} = 0 \tag{8-23}$$

These equations are identical to those of the Rouse model except that the relaxation times, $\{\hat{\lambda}_i\}$, are different. The effect of hydrodynamic interaction is to shorten the longer relaxation times and thereby ***compress the relaxation spectrum.*** If the number of beads is large, then the longest relaxation times depend on a single parameter, $h \equiv h^* N^{1/2}$. $h \to \infty$ is called the ***nondraining limit.*** In the nondraining limit, the spacing of the long relaxation times changes from $\lambda_p \sim p^{-2}$ of Rouse to $\lambda_p \sim p^{-3/2}$ of Zimm. At intermediate frequencies, G′ and G″ therefore scale as $\omega^{2/3}$ rather than $\omega^{1/2}$ of the Rouse theory:

$$G'' - \eta_s\omega = \sqrt{3}\,G' \propto \omega^{2/3} \tag{8-24}$$

Both the scaling with ω and the proportionality constant, $\sqrt{3}$, are confirmed by experimental data, as in Figure 8-3. Figure 8-3 compares G′ and $G'' - \eta_s\omega$ in the

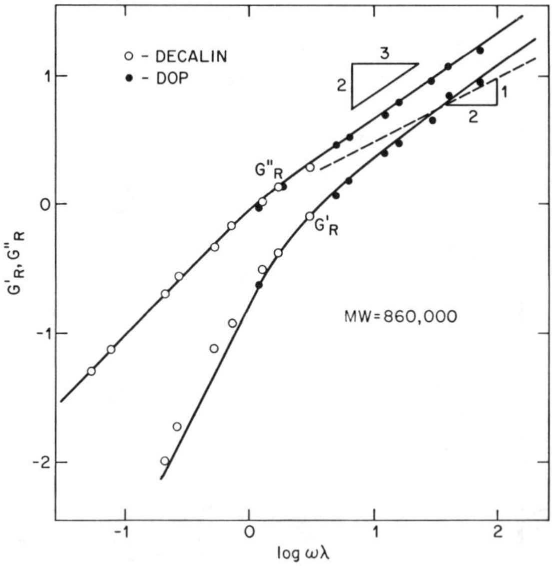

Fig. 8-3 Linear viscoelastic data (symbols) for polystyrene in two theta solvents compared to the predictions (lines) of the Zimm theory with dominant hydrodynamic interaction. For frequencies $\lambda\omega$ greater than 10, G′ and G″ scale as $\omega^{2/3}$, in agreement with the Zimm theory (from Johnson et al.,[4] reproduced with permission from Polym. J.).

nondraining limit with data for a polystyrene of high molecular weight, M = 860,000, in theta solvents. The agreement is excellent. For low-molecular-weight polymers, h cannot be considered infinite, and the relaxation spectrum is governed by both h^* and N. This is *partial draining*. The sets of relaxation times, $\{\hat{\lambda}_i\}$, for various values of N and h^* have been computed and tabulated by Lodge and Wu.[15,11] Values of h^* and N for a given polymer solution are usually obtained by fitting G′ and G″ calculated from the relaxation spectrum to the data. For polystyrene in a good solvent, the best fits[16] are usually obtained when the molecular weight of a single submolecule, M/N, is 5,000—10,000 and $h^* \doteq 0.15$.

8.2.2.4 Good Solvents The above theory has been extended to nontheta solvents. An increase in the solvent quality causes the relaxation spectrum to become more Rouse-like. This can also be seen in the molecular-weight asymptotes in Table 8-1. In practice, G′ and G″ for polymers in a good solvent can be fit to the Zimm theory[14] (which was derived for theta solvents) simply by adjusting h^*. For finite N, a reduction in h^* produces linear viscoelastic behavior that is more Rouse-like and thus can be made to fit linear spectra measured in good solvents.

8.2.3 High Frequency Behavior

The comparison in Figure 8-3 extends only up through the intermediate-frequency regime where $\omega << 1/\lambda_N$. At still higher frequency, $\omega \sim 1/\lambda_N$,

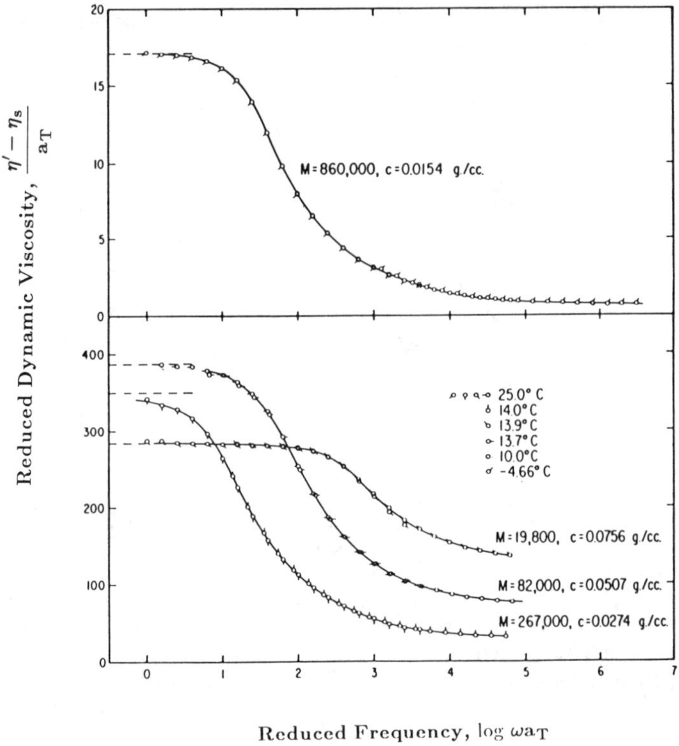

Fig. 8-4 Data for polystyrenes showing that η' approaches a high-frequency asymptote that is *greater* than the solvent viscosity η_s. The data have been time-temperature superposed using a shift factor a_T (from Massa et al.,[17] reprinted with permission from Macromolecules © 1971 American Chemical Society)

individual submolecules are stretched, according to the beads-and-springs model. A crossover from $G'' \sim \omega^{2/3}$ to $G'' = \eta_s\omega$ is therefore predicted. Equivalently, the polymer contribution to the ***dynamic viscosity***, which is defined as

$$\eta'_p \equiv \frac{G''}{\omega} - \eta_s \tag{8-25}$$

is predicted to approach zero at high frequency. In the beads-and-springs description, a polymer molecule contributes nothing to the viscous dissipation if the deformation is too fast for even the submolecules to relax. Experimentally, $\eta' = G''/\omega$ does approach a constant,[14] η_∞, at high ω, but η_∞ does not equal η_s. Usually η_∞ is greater than η_s. Data from Massa, Schrag, and Ferry in Figure 8-4 show that $\eta' - \eta_s$ approaches a nonzero constant at high ω; this constant, $\eta_\infty - \eta_s$, increases with polymer concentration. This phenomenon may be caused by a resistance of short segments of polymer to short-range distortions that would otherwise occur at high frequency.[14] As ω increases, all elastic motions available to the molecule are eventually exhausted, and G' reaches a plateau. Unable to deform, short segments of polymer act like rigid structures past which solvent must flow, thereby exerting drag and dissipating energy. In this high frequency limit, the dilute solution then acts like a suspension of rigid particles, and $\eta' \rightarrow \eta_\infty$, a constant.

The true situation is no doubt more complicated than this, with a role being played by specific steric and energetic interactions between polymer and solvent. Such interactions are probably responsible for an anomalous case in which $\eta_\infty < \eta_s$, the presence of the polymer actually ***reducing*** dissipation in the solvent.[18,16] Depending on its interactions with the solvent, a polymer can either reduce or enhance solvent structure, thereby affecting its ability to dissipate energy.

Polymer-solvent interactions are also probed in ***high-frequency birefringence*** measurements[16] of S'_∞, the optical counterpart to η_∞. S' is the component of the birefringence that is in phase with the velocity gradient, divided by the amplitude of the oscillation and the frequency ω; S'_∞ is the high-frequency limit of this quantity. When the high-frequency asymptotes are subtracted from η' and S', it is found that[16]

$$\eta' - \eta_\infty \propto S' - S'_\infty \tag{8-26}$$

which is a ***stress-optic law***. As discussed in Section 4.2.3, the stress-optic law supports the assumption that the viscoelastic contribution to the stress tensor, $\underset{\approx}{\tau} - 2\eta_\infty \underset{\approx}{D}$, is proportional to $\sum_i \langle \underset{\sim}{R}_i \underset{\sim}{R}_i \rangle$. Thus, except for the high-frequency component, the linear viscoelastic stresses in dilute solution are a direct reflection of the distribution of polymer conformations. For dilute solutions, it is not known if the stress-optic law is valid outside the linear limit.

8.3 NON-NEWTONIAN VISCOSITY

8.3.1 Experiments

Theoretical understanding of the nonlinear viscoelastic properties of dilute solutions is much less advanced than that of the linear properties. The equilibrium-averaging approximation, which works well in the predictions of linear properties, is not accurate in the nonlinear regime, where coil distortions are not small. Thus if one multiplies by $\underset{\sim}{R}_i' \underset{\sim}{R}_i'$ the Smoluchowski equation obtained from the equilibrium-averaged Oseen tensor, Equation 8-23, and integrates it over dR^3, the result is an UCM equation,

$$\hat{\lambda}_i \overset{\nabla}{\underset{\approx}{\tau_i}} + \underset{\approx}{\tau_i} = nkT \underset{\approx}{\delta} \tag{8-27}$$

for each mode. Here $\underset{\approx}{\tau_i} = 2nkT\beta^2 \langle \underset{\sim}{R}_i' \underset{\sim}{R}_i' \rangle$. This result differs from the Rouse result only in the spectrum of relaxation times. The predictions of the Rouse theory at high shear rates, such as the absence of shear thinning, are therefore unchanged.

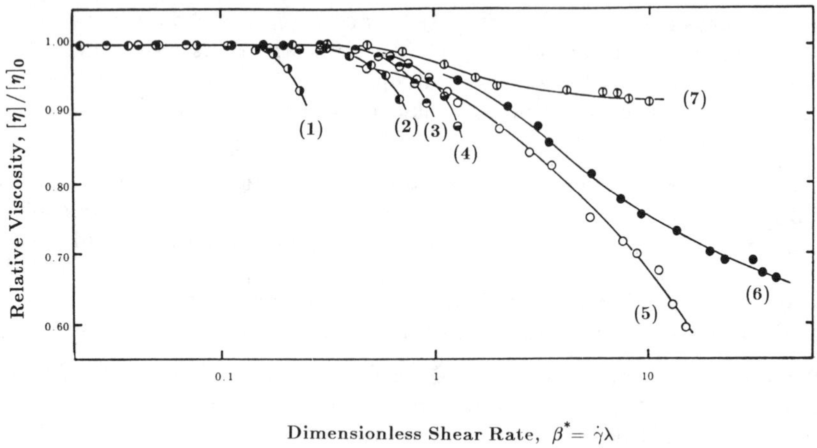

Fig. 8-5 Shear thinning in dilute solution: poly(α-methylstyrene) with molecular weights of (1) 690,000, (2) 1,240,000, (3) 1,460,000, (4) 1,820,000, and (5) 7,500,000 in toluene (good solvent). Curves (6) and (7) are for polystyrene of molecular weight 13,600,000 in toluene (good solvent) and decalin (theta solvent), respectively (from Noda et al.,[19] reprinted with permission from J. Phys. Chem. ©1968 American Chemical Society)

8.3.1.1 Shear Thinning Real polymer solutions show shear thinning. Figure 8-5, for example, shows the ***relative*** viscosity, $[\eta]/[\eta]_0$, plotted against a dimensionless shear rate,

$$\beta^* \equiv \lambda\dot{\gamma} \tag{8-28}$$

where $\lambda = [\eta]_0 M\eta_s/N_A kT$ is the characteristic relaxation time. β^* is the only dimensionless shear rate that can be formed from the quantities $[\eta]_0$, η_s, M, $\dot{\gamma}$, and $N_A kT$. Figure 8-5 includes data for polymers in both good and theta solvents. Although present in both types of solvent, the degree of shear thinning is much greater in the good solvents than in the theta solvent.

The degree of shear thinning, even in a good solvent, is much less in dilute solution than occurs in concentrated solutions or melts. For a melt, the steady shear viscosity can fall by three orders of magnitude;[20] see Figure 3-6 for example. η falls off steeply, scaling as $\dot{\gamma}^{-0.8}$ at high $\dot{\gamma}$ in melts of low polydispersity.[21] In truly dilute solutions, a decrease of about 40% is more typical; the decrease can be even less in theta solvents. Figure 8-5 shows at most a 40% drop in one and a half decades of shear rate for good solvents.

8.3.1.2 Causes of Shear Thinning Several causes of shear thinning in dilute solution have been proposed. These include *finite extensibility*, nonequilibrium *hydrodynamic interaction*, *excluded volume*, and *internal viscosity*. Experimental results, such as those depicted in Figure 8-5 suggest that *excluded volume* perhaps acting synergistically with other factors, is a significant source of shear thinning.

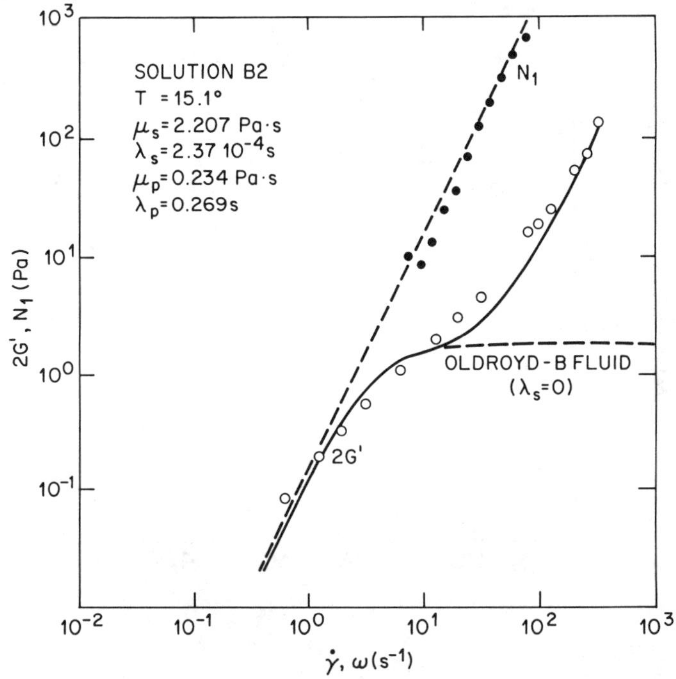

Fig. 8-6 2 G′ versus ω and N_1 versus $\dot{\gamma}$ for a Boger fluid, in this case a high-molecular-weight polyisobutylene in a solvent consisting of poly(1-butene) of low molecular weight and kerosene. η_p and λ_p are the viscosity and relaxation time of the high molecular-weight polymer; η_s and λ_s are the viscosity and relaxation time of the solvent. The solid line gives the best fit to G′ data of the upper-convected Maxwell equation with two relaxation times; the dashed lines give the best fit of the Oldroyd-B equation (from Mackey and Boger,[24] reproduced with permission from J. Non-Newtonian Fluid Mech.).

Some shear thinning occurs even in theta solvents, however.

The causes of shear thinning can be sought indirectly by looking for conditions under which it is absent or minimal. Specially concocted mixtures of high-molecular-weight polymers in viscous solvents are thought to be nearly free of shear thinning.[22] These solutions are nowadays called *Boger fluids*, although Peterlin had studied similar fluids.[23] Boger fluids are reasonably well described, out to high frequency, by a two-mode version of the UCM equation;[24] see Figure 8-6. At low and modest frequencies, the faster of these modes can be collapsed into a retardation term, and a single-mode Oldroyd-B equation (see Section 3.4.3) is then a sufficient rheological description. Experimental flows of Boger fluids have compared favorably to numerical simulations of these flows with the Oldroyd-B equation.[25]

For Boger fluids, $\eta = \eta_p + \eta_s$ shows little shear thinning and in this sense is close to the UCM or Oldroyd-B prediction of no shear thinning. This is no doubt true in part because the polymeric contribution to η is swamped by the nearly

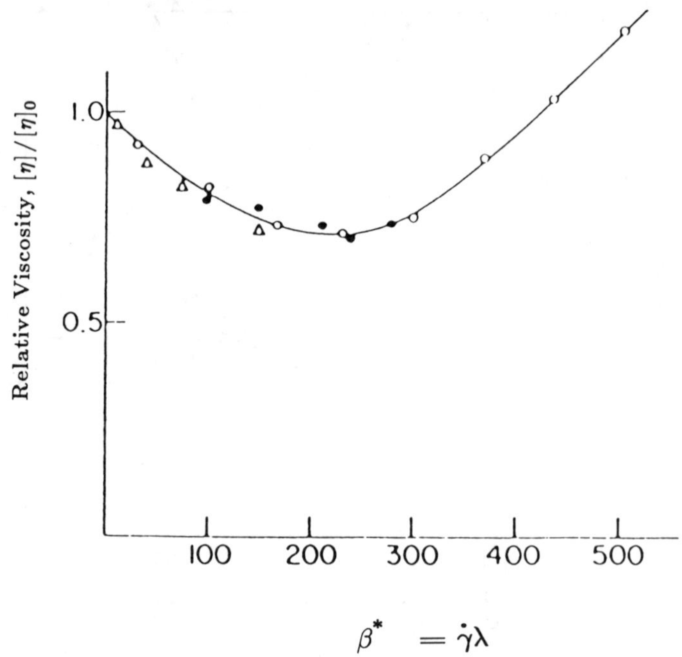

Fig. 8-7 Shear thinning and thickening in high molecular-weight polyisobutylene ($M_n = 7 \times 10^6$) dissolved in low molecular-weight polybutene. The different symbols correspond to different ways of extrapolating to zero concentration (from Bianchi and Peterlin,[23] reprinted from the Journal of Polym. Sci. ©1968 by permission of John Wiley & Sons, Inc.).

Newtonian solvent contribution, η_s, which is typically 90% of the total. Thus the *intrinsic* viscosity of a Boger fluid may not be any less shear thinning than other dilute solutions. However, the first normal stress difference, N_1, of a Boger fluid, to which there is no direct solvent contribution, also follows the UCM prediction; it is quadratic in shear rate up to moderately high $\dot{\gamma}$. The dimensionless ratio, N_1/τ_p, of the first normal stress difference to the polymeric part of the shear stress, can be considered a dimensionless *Weissenberg number* characterizing the ratio of elastic to viscous forces for the polymer. The magnitude that this quantity attains before the occurrence of significant deviations from the UCM equation — that is, either shear thinning or nonquadratic N_1 — can be considered a measure of how close the fluid is to being an UCM fluid. For the Boger fluid depicted in Figure 8-6, N_1 remains roughly quadratic in $\dot{\gamma}$ even at values of N_1/τ_p as high as 40, whereas for other solutions or melts, highly nonquadratic behavior in N_1 usually occurs when this ratio is near unity.[20] Since the ratio N_1/τ_p is a rough measure of the degree of stretch or aspect ratio of the macromolecule in the flow field, it is apparent that in Boger fluids coils can become highly stretched before significant deviations from the UCM equation — that is, from Rouse-like non-Newtonian behavior — occurs.

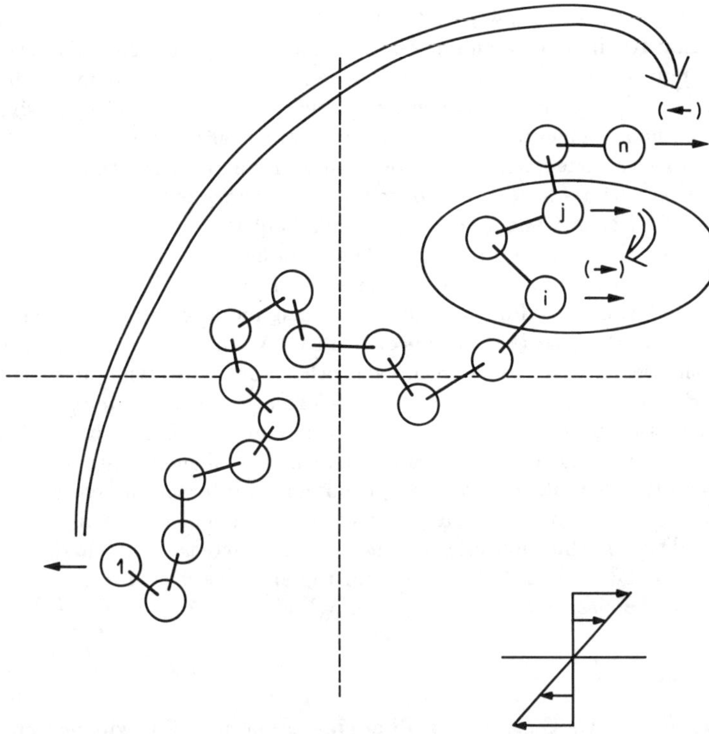

Fig. 8-8 Hydrodynamic interaction between beads 1 and n at opposite ends of the chain leads to shear thinning, while hydrodynamic interaction between neighboring beads i and j leads to shear thickening. See text for details.

The important characteristics of Boger fluids that seem to be responsible for for their Rouse-like behavior are that the polymeric molecular weight is high (in the millions) and that the solvent is very viscous. Under these conditions, shear thinning or any other deviations from the UCM predictions that result from internal viscosity and finite extensibility are expected to be suppressed. In addition, the solvents in Boger fluids are polymeric, although of much lower molecular weight than the dilute polymeric solute. The polymeric nature of the solvent may partially screen hydrodynamic interaction and excluded volume.

8.3.1.3 Shear Thickening When the molecular weight, M, of the polymer is very high, ***shear thickening*** is sometimes seen, most often in viscous solvents, at very large values of β^*. Figure 8-7, for example, shows the shear rate dependence of the intrinsic viscosity of a dilute solution of polyisobutylene, with molecular weight of 7×10^6, in polybutene, measured by Bianchi and Peterlin.[23] $[\eta]$ shows a minimum at $\beta^* \approx 200$, followed by an upturn. Layec-Raphalen and Wolff[26] report shear thickening for polystyrene in decalin. A slight upturn in Ψ_1, the first normal stress coefficient, has also been reported for a Boger fluid, again a high-molecular-weight polyisobutylene in polybutene, by Mackay and Boger.[24]

Both the decrease in $[\eta]$ at low β^* and the subsequent upturn can be accounted for by the changes in hydrodynamic interaction induced by coil distortion, as Peterlin has pointed out.[27] Hydrodynamic interaction can account for both shear thinning at low β^* and shear thickening at high β^* because the hydrodynamic interaction of chain segments that are close to each other is qualitatively different from the interaction of segments that are on opposite ends of the chain. Consider a chain in a shear field whose center of mass is stationary with respect to the solvent, Figure 8-8. Beads 1 and n, on opposite ends of the chain, move in opposite directions. Thus the extra hydrodynamic drag induced by the motion of bead 1 tends to *resist* the motion of bead n. Beads that are near each other move in the same direction, however. Thus the immediate neighbors of bead i induce hydrodynamic effects that *assist* the motion of bead i. As the coil is deformed in the shear field, the beads move away from each other and hydrodynamic interaction is reduced. But at low β^*, the predominate deformation is in the *longest mode* — that is, the separation between beads on opposite ends of the molecule. Thus the hydrodynamic interaction between beads 1 and n is weakened first; only at high β^* is there significant weakening of the hydrodynamic interaction between beads i and j. Therefore at low β^*, only the gross shape of the molecule is distorted and the effective drag on the molecule *decreases* as β^* increases. At high β^*, even small sections of the chain are stretched and the effective drag *increases*, leading to an upturn in the viscosity.

8.3.2 Theories

8.3.2.1 Fixman's and Öttinger's Theories Experimental evidence suggests hydrodynamic interaction and especially excluded volume as sources of shear thinning for high-molecular-weight polymers in dilute nonpolymeric solvents. With equilibrium averaging of hydrodynamic interaction, no shear thinning is predicted. Thus a more precise treatment of hydrodynamic interaction is required if predictions of shear thinning are to be derived from it. For the beads-and-springs model, the most detailed theories of shear thinning are those of Fixman[28] (1966) and Öttinger[29,30] (1987). Öttinger deals only with hydrodynamic interaction; whereas Fixman considers both hydrodynamic interaction and excluded volume. Fixman and Öttinger, like Zimm, *preaverage* the Oseen tensor and therefore *neglect fluctuations in the hydrodynamic interaction.* But Fixman and Öttinger obtain results that are more accurate than Zimm's, because they preaverage with a *nonequilibrium* distribution function rather than with the equilibrium distribution function.

In Öttinger's method, the nonequilibrium distribution function used to preaverage the Oseen tensor is ψ, the distribution function obtained by solving the Smoluchowski equation. The Oseen tensor appearing in this Smoluchowski equation is replaced by $\langle \underset{\approx}{\Omega}_{ij} \rangle$, the preaveraged Oseen tensor, which is not known until ψ is computed. But the Smoluchowski equation cannot be solved for ψ until $\langle \underset{\approx}{\Omega}_{ij} \rangle$ is known. In Öttinger's method, both ψ and $\underset{\approx}{\Omega}_{ij}$ are determined simultaneously in a *self-consistent* way.

Fixman, on the other hand, solved the problem by the following *second-order* iterative scheme:

1. Start with the equilibrium distribution function, ψ_0.

2. Average $\underset{\approx}{\Omega}_{ij}$ over $\psi_0 \rightarrow <\underset{\approx}{\Omega}_{ij}>_0$.

3. Replace $\underset{\approx}{\Omega}_{ij}$ by $<\underset{\approx}{\Omega}_{ij}>_0$ in the Smoluchowski equation and solve it to obtain *first-order* approximations to the distribution function ψ_1 and the stress $\underset{\approx}{\tau}_1 = 2nkT\beta^2 \sum_i \int \underset{\sim}{R}'_i \underset{\sim}{R}'_i \psi_1 \, dR^3$ (Zimm stopped here).

4. Average $\underset{\approx}{\Omega}_{ij}$ over $\psi_1 \rightarrow <\underset{\approx}{\Omega}_{ij}>_1$.

5. Replace $\underset{\approx}{\Omega}_{ij}$ by $<\underset{\approx}{\Omega}_{ij}>_1$ in the Smoluchowski equation and solve it to obtain *second-order* approximations to the distribution function ψ_2 and the stress $\underset{\approx}{\tau}_2 = 2nkT\beta^2 \sum_i \int \underset{\sim}{R}'_i \underset{\sim}{R}'_i \psi_2 \, dR^3$ (Fixman stopped here).

Thus Fixman's scheme is one step higher in order than Zimm's. Öttinger's result could be obtained by continuing the iteration out to infinite order. Zimm's technique is adequate to obtain exact* values of the first-order coefficient η_0, as well as the zero-shear value of the first normal stress coefficient** $\Psi_{1,0}$. Fixman's technique in principle supplies exact* values for the low shear-rate values of the first- *and* second-order coefficients —that is, $\Psi_{2,0}$ in addition to η_0 and $\Psi_{1,0}$. At high shear rates, Fixman's method, which is computationally much more economical than Öttinger's, continues to give nearly the same results as Öttinger's more exact method.[31]

For good solvents, Fixman[28] introduced an additional approximation for the effect of excluded volume. He represented the excluded volume by a *potential,* E, that drives beads away from each other. Using a perturbation expansion, he approximated this potential by a quadratic in the bead separation distance:

$$E \equiv kTX \sum_{i>j} (A_{ij} - B_{ij} R_{ij}^2) \tag{8-29}$$

Here R_{ij} is the distance of separation between beads i and j. X is an effective excluded volume per bead. The excluded volume potential is repulsive when the beads are closely spaced and diminishes in magnitude as the beads move apart. The quadratic approximation is required in order that the Smoluchowski equation have an analytic solution, that of an anisotropic Gaussian, for the configuration distribution function. The coefficients A_{ij} and B_{ij} depend on the average chain configuration and are therefore sensitive to chain deformation. This potential, for the case of the dumbbell, is discussed later in more detail.

---

* The technique is not completely exact because of the neglect of fluctuations in hydrodynamic interaction. In Fixman's method there are additional sources of error resulting, for example, from approximations involved in diagonalizing the Smoluchowski equation into normal modes.

** Although it multiplies the square of the shear rate, $\dot{\gamma}^2$, and is in that sense a second-order term, $\Psi_{1,0}$ can be predicted from a first-order perturbation, that is, from linear viscoelasticity; see Section 3.2.1.

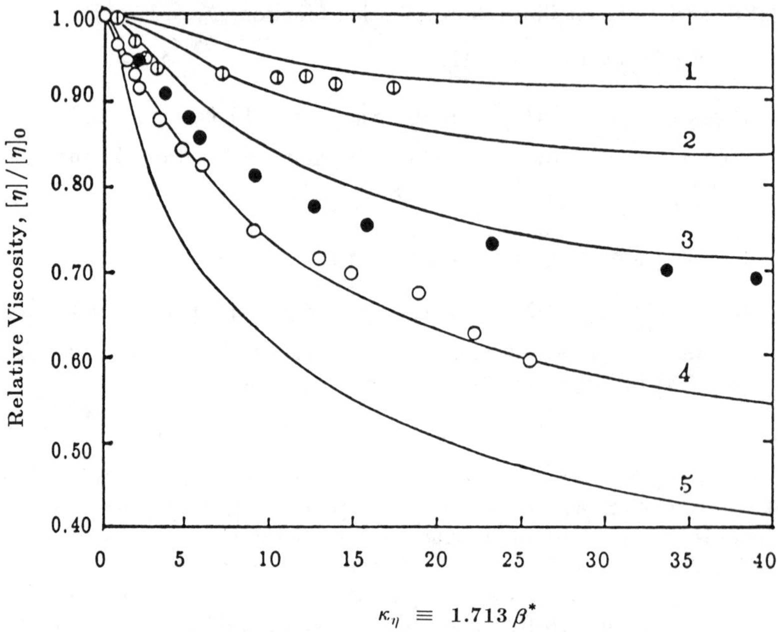

Fig. 8-9 Fixman's theory for shear thinning fitted to the data of Noda et al. The top and middle sets of data are respectively for polystyrene (MW = 13,600,000) in decalin (theta solvent), and in toluene (good solvent); the bottom set is for poly(α-methylstyrene), MW = 7,500,000 in toluene. Theoretical curves: (1) z=0, (2) 0.243, (3) 1.764, (4) 14.66, (5) 147.7 (from Noda et al.,[19] reprinted with permission from J. Phys. Chem. ©1968 American Chemical Society)

8.3.2.2 Predictions Noda et al. compared their experimental measurements of shear thinning with Fixman's theory; see Figure 8-9. The agreement is good for the theta solvent and is also reasonable for the good solvents if the excluded volume parameter

$$z \equiv (3/2\pi\ell_0^2)^{3/2}\mathrm{X}\mathrm{N}^{1/2} \tag{8-30}$$

is adjusted to obtain a fit. Here ℓ_0 is the average neighboring bead separation in the absence of excluded volume, and N is the number of springs. Because $z \propto \mathrm{N}^{1/2}$, the effect of excluded volume is predicted to increase as the molecular weight increases. The opposite trend with molecular weight appears, however, in Figure 8-5; the data show that the lower the molecular weight, the greater the shear thinning.

Two of the suggested sources of shear thinning, namely finite extensibility and internal viscosity, agree with the experimental observation that on a plot of $[\eta]$ versus β^* shear thinning sets in at smaller β^* as molecular weight increases. However, the effects of finite extensibility are felt only when the chain becomes extended enough that the force law relating spring extension and spring tension become nonlinear. This effect is predicted to become significant when the polymer chain is stretched to almost half its completely extended length. Models of finitely extensible dumbbells predict shear thinning effects[11] that can be fit to the data of Figure 8-5, but only if the values of maximum chain length are set unrealistically low, considering the high molecular weights involved. Thus finite extensibility is probably not the dominant source of shear thinning for these fluids.

The Rouse beads-and-springs model assumes that polymer chains are completely *limp*, except for the elastic spring forces that resist stretching. In models with *internal viscosity*, first introduced by Kuhn and Kuhn[32] (1946), frictional barriers to short-range distortions, such as bond rotations, make the chain somewhat rigid and resistant to rapid deformation. For a given overall deformation of a macromolecule, the amount of distortion that must be distributed to each unit of the molecule decreases as the molecular weight increases. Thus the influence of internal viscosity on the slow relaxation modes is expected to diminish as molecular weight increases. The dimensionless shear rate, β^*, at which internal viscosity produces significant shear thinning should therefore increase as the molecular weight increases. This is consistent with the data of Figure 8-5.

Internal viscosity (IV) is introduced in the model of Kuhn and Kuhn by adding a term to the Rouse-Zimm balance of forces on a bead. For a dumbbell,

$$\underset{\text{(drag)}}{\varsigma(\dot{\underset{\sim}{R}} - \underset{\sim}{\nabla}\underset{\sim}{v})} + \underset{\text{(spring)}}{4kT\beta^2\underset{\sim}{R}} + \underset{\text{(Brownian)}}{2kT\frac{\partial \log \psi}{\partial \underset{\sim}{R}}} + \underset{\text{(internal viscosity)}}{\phi\frac{\dot{\underset{\sim}{R}}\cdot\underset{\sim}{R}\underset{\sim}{R}}{R^2}} = \underset{\approx}{0} \tag{8-31}$$

ϕ is the "internal viscosity coefficient." Although the exact dependence of ϕ on molecular weight is in doubt, it is agreed that ϕ should decrease as molecular weight increases. Thus Fixman's theory, which neglects internal viscosity, should become more accurate as molecular weight increases.[33a]

If ς in Equation 8-31 is large compared to ϕ, internal viscosity is eclipsed by bead-solvent viscous drag forces. Since the drag coefficient, ς, is proportional to the solvent viscosity, η_s, it might be possible to swamp the influence of internal viscosity by increasing the solvent viscosity.*

Manke and Williams[33] showed that the onset of shear thinning is indeed delayed to higher values of β^* if the solvent viscosity is increased while the solvent quality is held fixed at the theta point. The increase in solvent viscosity was achieved by adding a low molecular weight polymer to a low-viscosity solvent, as is done to make Boger fluids. The delay in shear thinning produced when the *molecular weight* is increased can, at least in principle, be explained by either internal viscosity or finite extensibility. The delay produced by an increase in *solvent viscosity*, however, cannot be explained by finite extensibility; internal viscosity seems to be the only known explanation.

---

* In some proposed models of internal viscosity, however, ϕ, like ς, is proportional to η_s.[33a]

Chain rigidity or internal viscosity is also the proposed explanation for an observed lack of large coil distortion at high shear rates. Light scattering measurements have shown that in good solvents of low viscosity, strongly sheared coils are only modestly nonspherical; aspect ratios of 1.5 were the maximum attained.[34]

Thus hydrodynamic interaction, excluded volume, and internal viscosity all seem to play a role in producing shear thinning at low and modest shear rates. When the shear rate is high, or when the molecular weight is low, finite extensibility may also important. Shear thinning is minimized in dilute solution when the molecular weight is high, the solvent viscosity is high, and theta conditions are approached.

At high shear rates, $\beta^* > 10$, hydrodynamic interaction can cause an upturn in the viscosity and first normal stress coefficient, as discussed earlier. Figure 8-10 shows the upturn predicted by Fixman's theory.[31] The shape of the curve is in qualitative agreement with the experimental data of Figure 8-7, although in Bianchi and Peterlin's experiments, the β^* value at which the upturn occurs is 10 times higher than predicted. Unless the molecular weight of the polymer is sufficiently high, and the solvent viscous, shear thinning from finite extensibility, internal viscosity, or excluded volume is likely to suppress this upturn.[35]

To illustrate how shear thinning is produced, we incorporate, one at a time, the mechanisms of hydrodynamic interaction, excluded volume, finite extensibility, and internal viscosity into the elastic dumbbell model. Though unrealistic, the dumbbell model illustrates the qualitative phenomena and avoids the tedious details involved in any consideration of the multiple beads-and-springs model.

8.4 EXPRESSIONS FOR THE STRESS TENSOR

8.4.1 Kirkwood-Riseman Expression

The expression

$$\underset{\approx}{\tau} = 2nkT\beta^2 \langle \underset{\sim}{R}\underset{\sim}{R} \rangle \tag{8-32}$$

that we have heretofore used for the polymer contribution to the stress tensor is not adequate when excluded volume, finite extensibility, or internal viscosity enter. Kirkwood and Riseman derived the more general formula[6]

$$\underset{\approx}{\tau} = -n \sum_{i=1}^{N+1} \langle \underset{\sim}{r_i}\, \underset{\sim}{f_i^d} \rangle \tag{8-33}$$

where $\underset{\sim}{r_i}$ is the position of the ith bead, $\underset{\sim}{f_i^d}$ is the drag force that bead i exerts on the solvent, and the summation is over all the beads on the chain. For the dumbbell, N=2, Equation 8-33 reduces to

$$\underset{\approx}{\tau} = -n \langle \underset{\sim}{R}\, \underset{\sim}{f_2^d} \rangle \tag{8-34}$$

where $\underset{\sim}{R} \equiv \underset{\sim}{r_2} - \underset{\sim}{r_1}$ is the end-to-end vector of the dumbbell. The Kirkwood-Riseman formula follows from a calculation of the momentum transported from

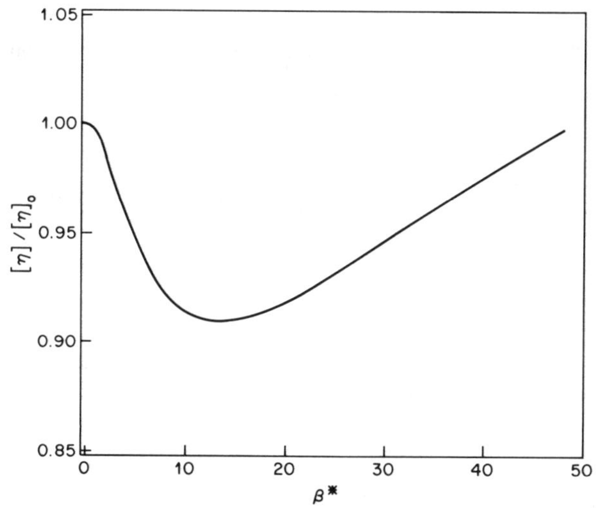

Fig. 8-10 Intrinsic viscosity versus dimensionless shear rate, $\beta^* = \lambda\dot{\gamma}$, predicted by Fixman's theory with 200 submolecules (from Magda et al.[31]).

the polymer molecules to the solvent and thence to the macroscopic boundaries. For the simple Hookean dumbbell, Equation 8-32 is recovered to within an isotropic constant. This can be seen by noting that for the Hookean dumbbell, the drag force exerted by bead "2" is

$$\underset{\sim}{f}_2^d = \frac{1}{2}\varsigma(\dot{\underset{\sim}{R}} - \underset{\sim}{R}\cdot\nabla\underset{\sim}{v}) = -2kT\beta^2\underset{\sim}{R} - kT\frac{\partial \log\psi}{\partial \underset{\sim}{R}} \qquad (8\text{-}35)$$

(drag on solvent) (spring force) (Brownian force)

Eqns. (34) and (35) imply that

$$\underset{\approx}{\tau} = 2nkT\beta^2 <\underset{\sim\sim}{RR}> + nkT<\underset{\sim}{R}\frac{\partial\log\psi}{\partial\underset{\sim}{R}}> = 2nkT\beta^2<\underset{\sim\sim}{RR}> - nkT\underset{\approx}{\delta}$$

(spring) (Brownian) (8-36)

The Brownian term in this case is isotropic and can be omitted for incompressible fluids. The elastic spring term is the same as was derived earlier from a virtual work argument in Section 2.3.2. As we shall see, consideration of excluded volume, finite extensibility, or internal viscosity alters the force balance of Equation 8-35 so that Equation 8-32 is no longer a valid formula for the stress tensor. In general, the polymeric part of the stress tensor has three contributions:

$$\underset{\approx}{\tau} = \underset{\approx}{\tau}^S + \underset{\approx}{\tau}^B + \underset{\approx}{\tau}^V$$

from the elastic *spring* forces, the *Brownian* forces, and a purely *viscous* contribution, $\underset{\approx}{\tau}^V$, that appears when internal viscosity is considered. Of course there is also the solvent contribution, $\underset{\approx}{\tau}_s = 2\eta_s \underset{\approx}{D}$.

8.4.2 Giesekus Expression

The *Giesekus* form[36,11] for the stress tensor can be derived if hydrodynamic interaction is neglected; for then the drag force is

$$\underset{\sim}{f}_2^d = \frac{1}{2}\varsigma(\dot{\underset{\sim}{R}} - \underset{\sim}{R}\cdot\underset{\sim}{\nabla}\underset{\sim}{v}) \tag{8-37}$$

Using this equation, the Kirkwood-Riseman expression, Equation 8-34, and symmetry of the stress tensor,

$$\underset{\approx}{\tau} = -n\langle\underset{\sim}{R}\underset{\sim}{f}_2^d\rangle = \frac{-n}{2}[\langle\underset{\sim}{R}\underset{\sim}{f}_2^d\rangle + \langle\underset{\sim}{f}_2^d\underset{\sim}{R}\rangle]$$

$$= -\frac{n\varsigma}{4}[\frac{\partial}{\partial t}\langle\underset{\sim\sim}{RR}\rangle - \underset{\sim}{\nabla}\underset{\sim}{v}^T\cdot\langle\underset{\sim\sim}{RR}\rangle - \langle\underset{\sim\sim}{RR}\rangle\cdot\underset{\sim}{\nabla}\underset{\sim}{v}]$$

Therefore

$$\underset{\approx}{\tau} = -\frac{n\varsigma}{2}\overset{\nabla}{\langle\underset{\sim\sim}{RR}\rangle} \quad \text{(Giesekus expression)} \tag{8-38}$$

where ∇ denotes the upper-convected time derivative

$$\overset{\nabla}{\langle\underset{\sim\sim}{RR}\rangle} \equiv \frac{\partial}{\partial t}\langle\underset{\sim\sim}{RR}\rangle - \underset{\sim}{\nabla}\underset{\sim}{v}^T\cdot\langle\underset{\sim\sim}{RR}\rangle - \langle\underset{\sim\sim}{RR}\rangle\cdot\underset{\sim}{\nabla}\underset{\sim}{v}$$

8.5 DUMBBELLS WITH SHEAR THINNING

We now consider the elastic dumbbell generalized to allow hydrodynamic interaction, excluded volume, internal viscosity, and finite extensibility. Each mechanism is considered one at a time.

8.5.1 Dumbbells with Hydrodynamic Interaction

When there are only two beads, the Smoluchowski equation with hydrodynamic interaction reduces to

$$\frac{\partial\psi}{\partial t} = -\frac{\partial}{\partial\underset{\sim}{R}}\cdot[\underset{\sim}{R}\cdot\underset{\sim}{\nabla}\underset{\sim}{v}\psi - \frac{2kT}{\varsigma}(\underset{\approx}{\delta} - \varsigma\underset{\approx}{\Omega})\cdot(2\beta^2\underset{\sim}{R}\psi + \frac{\partial\psi}{\partial\underset{\sim}{R}})] \tag{8-39}$$

The Oseen tensor, $\underset{\approx}{\Omega}$, is given by Equation 8-11. Öttinger solved[29] this equation numerically and by a perturbation expansion with a single approximation: the Oseen tensor is replaced by its preaveraged (or consistently averaged) value. Thus fluctuations of the hydrodynamic interaction are neglected. For large coil distortions that occur at high strain rates, this approximation is much better than the equilibrium-averaging approximation of Kirkwood and Riseman and of Zimm.

If Equation 8-39 is multiplied by $\underset{\sim\sim}{RR}$ and integrated over dR^3, the result is

$$\overset{\nabla}{\underset{\approx}{S}} + \frac{8kT}{\varsigma}(\underset{\approx}{\delta} - \varsigma<\underset{\approx}{\Omega}>)\cdot(\underset{\approx}{S} - \beta^{-2}\underset{\approx}{\delta}) = \underset{\approx}{0} \qquad (8\text{-}40)$$

with

$$\underset{\approx}{S} = \underset{\approx}{\tau}/nkT + \underset{\approx}{\delta} \qquad (8\text{-}41)$$

Equation 8-40 is a modified UCM equation, with $8kT\beta^2(\underset{\approx}{\delta} - \varsigma\underset{\approx}{\Omega})/\varsigma$ playing the role of a reciprocal relaxation time.

As the shear rate increases, and the dumbbell stretches out, the hydrodynamic interaction between the beads diminishes and $<\underset{\approx}{\Omega}>$ tends towards zero. Thus the effective relaxation time *decreases*. As the effective relaxation time decreases, so does the viscosity; the result is shear thinning. η reaches a high-shear-rate plateau η_∞, when $\dot\gamma$ is high enough that $<\underset{\approx}{\Omega}>$ is nearly zero. Öttinger showed that the ratio of the high-shear-rate to the zero-shear-rate values of η and Ψ_1 are

$$\frac{\eta_\infty}{\eta_0} = 1-\sqrt{2}h^* = 0.65 \quad \text{for } h^* = 0.25$$

$$\frac{\Psi_{1,\infty}}{\Psi_{1,0}} = (1-\sqrt{2}h^*)^2 = 0.42 \quad \text{for } h^* = 0.25$$

Ψ_2 goes to zero at high shear rates.

At low shear rates,

$$\frac{\Psi_{2,0}}{\Psi_{1,0}} = \frac{1}{20}\,\frac{\sqrt{2}h^*}{1-\sqrt{2}h^*} = 0.027 \quad \text{for } h^* = 0.25$$

The second normal stress difference is very small and *positive*. For a large number of beads, Öttinger's[30] and Fixman's[31] theories show that $\Psi_{2,0}/\Psi_{1,0} = 0.011$. Unfortunately Ψ_2 is almost never measured in dilute solutions. However, Keentok et al.[37] found Ψ_2 to be near zero in a Boger fluid, which is consistent with Öttinger's and Fixman's calculations. In contrast, for concentrated solutions and melts, Ψ_2 is usually found to be –0.05 to –0.30 times Ψ_1.

For a large number of beads with hydrodynamic interaction, shear thinning gives way to shear thickening at sufficiently high $\dot\gamma$ (see Section 8.3.2.2); this is *not* predicted by the dumbbell model. As illustrated in Figure 8-8, this shear thickening is caused by hydrodynamic interactions of *interior* beads, which are not present in the dumbbell.

8.5.2 Dumbbells with Excluded Volume

Fixman's *excluded volume potential,* E, for the dumbbell problem can be obtained by considering the probability that the two beads come near to overlap.[28] If there were no excluded volume, the probability distribution function at equilibrium would be $\psi_0 = \pi^{-3/2}\beta^3 \exp(-\beta^2 R^2)$. Fixman assumed that the distribution function for the coil with excluded volume is the same as without, except that the linear dimensions of the coil are increased by an as yet undetermined factor α. That is,

$$\psi_0 = \pi^{-3/2}\beta^3\alpha^{-3}\exp(-\beta^2 R^2/\alpha^2) \tag{8-42}$$

Using a perturbation expansion, Fixman approximated the excluded volume potential at equilibrium, E_0, to be

$$E_0 \approx kTx\pi^{-3/2}\beta^3\alpha^{-3}(1 - \beta^2\alpha^{-2}R^2) \tag{8-43}$$

Here x is the effective excluded volume of a bead. E_0 is a *repulsive* potential that becomes weaker as the bead separation distance increases. x must scale as the *square* of the molecular weight of the chain, since excluded volume is a pairwise interaction. For the many-bead representation of the chain, the square of the molecular weight enters indirectly through the double sum over i and j in Equation 8-29. Thus we put $x = XN_K^2$, where N_K is the number of Kuhn steps in the chain. The total potential, V, experienced by the beads of the dumbbell is the sum of the spring potential, S, and the excluded volume potential, E,

$$V = E + S \tag{8-44}$$

where $S = kT\beta^2R^2$. The Smoluchowski equation for a dumbbell in a potential V is

$$\frac{\partial\psi}{\partial t} = -\frac{\partial}{\partial \underset{\sim}{R}}\cdot[\underset{\sim}{R}\cdot\underset{\sim}{\nabla}\underset{\sim}{v}\psi - \frac{2kT}{\varsigma}(\frac{1}{kT}\frac{\partial V}{\partial \underset{\sim}{R}}\psi + \frac{\partial\psi}{\partial \underset{\sim}{R}})] \tag{8-45}$$

At equilibrium, the solution is

$$\psi_0 = c\exp(-V_0/kT) \tag{8-46}$$

where V_0 is the potential at equilibrium and c is a normalization constant.

The constant part of the potential can be absorbed into the prefactor, and because the distribution is normalized to unity it does not have any affect and can be dropped. Thus at equilibrium,

$$V_0 = E_0 + S = \underbrace{-kTXN_K^2\pi^{-3/2}\beta^5\alpha^{-5}R^2}_{\text{(excluded volume)}} + \underbrace{kT\beta^2R^2}_{\text{(spring)}} \tag{8-47}$$

The unknown factor α can now be calculated by a self-consistency argument. Equating the exponentials in Equations 8-42 and 8-46, and using Equation 8-47, we obtain

$$\beta^2 R^2 \alpha^{-2} = \frac{E_0 + S}{kT} = -X N_K^2 \pi^{-3/2} \beta^5 R^2 \alpha^{-5} + \beta^2 R^2 \tag{8-48}$$

Now $\beta^2 = 3/2b^2 N_K$, where b is the length of a Kuhn step. Defining

$$z \equiv (3/2\pi b^2)^{3/2} X N_K^{1/2} , \tag{8-49}$$

we can rewrite Equation 8-48 as

$$\alpha^3 = -z + \alpha^5 \tag{8-50}$$

This equation can be solved for α.

For large z, that is, large X or large molecular weight,

$$\alpha \approx z^{1/5} \propto N_K^{0.1} \tag{8-51}$$

The mean square coil dimension therefore scales as

$$<R^2>_0 \propto \alpha^2 N_K \propto N_K^{6/5} \equiv N_K^{2\nu} \tag{8-52}$$

with ν equal to 3/5, which is the Flory exponent expected for a good solvent. Thus the equilibrium distribution function in Equation 8-42 with α given by Equation 8-50 yields the correct scaling with molecular weight.

Away from equilibrium, Fixman found E by a perturbation expansion for the multiple-spring chain. For the two-bead problem, the analogous expression is

$$E = kTXN_K^2 \pi^{-3/2} [\det(\underset{\approx}{S})]^{-1/2} (1 - \beta^2 \alpha^{-2} R^2) \tag{8-53}$$

where $\underset{\approx}{S} \equiv 2<\underset{\sim}{R}\underset{\sim}{R}>$. According to Equation 8-53, the excluded volume potential weakens as the average coil volume, given by det $\underset{\approx}{S}$, is expanded in a flow field. Thus

$$\frac{\partial V}{\partial \underset{\sim}{R}} = \frac{\partial S}{\partial \underset{\sim}{R}} + \frac{\partial E}{\partial \underset{\sim}{R}} = 2kT\beta^2\alpha^{-2}(\alpha^2 - z\beta^{-3}[\det \underset{\approx}{S}]^{-1/2}) \underset{\sim}{R}$$

$$= 2kT\beta^2 \alpha^{-2}(1 + g) \underset{\sim}{R} \tag{8-54}$$

where

$$g \equiv \alpha^2 - z\beta^{-3}[\det \underset{\approx}{S}]^{-1/2} - 1 \tag{8-55}$$

With this definition, g = 0 at equilibrium. Multiplying the Smoluchowski equation, Equation 8-45, by $2\beta^{-2}\underset{\sim}{R}\underset{\sim}{R}$ and integrating over configuration space gives

$$\overset{\nabla}{\underset{\approx}{S}} = \frac{1}{\lambda_R}\alpha^{-2}(1+g)\underset{\approx}{S} - \frac{1}{\lambda_R}\beta^{-2}\underset{\approx}{\delta} \tag{8-56}$$

where $\lambda_R \equiv \varsigma/8kT\beta^2$ is the Rouse relaxation time. To within an isotropic constant, the Kirkwood-Riseman stress tensor is

$$\underset{\approx}{\tau} = n < \underset{\sim}{R}\frac{\partial V}{\partial \underset{\sim}{R}} > = nkT\beta^2\alpha^{-2}(1+g)\underset{\approx}{S} \tag{8-57}$$

Equations 8-56 and 8-57 define a UCM-type of equation with an *effective relaxation time* of $\lambda_R\,\alpha^2/(1+g)$, and an *effective modulus* of nkT; the product of these is an *effective viscosity*, $nkT\lambda_R\,\alpha^2/(1+g)$. At high shear rate, the dumbbell stretches out, and excluded volume forces vanish; $1+g \rightarrow \alpha^2$, according to Equation 8-55. Thus the dumbbell behaves like a Hookean dumbbell in the limit of high strain rates. We found similar behavior for the dumbbell with hydrodynamic interaction. The asymptotic values of η and of Ψ_1 at high shear rates are

$$\frac{\eta_\infty}{\eta_0} = \frac{1}{\alpha^2}\,; \quad \frac{\Psi_{1,\infty}}{\Psi_{1,0}} = \frac{1}{\alpha^4}\,; \quad \alpha \geq 1$$

The second normal stress difference is zero for all shear rates.

Shear thinning occurs when excluded volume enters because the force law for the bead to bead end-to-end vector becomes nonlinear. Based on a "blob" theory for the polymer chain, Pincus[38,2] argues that the force law for a chain with excluded volume should be $f \propto R^{3/2}$. This force law, like the force law used for finitely-extensible polymer chains (see below), produces shear thinning.

8.5.3 Dumbbells with Finite Extensibility

At high molecular extensions, the relation between the spring force, f^s, and the molecular extension, R, is nonlinear, the force becoming very large as the molecular length approaches full extension, R_0. For a freely jointed chain with many bonds, the force-extension relationship has been computed; it is an *inverse Langevin function,* a mathematically complex function, but one that can be roughly approximated by a simple law proposed by Warner:[39]

$$\underset{\sim}{f^s} = \frac{2\beta^2 kT}{1-(R/R_0)^2}\underset{\sim}{R} \equiv H(R^2)\underset{\sim}{R} \tag{8-58}$$

Figure 8-11 compares the Warner force law with the inverse Langevin function. Large departures from linear behavior occur when the chain reaches half or more of its full extension. For the Warner force law, the equilibrium distribution function is[39,11]

$$\psi_0 = \exp[-\frac{1}{kT}\int_0^R \underset{\sim}{f^s}\, dR] = [1-(\frac{R}{R_0})^2]^{b/2} \tag{8-59}$$

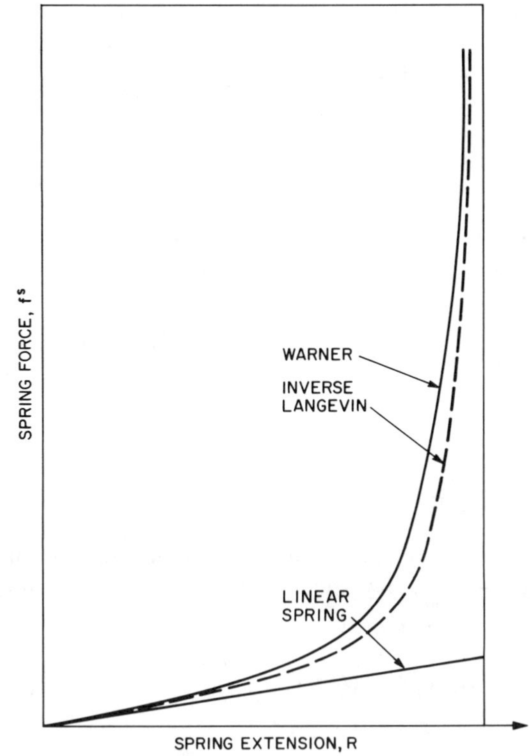

Fig. 8-11 Force law for the Warner spring, Equation 8-58 compared to the inverse Langevin and the Gaussian springs.

where $b \equiv 2\beta^2 R_0^2$.

The Smoluchowski equation for a nonlinear dumbbell is changed from that of the Gaussian dumbbell only in the term for the spring force:

$$\frac{\partial}{\partial t}\psi = -\frac{\partial}{\partial \underset{\sim}{R}}\cdot\left[\underset{\sim}{R}\cdot\underset{\sim\sim}{\nabla v} - \frac{2kT}{\zeta}\left(\frac{1}{kT}H(R^2)\underset{\sim}{R} - \frac{\partial\psi}{\partial \underset{\sim}{R}}\right)\right] \qquad (8\text{-}60)$$

Because $H(R^2)$ is not a constant, this equation lacks a general analytic solution, and a closed-form constitutive equation for the stress is not obtained unless an approximation is made. Following Peterlin,[40] Tanner[41] replaced $H(R^2)$ by the *preaveraged*[*] quantity, $H(<R^2>)$. Equation 8-60 can then be multiplied by $\underset{\sim\sim}{RR}$

---

* Preaveraging can be dangerous if there are large fluctuations about the mean value of the preaveraged quantity. For the Warner dumbbell, however, preaveraging seems to be a good approximation.[41]

and integrated over configuration space in the usual way:

$$\overset{\nabla}{\underset{\approx}{S}} + \frac{1}{\lambda}\left(\frac{\underset{\approx}{S}}{1-(\mathrm{tr}\underset{\approx}{S})/2R_0^2} - \beta^{-2}\underset{\approx}{\delta}\right) = \underset{\approx}{0} \qquad (8\text{-}61)$$

where $\underset{\approx}{S} \equiv 2\langle\underset{\sim}{R}\underset{\sim}{R}\rangle$, as usual, and $\lambda = \varsigma/8kT\beta^2$. Because the spring force law is not linear and the equilibrium distribution function is not Gaussian, the stress tensor is not proportional to $\langle\underset{\sim}{R}\underset{\sim}{R}\rangle$. The Giesekus expression for the stress tensor, Equation 8-38, gives

$$\underset{\approx}{\tau} = -\frac{n\varsigma}{4}\overset{\nabla}{\langle\underset{\sim}{R}\underset{\sim}{R}\rangle} = \frac{-n}{8\varsigma}\overset{\nabla}{\underset{\approx}{S}} = nkT\beta^2\left((1-\frac{\mathrm{tr}\underset{\approx}{S}}{2R_0^2})^{-1}\underset{\approx}{S} - \beta^{-2}\underset{\approx}{\delta}\right) \qquad (8\text{-}62)$$

where the last equality is obtained by substituting for $\overset{\nabla}{\underset{\approx}{S}}$ from Equation 8-61.

The high shear-rate asymptotics of the Warner dumbbell model can be calculated by letting $1 - \mathrm{tr}\underset{\approx}{S}/2R_0^2$ equal ϵ, a small quantity, and then solving the equations to lowest order in ϵ. We find

$$\frac{\eta_\infty}{\eta_0} = (\beta R_0)^{2/3}(\lambda\dot{\gamma})^{-2/3}\,; \qquad \frac{\Psi_{1,\infty}}{\Psi_{1,0}} = (\beta R_0)^{4/3}(\lambda\dot{\gamma})^{-4/3}$$

The second normal stress difference is zero at all shear rates.

8.5.4 Dumbbells with Internal Viscosity

In the original formulation of internal viscosity by Kuhn and Kuhn,[32] the force on a bead arising from internal viscosity was given by

$$\underset{\sim}{f}^{iv} = -\phi\frac{\dot{\underset{\sim}{R}}\cdot\underset{\sim}{R}}{R^2}\underset{\sim}{R} \qquad (8\text{-}63)$$

This force is proportional to the rate at which the end-to-end vector distance between the beads changes, and is *directed along the orientation of the spring.* In Cerf's later formulation,[42] $\underset{\sim}{f}^{iv}$ is given by

$$\underset{\sim}{f}^{iv} = -\phi(\dot{\underset{\sim}{R}} - \underset{\sim}{L}\times\underset{\sim}{R}) \qquad (8\text{-}64)$$

where $\underset{\sim}{L}$ is the *angular velocity* of the dumbbell, and $\underset{\sim}{L}\times\underset{\sim}{R}$ is the cross product of $\underset{\sim}{L}$ and $\underset{\sim}{R}$. Here we consider only the formulation of Kuhn and Kuhn, Equation 8-63.

From the force balance for a Hookean dumbbell with internal viscosity, Equation 8-31, we then obtain

$$\dot{\underset{\sim}{R}} - \underset{\sim}{R} \cdot \underset{\sim}{\nabla} \underset{\sim}{\mathbf{v}} + \frac{4kT\beta^2}{\varsigma} \underset{\sim}{R} + \frac{2kT}{\varsigma} \frac{\partial \log \psi}{\partial \underset{\sim}{R}} + \epsilon \frac{\dot{\underset{\sim}{R}} \cdot \underset{\sim}{R} \underset{\sim}{R}}{R^2} = \underset{\sim}{0} \qquad (8\text{-}65)$$

where $\epsilon \equiv \phi/\varsigma$ is the internal viscosity parameter divided by the friction coefficient. $\dot{\underset{\sim}{R}}$ appears on both the left and right sides of this expression.* However, multiplying Equation 8-65 by $\underset{\sim}{R}$ and regrouping gives an explicit expression for $\dot{\underset{\sim}{R}} \cdot \underset{\sim}{R}$:

$$\dot{\underset{\sim}{R}} \cdot \underset{\sim}{R} = \frac{1}{1+\epsilon} \underset{\sim}{R} \underset{\sim}{R} : \underset{\sim}{\nabla} \underset{\sim}{\mathbf{v}} - \frac{4kT\beta^2 R^2}{\varsigma(1+\epsilon)} - \frac{2kT}{\varsigma(1+\epsilon)} \underset{\sim}{R} \cdot \frac{\partial \log \psi}{\partial \underset{\sim}{R}} \qquad (8\text{-}66)$$

This equation, when substituted into Equation 8-65, gives an equation that can be rearranged so that $\dot{\underset{\sim}{R}}$ appears only on the left side:[43]

$$\dot{\underset{\sim}{R}} = (\underset{\approx}{\delta} - \frac{\epsilon}{1+\epsilon} \underset{\sim}{u} \underset{\sim}{u}) \cdot (\underset{\sim}{R} \cdot \underset{\sim}{\nabla} \underset{\sim}{\mathbf{v}} - \frac{4kT\beta^2}{\varsigma} \underset{\sim}{R} - \frac{2kT}{\varsigma} \frac{\partial \log \psi}{\partial \underset{\sim}{R}}) \qquad (8\text{-}67)$$

For rapid deformations, that is, for large $\underset{\sim}{\nabla} \underset{\sim}{\mathbf{v}}$, the above equation simplifies to

$$\dot{\underset{\sim}{R}} = \underset{\sim}{R} \cdot \underset{\sim}{\nabla} \underset{\sim}{\mathbf{v}} - \frac{\epsilon}{1+\epsilon} \underset{\sim}{R} \underset{\sim}{u} \underset{\sim}{u} : \underset{\approx}{\mathbf{D}} \qquad (8\text{-}68)$$

This expression represents *nonaffine motion* and is identical to a nonaffine model for melts in which polymer strands are *partially extensible*; see Section 5.4.

In the limit, $\epsilon \to \infty$, Equation 8-67 describes the dynamics and stress in solutions of *rigid dumbbells*; Thus Kuhn's internal viscosity model encompasses both the *flexible and rigid* dumbbell limits. Theories for rigid dumbbells are discussed in Section 8.7.1.

Equation 8-67 can be inserted into the probability conservation equation

$$\frac{\partial \psi}{\partial t} + \frac{\partial}{\partial \underset{\sim}{R}} \cdot (\dot{\underset{\sim}{R}} \psi) = 0$$

The result can be rearranged into

* Equation 8-65 is somewhat sloppy in that the Brownian term, $(2kT/\varsigma)(\partial \log\psi/\partial \underset{\sim}{R})$, has been averaged over the velocity distribution and $\dot{\underset{\sim}{R}}$ has not been. A more careful treatment would avoid this inconsistency.

$$
\underset{\approx}{\tau} = -\frac{n\zeta}{4} \overset{\nabla}{\underset{\sim\sim}{<RR>}} \qquad \text{(Giesekus expression)}
$$

$$
= \frac{2nkT}{1+\epsilon}\beta^2 \underset{\sim\sim}{<RR>} \; - \; nkT[\underset{\approx}{\delta} - 3\frac{\epsilon}{1+\epsilon}\underset{\sim\sim}{<uu>}] + \frac{n\zeta}{2}\frac{\epsilon}{1+\epsilon}\underset{\sim\sim\sim\sim}{<uuRR>} : \underset{\approx}{D}
$$

$$
\text{(spring)} \qquad\qquad \text{(Brownian)} \qquad\qquad \text{(viscous)} \qquad (8\text{-}69)
$$

where the first equality is obtained from Equation 8-38. Note that internal viscosity reduces the spring contribution to the stress tensor by the factor $1/(1+\epsilon)$. Also because of internal viscosity, the Brownian stress contribution is *anisotropic*. In Section 5.2.1, we argued that for melts, nonaffine motion or "slip," reduces the spring contribution and can introduce a Brownian term proportional to $\underset{\sim\sim}{<uu>}$; see Equation 5-38. The viscous term, $\underset{\sim\sim\sim\sim}{<uuRR>} : \underset{\approx}{D}$, however, is absent or very small in the melt. It represents the *viscous dissipation* introduced when solvent flows around a rigid or a semirigid object. Because it is proportional to $\underset{\approx}{D}$, it produces an instantaneous *stress jump* whenever flow is started; when flow ceases, this contribution to the stress also disappears instantaneously. Jumps in the *polymeric* contribution to the stress on start-up or cessation of flow have not yet been observed.

In a small-amplitude oscillatory shear flow, the viscous term prevents the polymeric contribution to η' from approaching zero as the frequency becomes very high. Instead, $\eta' \rightarrow \eta_\infty$, a constant. For small values of ϵ, Booij and van Wiechen found that

$$
G' = \frac{nkT\omega^2\lambda^2}{1+\omega^2\lambda^2}\left(1 - \frac{0.8\epsilon\omega^2\lambda^2}{1+\omega^2\lambda^2}\right)
$$

$$
G'' = \frac{nkT\omega\lambda}{1+\omega^2\lambda^2}\left(1 - \frac{0.4\epsilon(1-\omega^2\lambda^2)\omega^2\lambda^2}{1+\omega^2\lambda^2}\right)
$$

At high frequency, $\eta' \equiv G''/\omega$ approaches a nonzero constant, $\eta'_\infty = 0.4\epsilon\,\eta_0 + \eta_s$. Values of η'_∞ higher than η_s at high frequency have often been found in dilute solutions (see Section 8.2.3).

Again for low values of ϵ, in steady-state shear,*

$$
\frac{\eta_\infty}{\eta_0} = 1 - \epsilon \; ; \qquad \frac{\Psi_{1,\infty}}{\Psi_{1,0}} = 1 - 2\epsilon
$$

At all shear rates, the ratio N_2/N_1 is zero[44] and the optical counterpart, $<R_2^2 - R_3^2>/<R_1^2 - R_2^2>$, is $-\epsilon/5$ at low shear rates and zero at high shear rates.[43]

---

* $\Psi_{1,\infty}$ was computed by the author with the assistance of a table in Appendix B of Booij and van Weichen.[43]

8.5.5 Summary of Dumbbell Models

Table 8-2 Mechanisms of Shear Thinning

Dumbbell	Asymptotics			Shear Thinning	Stress-Optic Law
	$\frac{\eta_\infty}{\eta_0}$	$\frac{\Psi_{1,\infty}}{\Psi_{1,0}}$	$\frac{\Psi_{2,0}}{\Psi_{1,0}}$		
HI	$1-\sqrt{2}h^*$	$(1-\sqrt{2}h^*)^2$	$\frac{0.07h^*}{1-\sqrt{2}h^*}$		holds
EV	$\frac{1}{\alpha^2}$	$\frac{1}{\alpha^4}$	0	increases with MW	fails at high $\dot{\gamma}$
FE	$(\lambda\dot{\gamma})^{-2/3}$	$(\lambda\dot{\gamma})^{-4/3}$	0	decreases with MW	fails at high $\dot{\gamma}$
IV	$1 - \epsilon$	$1 - 2\epsilon$	0	decreases with MW decreases with η_s	fails at high $\dot{\gamma}$

Table 8-2 includes shear thinning predictions for the various dumbbell models. Only *hydrodynamic interaction* does not lead to a failure of the stress-optic law. The stress-optic law fails whenever the stress tensor is not proportional to $<\underset{\sim}{R}\underset{\sim}{R}>$. The calculations of Öttinger and of Fixman for the multiple-spring chain, as well as the experimental data of Noda et al. suggest that hydrodynamic interaction, though important in its effect on the relaxation spectrum, is a minor contributer to shear thinning. *Excluded volume* could be an important factor in shear thinning, but models with excluded volume predict that, on a plot of $[\eta]$ versus β, shear thinning should become more pronounced as the molecular weight increases; the opposite is seen experimentally. *Finite extensibility* does predict that shear thinning will decrease as molecular weight increases, but the value of β^* at which shear thinning becomes evident in the experimental data is often too low to be consistent with predictions that are based on the known fully-extended length of the molecule. Of the four mechanisms, only Kuhn's *internal viscosity* predicts the dependences of shear thinning on both molecular weight and with solvent viscosity that are actually observed experimentally. Internal viscosity also seems to be the only available explanation for the observation that polymer coils are only minimally deformed in a nonviscous solvent at high shear rates. Confidence in the significance of internal viscosity is limited, however, because the description of it is often vague and somewhat arbitrary, the parameter (ϕ) that quantifies it cannot be estimated readily from molecular structure, and computations that involve internal viscosity are difficult, even for the dumbbell model and much more so for the multiple-spring model. Definitive conclusions regarding the sources of shear thinning can therefore only be made after more thorough study is made of the dependence of both stress and birefringence on molecular weight, solvent viscosity, and solvent quality.

8.6 EXTENSIONAL FLOW

In steady extensional flows, such as uniaxial extension, the Rouse or Zimm models predict that the steady-state extensional viscosity becomes infinite at a finite strain rate, $\dot{\epsilon}$. In the dumbbell model, for example, this occurs when the frictional drag force that stretches the dumbbell exceeds the contractile force of the spring. If z is the direction of extensional flow, the net tensile force on the spring in the z direction is

$$f_z \approx \underset{\text{(drag force)}}{\frac{1}{2}\varsigma R_z\dot{\epsilon}} - \underset{\text{(spring force)}}{HR_z} \tag{8-70}$$

where R_z is the z component of the spring end-to-end vector vector. For a linear spring, H is a constant, H_0, and the spring force exactly balances the drag force when

$$\dot{\epsilon} = \frac{2H_0}{\varsigma} = \frac{1}{2}\lambda \tag{8-71}$$

where λ is the relaxation time. For $\lambda\dot{\epsilon} < 1/2$, the spring force dominates and the dumbbell collapses. The mean square end-to-end vector, $<R^2>$, remains nonzero only because the Brownian diffusive force, neglected in the above force balance, opposes a complete collapse and in general acts to smear out the distribution of end-to-end vectors. For $\lambda\dot{\epsilon} > 1/2$, the drag force dominates, and because R_z cancels out of Equation 8-70, no extension of the chain can bring the forces back into balance. As R_z increases, the spring and drag forces remain in the same relative proportion and the dumbbell extends indefinitely. This is manifested in the UCM equation by a singularity in stress at $\dot{\epsilon}\lambda = 1/2$. Thus a transition from a coiled configuration to an infinitely stretched one is predicted to occur at $\dot{\epsilon}\lambda = 1/2$. Brownian forces smooth the transition, but do not shift the value of $\dot{\epsilon}$ at which it occurs.

A sudden *coil-stretch* transition is not expected to occur in steady shear flows. This is because the velocity gradient tensor in a shear flow is skew; therefore the drag and spring forces are proportional to different components of the end-to-end vector. If x is the flow direction and y is perpendicular to the shearing plane, then the x component of the force is

$$f_x \approx \frac{1}{2}\varsigma(\underset{\sim}{R}\cdot\underset{\sim\sim}{\nabla\mathbf{v}})_x - HR_x = \frac{1}{2}\varsigma R_y\dot{\gamma} - HR_x \tag{8-72}$$

As the shear rate, $\dot{\gamma}$, increases, R_x increases without an increase in R_y; thus the drag force can be balanced by the spring force at any shear rate.

Other strong flows —that is, flows with an extensional component— also have coil-stretch transitions. The smaller the extensional component is relative to the

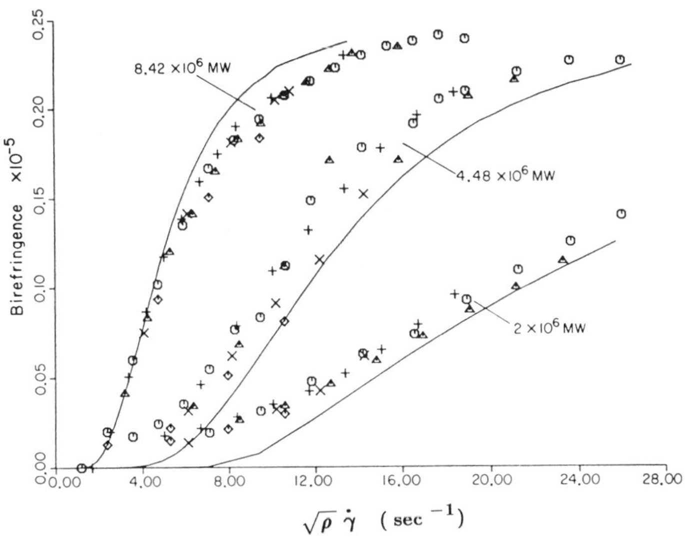

Fig. 8-12 Birefringence versus the eigenvalue of the velocity gradient tensor, $\sqrt{\rho}\,\dot{\gamma}$, for dilute polystyrene solutions in polychlorinated biphenyl solvent. The different symbols represent ρ values of 1.0 (○), 0.8 (Δ), 0.5 (+), 0.33 (×), and 0.25 (□). The curves are theoretical calculations that account for conformation-dependent friction and finite extensibility (from Fuller and Leal,[45] reproduced with permission from J. Non-Newt. Fluid Mech.)

overall strain rate, the higher is the overall strain rate at which the transition takes place.[36] In steady planar flows, for example, the velocity gradient tensor, $\underset{\sim}{\nabla}\underset{\sim}{\mathbf{v}}$, can be expressed as[45]

$$\underset{\sim}{\nabla}\underset{\sim}{\mathbf{v}} = \frac{1}{2}\dot{\gamma}\begin{pmatrix} 1+\rho & 1-\rho \\ -(1-\rho) & -(1+\rho) \end{pmatrix}$$

Here ρ is a parameter that determines the flow type. $\rho = 1$ is planar extension, a strong flow, and smaller values of ρ correspond to weaker flows, $\rho = 0$ being a simple shear. Such flows can be generated in a four-roller-mill device, as discussed by Fuller and Leal.[45] It is found that the birefringence of a dilute solution of flexible polymers measured in such a device is a nearly universal function of the combination variable $\dot{\gamma}\sqrt{\rho}$; see Figure 8-12. $\dot{\gamma}\sqrt{\rho}$ is the eigenvalue of the velocity gradient tensor. Thus for a given strain rate, $\dot{\gamma}$, as ρ increases from small values to unity, the flow strength increases and so does the steady-state distortion of the macromolecules. According to Fuller and Leal, the superposition of the data in Figure 8-12 when plotted agains $\dot{\gamma}\sqrt{\rho}$ implies that the influence of internal viscosity in these experiments must be small; the parameter ϵ in Equation 8-65 must be less than about 0.3.[45]

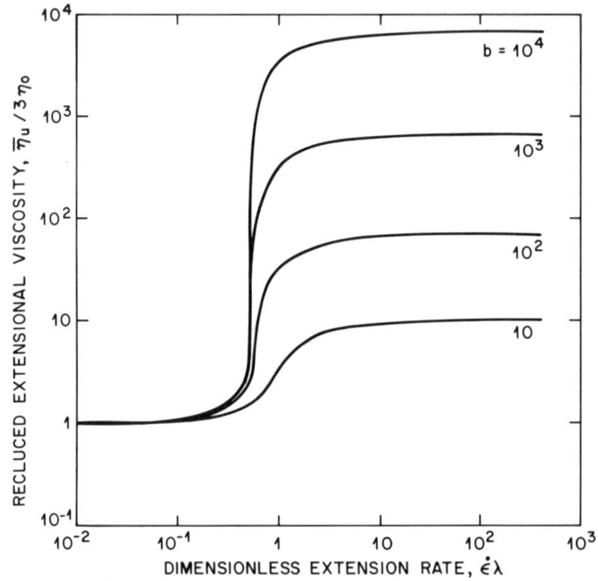

Fig. 8-13 Steady-state extensional viscosity for the dumbbell with finite extensibility (from Bird et al.,[11] reprinted from *Dynamics of Polymeric Liquids,* Vol. 2 ©1977 by permission of John Wiley & Sons, Inc.).

8.6.1 The Effect of Finite Extensibility

The unlimited extension and infinite stresses predicted for the Gaussian dumbbell in an extensional flow when $\dot{\epsilon}\lambda > 1/2$ are artifacts of the infinite extensibility of the spring. For finitely extensible dumbbells, H increases to infinity as R approaches R_0; thus drag force in Equation 8-70 can always be balanced at large $\dot{\epsilon}$ by the spring force. For the Warner spring, $H = H_0(1 - R^2/R_0^2)^{-1}$, and

$$\dot{\epsilon} = \frac{1}{2\lambda}[1 - \frac{R^2}{R_0^2}]^{-1} \tag{8-73}$$

from which we obtain

$$R = R_0(1 - \frac{1}{2\lambda\dot{\epsilon}})^{1/2} . \tag{8-74}$$

The polymer contribution to the stress tensor at high $\dot{\epsilon}$ is

$$\sigma_p \sim nRf^s = nR^2H = nR_0^2H_0(2\lambda\dot{\epsilon} - 1) \approx 2n\lambda R_0^2H_0\dot{\epsilon} \tag{8-75}$$

The polymer contribution to the elongational viscosity, $\sigma_p/\dot{\epsilon}$, at high $\dot{\epsilon}$ therefore approaches a constant:

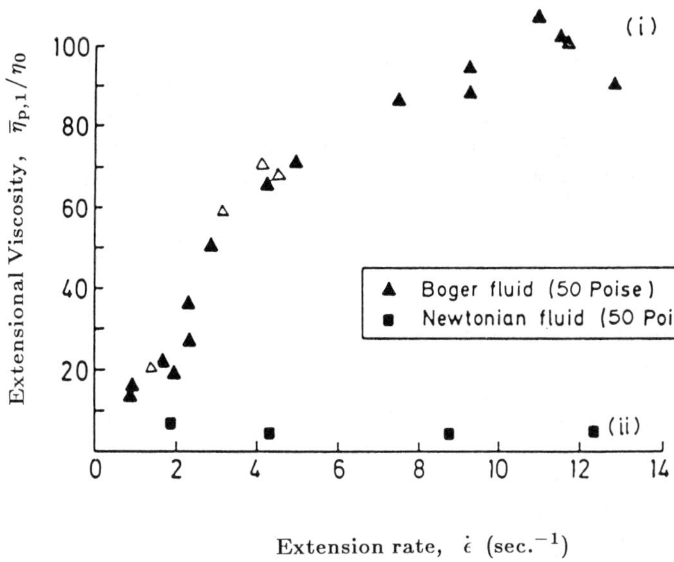

Fig. 8-14 Steady-state extensional viscosity versus extension rate for a Boger fluid (i) compared to that for a Newtonian fluid (ii) (from Williams and Williams,[48] reproduced with permission from J. Non-Newt. Fluid Mech.)

$$\bar{\eta}_p \rightarrow \bar{\eta}_{p,\infty} = 2nkT\,b\lambda \quad \text{as} \quad \dot{\epsilon} \rightarrow \infty \tag{8-76}$$

where $b = 2\beta^2 R_0^2$. Figure 8-13 shows the dependence of $\bar{\eta}$ on $\dot{\epsilon}$ for the finitely extensible dumbbell for various values of b.

As the molecular weight of the polymer increases at fixed mass concentration, the number of dumbbells per unit volume, n, scales as M^{-1}. R_0 scales as M and β^2 as M^{-1}. Hydrodynamic interaction seems to be important even in the extended state, because λ for extended molecules follows the scaling[46] $M^{3/2}$, rather than M^2 for the free-draining limit. Combining these scalings gives

$$[\bar{\eta}]_\infty \propto n\beta^2 R_0^2 \lambda \sim M^{3/2}$$

where $[\bar{\eta}]_\infty$ is polymer contribution to the elongational viscosity per unit mass of polymer in the limit of high extension rate. A cylinder is probably a better model of an extended coil than is a dumbbell; for the cylinder,[47] $[\bar{\eta}]_\infty$ is proportional to M^2. In either case, either $[\bar{\eta}]_\infty \propto M^{3/2}$ or $[\bar{\eta}]_\infty \propto M^2$, the extensional viscosity must become very large when $\lambda\dot{\epsilon} > 1$, if M is large. The corresponding zero-shear viscosity in a theta solvent scales only as $M^{1/2}$ (Table 8-1), and is usually a *decreasing* function of shear rate. Thus polymer solutions may have large extensional viscosities even if the polymer contribution to the shear viscosity is negligible.

Although extensional viscosities of dilute solutions are notoriously difficult to measure accurately, experimental data are unambiguous in showing a transition to very high extensional viscosities, much higher than shear viscosities for the same materials. Figure 8-14, for example, shows the dependence of the planar extensional viscosity $\overline{\eta}_{p,1}$ on the extension rate for a Boger fluid compared to the planar extensional viscosity for a Newtonian fluid having the same zero-shear viscosity. The Boger fluid shows a sudden transition, as strain rate increases, to a high extensional viscosity and small variations thereafter, in qualitative agreement with the theoretical results for a finitely extensible dumbbell.

Very-high-molecular-weight polymers may have knots or "self-entanglements," particularly in theta solvents,[49] that could prevent them from becoming completely stretched in a flow field. In spite of the knots, a typical segment of the molecule would become highly oriented, so the birefringence (which measures average segment orientation) of an extremely deformed knotted molecule would be the same as a completely stretched one. Although stress measurements should distinguish between the two conformations, sufficiently accurate measurements are difficult, and the presence or absence of effects from self-entanglements has not yet been settled.

8.6.2 Dumbbells with Variable Drag

Hysteresis in the coil-stretch transition could be produced by the effects of hydrodynamic interaction. Suppose that the chain is nearly aligned along the direction of stretch and has end-to-end distance R. For the dumbbell model, the frictional drag is proportional to R, and if R is large, hydrodynamic interaction is negligible (Section 8.5.1). This artificial result emerges because the dumbbell has only two points of frictional contact with the solvent, namely the beads at either end of the chain. A real chain, however, contacts the solvent all along its contour.

A more realistic hydrodynamic model of the chain therefore is a *cylinder* with length R and radius a. Recall that in the coiled state when hydrodynamic interaction is dominant, the frictional properties of the chain are accurately modeled by a single solid sphere with a radius equal to the coil radius (Section 8.2.2). In a partially stretched configuration, the analogous model is a solid cylinder.[2] Following de Gennes, we observe that the increment in drag df^d on a segment of the cylinder of length dx located a distance x from the center is

$$df^d = \eta_s \frac{2\pi}{\log(R/2a)} v\, dx \tag{8-77}$$

where $v = \dot{\epsilon}x$ is the velocity of the solvent relative to the cylinder at position x. The geometric factor $2\pi/\log(R/2a)$ as a rough approximation can be regarded as a constant, k. The total drag on the cylinder then is

$$f^d = \eta_s k \int_{-R/2}^{R/2} v\, dx = \eta_s k \int_{-R/2}^{R/2} \dot{\epsilon}x\, dx = \frac{1}{4}\dot{\epsilon}\eta_s k R^2 \tag{8-78}$$

Thus in the cylinder model the frictional force is *quadratic* in the end-to-end distance rather than linear as it is in the dumbbell model.

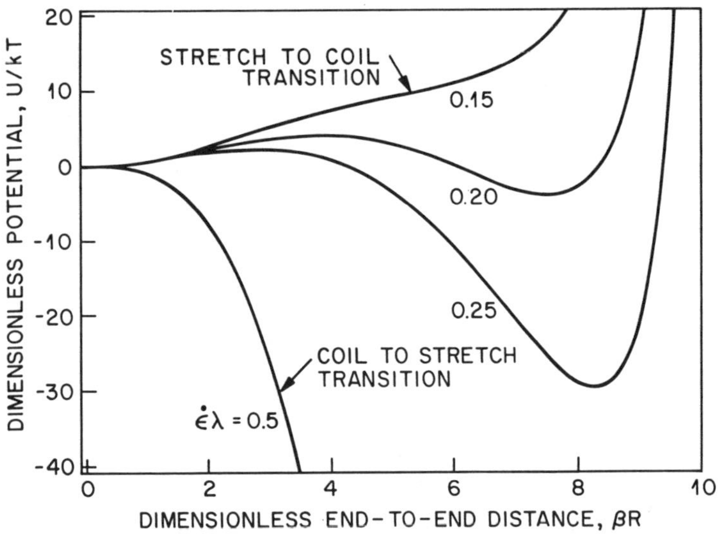

Fig. 8-15 The dumbbell potential from Equation 8-81. Here $\lambda = \varsigma_0/8\beta^2 kT$, $H_0 = 2\beta^2 kT$, $R_0 = 10$, and $2\alpha/3 = 5$.

Let us compare this result with the drag for a stretched chain without hydrodynamic interaction. The drag on one bead is $6\pi\eta_s av$, where $v = \dot{\epsilon}x$. If there are N beads of radius a spaced uniformly along an end-to-end vector distance R, then there are $(N/R)\,dx$ beads in a segment of length dx. The total drag is then

$$f^d = \int_{-R/2}^{R/2} 6\pi\eta_s a\,\dot{\epsilon}x\,\frac{N}{R}\,dx = \frac{3}{2}\pi\eta_s\,a\dot{\epsilon}\,NR \tag{8-79}$$

When the chain is nearly fully extended, $R \propto N$. Thus without hydrodynamic interaction the drag at high extensions scales as N^2. The same scaling is obtained in the cylinder model with hydrodynamic interaction, according to Equation 8-78, when $R \propto N$ at high extensions. The explanation for the convergence of these results is that as the chain is stretched, the cylinder representing the chain presents a greater surface area to the solvent, and hence experiences more drag. When the chain is fully stretched, the surface area of the cylinder is proportional to the surface area of all the beads of the chain, and the cylinder model is roughly equivalent to a freely draining chain. Thus the cylinder model describes the process whereby hydrodynamic "shielding" of interior beads from solvent drag is stripped away as the chain is uncoiled. Because of the extra drag produced by this stripping action, *shear thickening* in shearing flows is predicted to occur at high shear rates; see Section 8.3.1.3.

This phenomenon has been incorporated into the dumbbell model by making the friction coefficient, ς, depend on chain extension:[50,51]

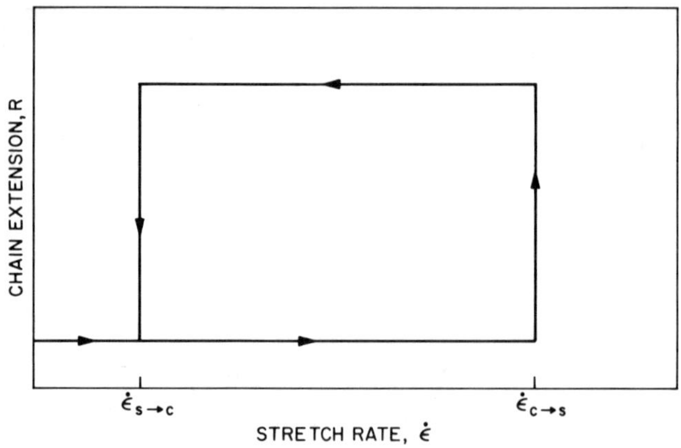

Fig. 8-16 Hysteresis in the coil to stretch transition.

$$\varsigma = \varsigma_0 (1 + \alpha\beta R) \qquad (8\text{-}80)$$

Here α is a dimensionless parameter. It scales as $N_K^{1/2}$. Since $f^d = \frac{1}{2}\varsigma R\dot{\epsilon}$, the scaling, $f^d \propto R^2$ that one obtains for the cylinder is then also obtained at high extensions for the *dumbbell with configuration-dependent friction,* described by Equation 8-80. Adding the drag force to the finitely extensible spring force and integrating, one obtains an effective *dumbbell potential,* U:[2]

$$U = -\int (f^d + f^s)\, dR = -\frac{1}{4}\varsigma_0\dot{\epsilon}\,(R^2 + \frac{2}{3}\alpha\beta R^3) - \frac{1}{2}H_0R_0^2 \log[1 - (\frac{R}{R_0})^2] \qquad (8\text{-}81)$$

The dumbbell potential is plotted for various strain rates, $\dot{\epsilon}$, in Figure 8-15. Dumbbell configurations tend to concentrate at minima in the potential where $f = -\partial U/\partial R = 0$. When $\dot{\epsilon}\lambda << 1/2$, the only minimum is in the coil state at $R = 0$. But as $\lambda\dot{\epsilon}$ increases (to 0.20 in Figure 8-15), a *second minimum* appears at high R, corresponding to a stretched state. The potential barrier, ΔU, between the two minima can be large compared to kT when the molecular weight is high. Thus though the second minimum becomes much deeper than the first, Figure 8-15, coiled molecules require a very long time, on the order of $\lambda\exp(\Delta U/kT)$, to *diffuse* by Brownian processes over the barrier into the stretched state. As $\dot{\epsilon}$ is further increased, the barrier disappears at a strain rate $\dot{\epsilon}_{c\to s} = 1/2\lambda$, and the transition to the stretched state rapidly occurs, Figure 8-16.

Starting at high $\dot{\epsilon}$, with the molecule in the stretched state, if $\dot{\epsilon}$ is decreased to $\dot{\epsilon}_{c\to s}$, the dumbbell nevertheless stays in the stretched state where the potential is lowest. The transition back to the coiled state occurs when the strain rate is further reduced to $\dot{\epsilon}_{s\to c}$ ($= 0.15/\lambda$ in Figure 8-15); here the coiled state is the only solution. Thus the transitions between coiled and stretched states have hysteresis,[51,45,52] Figure 8-16. For strain rates between $\dot{\epsilon}_{s\to c}$ and $\dot{\epsilon}_{c\to s}$, there are *two* chain states, one coiled and one stretched. In this regime, the configuration of a chain depends on its stretch history.* Because of diffusion, there is actually a distribution of chain configurations, and at any steady $\dot{\epsilon}$ there can be only one truly steady distribution;[53] this distribution is independent of the initial distribution. Thus if one waits long enough at a given $\dot{\epsilon}$ between $\dot{\epsilon}_{s\rightarrow c}$ and $\dot{\epsilon}_{c\rightarrow s}$, the configuration with the lowest potential, be it coiled or stretched, will eventually predominate no matter what the initial distribution of configurations is. However, the time required to reach the unique steady-state distribution can be extremely long, much longer than the nominal "longest relaxation time," λ_1, of the chain. The situation is analogous to a classical first-order phase transition, where there is a unique equilibrium state, but, in addition, there can be long-lived metastable states.

8.7 SUSPENSIONS OF RIGID PARTICLES

In the dynamics of a polymer coil in dilute solution, complex hydrodynamic and statistical effects are coupled. Treating the combined statistical-hydrodynamic problem realistically is difficult, as Fixman's complicated theory for hydrodynamically interacting beads illustrates. The Rouse theory is much more tractable because it greatly oversimplifies the hydrodynamic part of the problem, treating the beads as hydrodynamically independent spheres while maintaining a sophisticated statistical treatment with multiple beads and springs. An alternative approach to gaining tractability is to oversimplify the statistical part of the problem and to treat the hydrodynamics realistically. One simple approach is to model the polymer molecule as a single rigid ellipsoid or as a rigid rod. Even the most flexible macromolecules have some stiffness, as suggested by the high-frequency polymer contribution to η' (Section 8.2.3), and the evidence for internal viscosity.

Theories for rigid particles are useful as crude models of flexible polymer molecules, but more so as realistic models of truly rigid rodlike polymer molecules, such as polysaccharides, tobacco mosaic viruses, and DNA and proteins in their helical conformations.[2]

8.7.1 Rigid Dumbbells

A simple model of a rigid molecule is the *rigid dumbbell*. It has spheres on each end of an infinitely thin rigid connector. For simplicity, we neglect consideration of hydrodynamic interaction between the beads. The influence of hydrodynamic interaction is discussed in Bird et al.[2] The rigid dumbbell without hydrodynamic interaction is obtained in the limit of $\epsilon \rightarrow \infty$ in Kuhn's model of internal viscosity; see Section 8.5.4.

8.7.1.1 Dynamics Because of the connecter, the bead spacing, L, is fixed. Thus the end-to-end vector, $\underset{\sim}{r}_1 - \underset{\sim}{r}_2 = L\underset{\sim}{u}$, $\underset{\sim}{u}$ being a unit vector, cannot respond affinely to a rapid deformation. Instead the response is

* Strictly speaking, the configuration distribution of the polymer molecule is always history dependent; a fluid composed of such molecules "remembers" its past deformation history. But when $\dot{\epsilon}$ lies between $\dot{\epsilon}_{s\rightarrow c}$ and $\dot{\epsilon}_{c\rightarrow s}$, the memory of the fluid is profoundly increased.

$$\dot{\underset{\sim}{u}} = \underset{\sim}{u} \cdot \nabla \underset{\sim}{v} - \underset{\sim}{u}\underset{\sim}{u}\underset{\sim}{u} : \underset{\approx}{D} \tag{8-82}$$

The first term on the right side represents an affine response. The second is a vector parallel to $\underset{\sim}{u}$ and of a magnitude such that $\partial u^2/\partial t = 0$; $\underset{\sim}{u}$ therefore always remains a unit vector. Thus the second term subtracts the stretching from the affine motion of the first term. Equation 8-82 describes *affine rotation* with no stretching.

A Brownian dumbbell is also subject to *rotary* diffusion. The total flux of probability, $\underset{\sim}{J}$, produced by affine rotation and rotary diffusion of the dumbbell is[2]

$$\underset{\sim}{J} = \underbrace{(\underset{\sim}{u} \cdot \nabla \underset{\sim}{v} - \underset{\sim}{u}\underset{\sim}{u}\underset{\sim}{u} : \underset{\approx}{D})\psi}_{\text{(convection flux)}} \quad \underbrace{- D_r \frac{\partial \psi}{\partial \underset{\sim}{u}}}_{\text{(diffusion flux)}} \tag{8-83}$$

$\psi(\underset{\sim}{u})d\underset{\sim}{u}$ is the probability that the orientation of the dumbbell lies between $\underset{\sim}{u}$ and $\underset{\sim}{u} + d\underset{\sim}{u}$. At equilibrium, each orientation is equally likely; thus $\psi_0 = 1/4\pi$. D_r, the rotary diffusivity, is given by

$$D_r = \frac{2kT}{\varsigma L^2}$$

where ς is the bead friction coefficient. The continuity equation for probability is

$$\frac{\partial \psi}{\partial t} + \frac{\partial}{\partial \underset{\sim}{u}} \cdot \underset{\sim}{J} = 0 \tag{8-84}$$

When Equation 8-83 is substituted into the continuity equation, the result is the Smoluchowski equation that gives the time evolution of ψ in a flow field:

$$\frac{\partial \psi}{\partial t} + \frac{\partial}{\partial \underset{\sim}{u}} \cdot [(\underset{\sim}{u} \cdot \nabla \underset{\sim}{v} - \underset{\sim}{u}\underset{\sim}{u}\underset{\sim}{u} : \underset{\approx}{D})\psi - D_r \frac{\partial \psi}{\partial \underset{\sim}{u}}] = 0 \tag{8-85}$$

Multiplying by $\underset{\sim\sim}{RR} = L^2 \underset{\sim\sim}{uu}$ and integrating over du^2 gives

$$\overset{\nabla}{\langle \underset{\sim\sim}{RR} \rangle} + \frac{2}{L^2} \langle \underset{\sim\sim\sim\sim}{RRRR} \rangle : \underset{\approx}{D} + \frac{1}{\lambda} \langle \underset{\sim\sim}{RR} \rangle - \frac{L^2}{3\lambda} \underset{\approx}{\delta} = \underset{\approx}{0} \tag{8-86}$$

where $\lambda \equiv 1/(6D_r) = \varsigma L^2/12kT$.

8.7.1.2 The Stress Tensor For flexible or rigid molecules, the configurational entropy is given by (see Section 2.3.2)

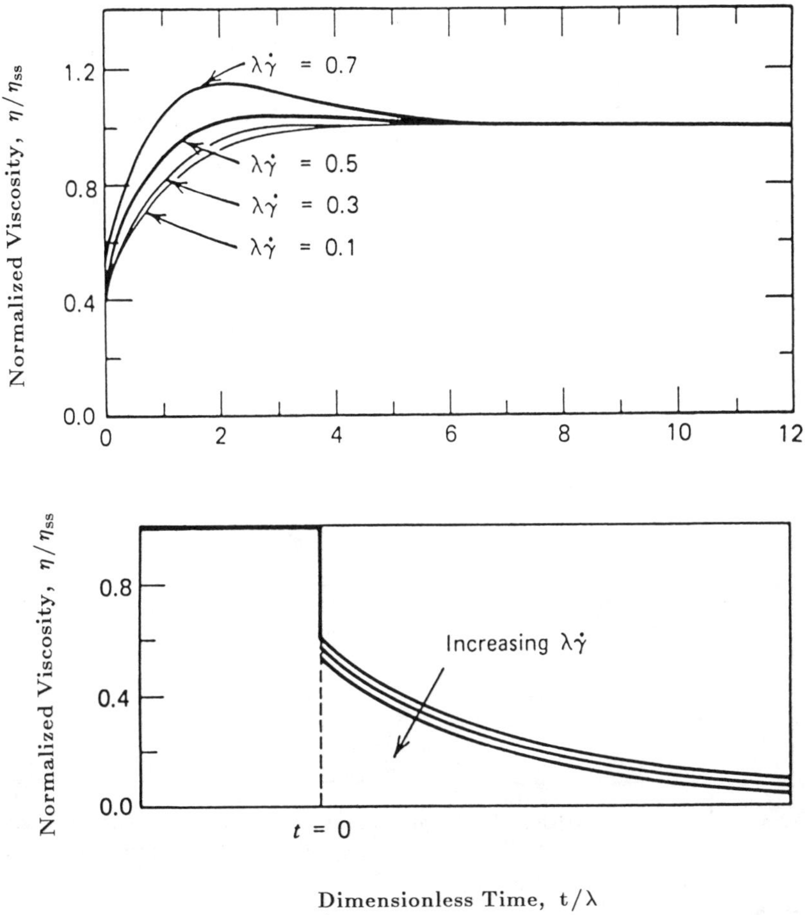

Fig. 8-17 Polymer contribution to the shear viscosity after start-up (top) and cessation (bottom) of steady shearing according to the rigid dumbbell model. The time-dependent viscosities are normalized by dividing by the steady-state viscosity, η_{ss} (from Bird et al.,[11] reprinted from *Dynamics of Polymeric Liquids,* Vol. 2 © 1977 by permission of John Wiley & Sons, Inc.).

$$S \;=\; \underset{\text{(orientation)}}{-\mathrm{k}\nu\int \psi \log \psi \, dR^3} \;-\; \underset{\text{(stretching)}}{-\nu \mathrm{k}\beta^2 \int \psi R^2 \, dR^3} \;+\; \text{constant} \qquad \text{(8-87a)}$$

The *entropic* contribution to the stress tensor $\underset{\approx}{\boldsymbol{\tau}}^{E}$ is obtained from the principal of virtual work by equating the power required to deform the fluid rapidly, $\underset{\approx}{\mathbf{D}} : \underset{\approx}{\boldsymbol{\tau}}^{E}$, with the temperature times the rate of entropy reduction,

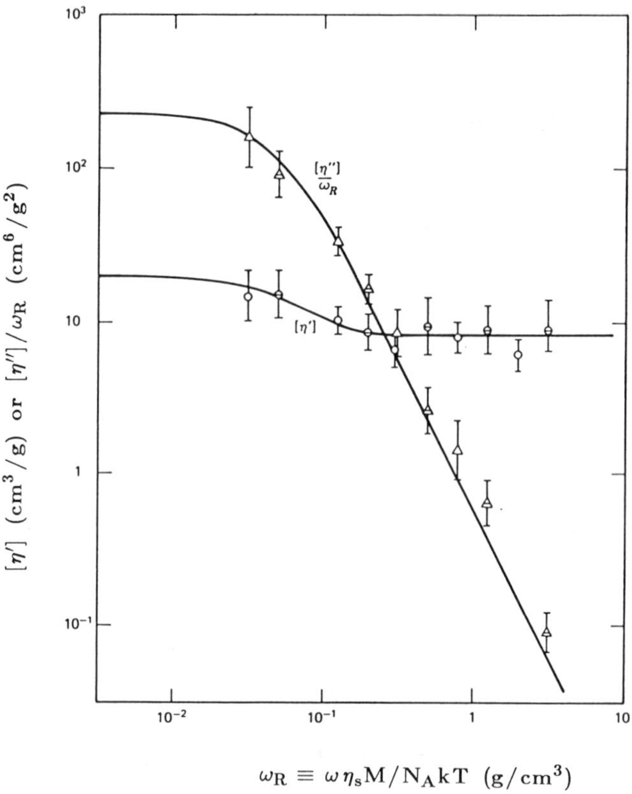

Fig. 8-18 Linear viscoelasticity of rigid dumbbells compared to data for solutions of tobacco mosaic virus (data obtained by Nemoto, et al.[56], figure from Bird et al.,[11] reprinted from *Dynamics of Polymeric Liquids,* Vol. 2 ©1977 by permission of John Wiley & Sons, Inc.).

$$\underset{\approx}{\mathbf{D}} : \underset{\approx}{\boldsymbol{\tau}}^{E} = -T\dot{S} = kT \frac{\partial}{\partial t} \int \psi \log \psi \, dR^3 + \nu k T \beta^2 \frac{\partial}{\partial t} \int \psi R^2 \, dR^3$$

(Brownian) (spring) (8-87b)

For rigid molecules the second term on the right side of Equation 8-87a is a constant and consequently the spring contribution in Equation 8-87b is zero. The entropic stress obtained from the first term is $\underset{\approx}{\boldsymbol{\tau}}^{E} = \underset{\approx}{\boldsymbol{\tau}}^{B} = 3nkT < \underset{\sim}{R}\underset{\sim}{R} > /L^2 - nkT\underset{\approx}{\boldsymbol{\delta}}$.

For completely flexible molecules, the entropic stress $\underset{\approx}{\boldsymbol{\tau}}^{E}$ is the only polymeric contribution to the stress; thus viscous dissipation occurs only by molecular diffusion. For the rigid dumbbell, on the other hand, the Giesekus expression for the stress tensor, Equation 8-38, yields

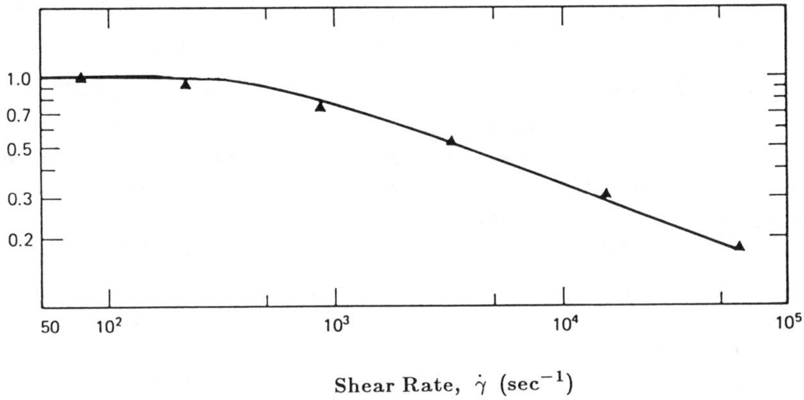

Fig. 8-19 Steady-state viscosity for the rigid dumbbell model compared to data for solutions of rigid poly(γ -benzyl-L-glutamate). The relaxation time λ was varied to obtain the best fit (data from Yang[57]; figure adapted from Bird et al.[55]).

$$\underset{\approx}{\tau} = \frac{-n\zeta}{4} \overset{\nabla}{\underset{\sim\sim}{<RR>}} = \underbrace{\frac{n\zeta}{2L^2} \underset{\sim\sim\sim\sim}{<RRRR>} : \underset{\approx}{\mathbf{D}}}_{\text{(viscous)}} + \underbrace{\frac{3nkT}{L^2} \underset{\sim\sim}{<RR>} - nkT\underset{\approx}{\delta}}_{\text{(entropic)}} \qquad (8\text{-}88)$$

In addition to the entropic term, $\underset{\approx}{\tau}^{E} = 3nkT\underset{\sim\sim}{<RR>}/L^2 - nkT\underset{\approx}{\delta}$, which was obtained from Equation 8-87b, this expression also contains a *purely viscous term*

$$\underset{\approx}{\tau}^{V} = \frac{n\zeta}{2L^2} \underset{\sim\sim\sim\sim}{<RRRR>} : \underset{\approx}{\mathbf{D}}$$

In the absence of Brownian motion, the dumbbell *rotates* freely in the flow field, but because the beads cannot change their distance of separation solvent is forced to flow around them, thus producing *viscous drag* and contributing $\underset{\approx}{\tau}^{V}$ to the stress tensor. When the dumbbells become oriented, Brownian motion produces a net disorienting motion of the beads which is resisted by the solvent; the solvent's resistance to this motion, which we call *Brownian drag*, contributes $\underset{\approx}{\tau}^{B} = \underset{\approx}{\tau}^{E}$ to the stress tensor. The Brownian drag is zero until the dumbbells are oriented; but once they are oriented and the flow is stopped, $\underset{\approx}{\tau}^{E}$ continues to contribute to the stress tensor until the dumbbells have again become disoriented. Thus $\underset{\approx}{\tau}^{E}$ relaxes gradually with time in the absence of flow and is therefore a *viscoelastic stress*. The viscous stress, $\underset{\approx}{\tau}^{V}$, in contrast, is nonzero if and only if there is a velocity gradient; it appears instantly after initiation of flow and disappears instantly after cessation of flow. Figure 8-17 shows the shear viscosity during start-up and cessation of steady shearing. Because of the viscous contribution, step changes in shear rate produce step changes in stress.

8.7.1.3 Predictions To obtain the stress, we must first solve Equation 8-85 for $\psi(\underset{\sim}{u})$ and then take second and fourth moments* of ψ so that we can be evaluate $\underset{\approx}{\tau}$ according to Equation 8-88. The predictions of the model are in qualitative and sometimes quantitative agreement with experimental data for rigid molecules. Figure 8-18 shows the components, $[\eta']$ and $[\eta'']$, of the complex intrinsic viscosity that are respectively in phase and out of phase with the shear rate as functions of frequency. The high-frequency limit in $[\eta']$ depicted in Figure 8-18 is produced by the viscous contribution to the stress tensor, $\underset{\approx}{\tau}^V$. Positive high-frequency asymptotes in $[\eta']$ are usually found with "flexible" polymers also (Section 8.2.3); this indicates that even apparently flexible polymers may have some rigidity. The steady-state intrinsic shear viscosity, $[\eta]$, of a solution of rigid dumbbells shows considerable shear thinning, Figure 8-19, in accord with data for rigid molecules.[55,11] According to the rigid-rod model, at large $\dot{\gamma}$, η scales as $\dot{\gamma}^{-2/3}$ and Ψ_1 scales as $\dot{\gamma}^{-4/3}$. Ψ_2/Ψ_1 at low shear rates is predicted to be zero,[11] the negative contribution to Ψ_2 from $\underset{\approx}{\tau}^E$ being exactly canceled by the positive contribution from $\underset{\approx}{\tau}^V$. When hydrodynamic interaction between the beads is accounted for, Ψ_2/Ψ_1 is found to be *positive*, but it becomes negative if more beads are added to the rigid rod to form a "shish kebab."[58] The steady-state uniaxial extensional viscosity increases monotonically with $\dot{\epsilon}$. For a thorough discussion of the rigid rod model, see Bird et al.[11]

8.7.2 Rigid Ellipsoids

Macromolecular coils become elliptic in shape in the presence of strong velocity gradients. While certainly not completely rigid, such elliptic coils are apparently not completely flexible either. Even if chain stiffness is negligible, hydrodynamic interaction causes the coil and the solvent in the coil to behave somewhat like an impenetrable object. Internal barriers to chain deformation could also stiffen the coil, in extreme cases making it act more as a bent wire than as a limp string. Chain rigidity is imperfectly represented by the rigid dumbbell (or any kind of dumbbell), because the dumbbell has an infinite aspect ratio, whereas the coil is spheroidal. An important difference between a rigid dumbbell and a rigid particle of finite aspect ratio is that in a shear field the solvent transmits no torque to the dumbbell when it becomes aligned in the shearing plane. Particles of finite aspect ratio, in contrast, experience torque in any orientation and thus *tumble* in a shear field.

8.7.2.1 Dynamics It is thus useful to consider *rigid ellipsoids* in dilute solution as a rough model or analog of polymer coils.[59] Let $\underset{\sim}{u}$ be a unit vector in the direction of the symmetry axis of an ellipsoid; let $\underset{\sim}{R} \equiv L\underset{\sim}{u}$, where L is the length of the symmetry axis. Then in the absence of Brownian motion, the response of the

---

* An alternative is to replace $\langle \underset{\sim\sim\sim\sim}{RRRR} \rangle : \underset{\approx}{D}$ with an *approximate* expression that contains at most a second moment of ψ, for then a closed-form constitutive equation for the stress tensor can be derived. The approximation must be chosen carefully, however, or large errors may be incurred. Leal and Hinch[54] present some useful closure approximations.

particle to a velocity gradient is[60]

$$\dot{\underset{\sim}{R}} = \underset{\sim}{R}\cdot\underset{\approx}{\omega} + \left(\frac{r^2-1}{r^2+1}\right)(\underset{\sim}{R}\cdot\underset{\approx}{D} - \underset{\sim\sim\sim}{Ruu}:\underset{\approx}{D}) \tag{8-89}$$

where r is the particle aspect ratio. In the limit of an infinite aspect ratio (rodlike particle), the expression reduces to

$$\dot{\underset{\sim}{u}} = \underset{\sim}{u}\cdot\underset{\sim\sim}{\nabla v} - \underset{\sim\sim\sim}{uuu}:\underset{\approx}{D} \tag{8-90}$$

which we used earlier to describe rigid dumbbell dynamics. When r=1, Equation 8-90 becomes

$$\underset{\sim}{R} = \underset{\sim}{R}\cdot\underset{\approx}{\omega} \tag{8-91}$$

which describes the tumbling motion of a sphere in a rotational flow, such as shear. As r increases, the period of the tumbling that occurs in shear steadily increases until at $r=\infty$ an infinite time is required for the particle (here a slender rod) to rotate through an angle of 2π. These tumbling motions are called *Jeffery orbits.*

8.7.2.2 The Stress Tensor When oriented, ellipsoidal particles produce an *entropic stress,* $\underset{\approx}{\tau}^E$, that is relieved by Brownian motion. The entropic stress can be computed using Equation 8-87b. It is

$$\underset{\approx}{\tau}^E = 3ankT<\underset{\sim\sim}{uu}> \tag{8-92}$$

where $a \equiv (r^2-1)/(r^2+1)$. As with the rigid dumbbell, there is also a purely *viscous* contribution to the stress given by Hinch and Leal as[60]

$$\underset{\approx}{\tau}^V = 2\eta_s n\left\{A<\underset{\sim\sim\sim\sim}{uuuu}>:\underset{\approx}{D} + B[<\underset{\sim\sim}{uu}>\cdot\underset{\approx}{D} + \underset{\approx}{D}\cdot<\underset{\sim\sim}{uu}>] + C\underset{\approx}{D}]\right\} \tag{8-93}$$

The coefficients A, B, and C, depend on the particle aspect ratio and shape. Values for these coefficients for axisymmetric ellipsoids are presented in Hinch and Leal.[61]

8.7.2.3 Predictions Hinch and Leal[60] have computed the stress components in simple shearing flows for suspensions of ellipsoidal particles that are rodlike ($r \rightarrow \infty$), nearly spherical ($r = 1 + \epsilon$, where ϵ is small), and disclike ($r \rightarrow 0$). They find that at high shear rates the shear viscosity reaches a constant value, η_∞, given by

$$\frac{\eta_\infty}{\eta_0} = \begin{cases} \dfrac{1.181}{r}, & r \rightarrow \infty, \quad \text{rodlike particles} \\ 1 - 0.24\epsilon^2, & r = 1+\epsilon,\ \epsilon \rightarrow 0, \quad \text{near spheres} \\ 4.61\,r, & r \rightarrow 0, \quad \text{disc-like particles} \end{cases}$$

For any of the particle shapes, the first normal stress difference reaches a plateau at high shear rates; thus

$$\frac{\Psi_{1,\infty}}{\Psi_{1,0}} \propto \dot{\gamma}^{-2}$$

At low shear rates, the second normal stress difference is *negative* and small;

$$\frac{\Psi_{2,0}}{\Psi_{1,0}} = \begin{cases} -1/7\,, & r \to \infty\,, \quad \text{rodlike particles} \\ -1/7\,, & r = 1+\epsilon\,,\ \epsilon \to 0\,, \quad \text{near spheres} \\ -2/7\,, & r \to 0\,, \quad \text{disc-like particles} \end{cases}$$

The uniaxial and planar extensional viscosities for ellipsoids show modest or small amounts of strain *thickening,* except for the planar extensional viscosity of near spheres, which shows a tiny amount of strain thinning.[62]

8.7.2.4 Gordon-Schowalter Nonaffine Motion There is a similarity between Equation 8-89 and the *nonaffine* motion proposed by Gordon and Schowalter[63] (Section 5.2):

$$\dot{\underset{\sim}{R}} = \underset{\sim}{R}\cdot\underset{\approx}{\omega} + a\underset{\sim}{R}\cdot\underset{\approx}{\mathbf{D}} \qquad \text{(Gordon and Schowalter)} \tag{8-94}$$

This equation can be obtained from Equation 8-89 by dropping the term $\underset{\sim\sim\sim}{Ruu}:\underset{\approx}{\mathbf{D}}$ and by identifying a with $(r^2-1)/(r^2+1)$. The term $\underset{\sim\sim\sim}{Ruu}:\underset{\approx}{\mathbf{D}}$ keeps the major axis fixed in length, $|\underset{\sim}{R}| = L$, in any deformation history. Dropping this term therefore allows stretching of the ellipsoid. Thus *the equation of Gordon and Schowalter can be used to model molecular rigidity with a finite and variable aspect ratio.* The stress tensor should then contain not only a viscoelastic component that is proportional to $\langle\underset{\sim\sim}{RR}\rangle$ but also a viscous component. If the end-to-end vector $\underset{\sim}{R}$ follows the evolution equation

$$\dot{\underset{\sim}{R}} = \underset{\sim}{R}\cdot\underset{\sim\sim}{\nabla \mathbf{v}} - \xi\underset{\sim}{R}\cdot\underset{\approx}{\mathbf{D}} - \frac{4kT\beta^2}{\varsigma}\underset{\sim}{R} - \frac{2kT}{\varsigma}\frac{\partial \log\psi}{\partial \underset{\sim}{R}}\,, \tag{8-95}$$

and if the Giesekus expression for the stress tensor is valid, we have

$$\underset{\approx}{\tau} = \frac{-n\varsigma}{4}\overset{\nabla}{\langle\underset{\sim\sim}{RR}\rangle} = \frac{\xi n\varsigma}{4}\left(\langle\underset{\sim\sim}{RR}\rangle\cdot\underset{\approx}{\mathbf{D}} + \underset{\approx}{\mathbf{D}}\cdot\langle\underset{\sim\sim}{RR}\rangle\right) + 2nkT\beta^2\langle\underset{\sim\sim}{RR}\rangle - nkT\underset{\approx}{\delta}$$

$$\text{(viscous)} \qquad\qquad \text{(viscoelastic)} \tag{8-96}$$

The stress tensor then contains both an elastic term and a purely viscous one. In the corotational limit, $\xi = 1$, $\langle\underset{\sim\sim}{RR}\rangle$ remains equal to $\frac{1}{2}\beta^{-2}\underset{\approx}{\delta}$ in any deformation history. Thus from Equation 8-96,

$$\underset{\approx}{\tau} = \frac{n\beta^{-2}\zeta}{4}\underset{\approx}{D} \tag{8-97}$$

The polymeric contribution to the stress tensor is then a Newtonian viscous term. Thus in the corotational limit, polymer coils are completely rigid and their contribution to the stress tensor is analogous to that of suspended rigid spheres.

8.8 SUMMARY

The viscoelasticity of dilute solutions is influenced by hydrodynamic interaction, excluded volume, finite extensibility, and internal viscosity. Hydrodynamic interaction is the indirect drag that one part of the polymer chain exerts on another; it is transmitted through the solvent. Excluded volume is the effective repulsive force that one part of the polymer chain feels for another in good solvents. The excluded volume force is canceled out in theta solvents. When coils are highly stretched, their dynamics are influenced by their finite extensibility, which produces a nonlinear relationship between the elastic force and molecular extension.

There are two important aspects of internal viscosity. The first is a *resistance to chain stretching*, proposed by Kuhn. As this resistance increases, a flexible dumbbell becomes a rigid one. The second element of internal viscosity is an apparent *finite aspect ratio* evidenced by the ability of the solvent to transmit torque to the coil in a shear flow even if the major axis of the coil is perfectly aligned with the flow. Thus the coil can pass through the shearing plane in a tumbling motion. This has been modeled by the nonaffine motion proposed by Gordon and Schowalter.

The Rouse predictions for the linear viscoelasticity of polymer coils in dilute solution are modified by consideration of hydrodynamic interaction, excluded volume, and internal viscosity. Hydrodynamic interaction changes the scaling of the zero-shear intrinsic viscosity with molecular weight from the Rouse prediction of M^1 to $M^{1/2}$; consideration of both hydrodynamic interaction and excluded volume changes it to $M^{4/5}$. These scalings are seen experimentally in theta and very good solvents, respectively. Hydrodynamic interaction shortens the longest Rouse relaxation times and thereby compresses the relaxation spectrum. Excluded volume has the reverse effect. Internal viscosity causes the high frequency dynamic viscosity to approach a constant that is greater than the solvent contribution.

In the nonlinear regime, hydrodynamic interaction, excluded volume, finite extensibility, and internal viscosity can all produce shear thinning. More general expressions for the stress tensor must be used, and the stress-optic law is predicted to fail at high shear rates when excluded volume, finite extensibility, or internal viscosity are significant. Experimental data and multiple beads-and-springs and dumbbell theories suggest that excluded volume and internal viscosity contribute significantly to shear thinning at modest shear rates in low-viscosity solvents. Shear thinning is minimized in viscous theta solvents when the dilute polymer molecule is high in molecular weight.

In extensional flows, the polymer molecule becomes highly extended at a critical extension rate. The transition from the coiled to the stretched state is apparently sharpened by the diminution of hydrodynamic interaction and consequent increase in drag forces as the molecule is stretched. It is possible that long-lived metastable states of stress exist in extensional flows.

References

1. P.J. Flory, *Principles of Polymer Chemistry*, Cornell University Press, Ithaca, New York (1953).
2. P. G. de Gennes, *Scaling Concepts in Polymer Physics*, Cornell University Press, Ithaca, New York (1979).
3. J.D. Ferry, *Viscoelastic Properties of Polymers*, 3rd ed., John Wiley & Sons, New York (1980), p. 117.
4. R.M. Johnson, J.L. Schrag, and J.D. Ferry, Polym. J. *1*, 742 (1970).
5. P. G. de Gennes, *Scaling Concepts in Polymer Physics*, Cornell University Press, Ithaca, New York (1979), p. 183.
6. J. G. Kirkwood and J. Riseman, J.Chem. Phys., *16*, 565 (1948).
7. A. Einstein, Ann. d. Physik, *17*, 549 (1905); *19* 371 (1906).
8. P.J. Flory, *Statistical Mechanics of Chain Molecules*, John Wiley & Sons, New York (1969), p. 40.
9. J.D. Ferry, *Viscoelastic Properties of Polymers*, 3rd ed., John Wiley & Sons, New York (1980), p. 113.
10. B.H. Zimm, J. Chem. Phys., *24*, 269 (1956).
11. R.B. Bird, O. Hassager, R.C. Armstrong, and C.F. Curtiss, *Dynamics of Polymeric Liquids*, Vol. 2, John Wiley & Sons, New York (1977).
12. C.W. Oseen, Arf. f. Mat. Astr. og Fys., *6*, No. 29, 1 (1910).
13. J. Rotne and S. Prager, J. Chem. Phys. *50*, 4831 (1969).
14. J.D. Ferry, *Viscoelastic Properties of Polymers*, 3rd ed., John Wiley & Sons, New York (1980, Chapter 9.
15. A.S. Lodge and Y. Wu, University of Wisconsin Rheology Research Center Report No. 19 (1972).
16. C.J. Landry, Ph.D. thesis, Univ. of Wisc. (1985).
17. D.J. Massa, J.L. Schrag, and J.D. Ferry, Macromolecules, *4*, 210 (1971).
18. V. Man, Ph.D. Thesis, Univ. of Wisc. (1984).
19. I. Noda, Y.Yamada, and M. Nagasawa, J. Phy. Chem., *72*, 2890 (1968).
20. H.M. Laun, Rheol. Acta, *17*, 1 (1978).
21. J.D. Ferry, *Viscoelastic Properties of Polymers*, 3rd ed., John Wiley & Sons, New York (1980), p. 382.
22. D.V. Boger and R. Binnington, Trans. Soc. Rheol., *21*, 515 (1977).

23. U. Bianchi and A. Peterlin, J. Polym. Sci. A-2, *6*, 1011 (1968).

24. M.E. Mackay and D.V. Boger, J. Non-Newtonian Fluid Mech., *22*, 235 (1987).

25. T. Sridhar, R.K. Gupta, D.V. Boger, and R. Binnington, J. Non-Newtonian Fluid Mech., *21*, 115 (1986).

26. M.-N. Layec-Raphalen and C. Wolff, J. Non-Newtonian Fluid Mech., *1*, 159 (1976).

27. A. Peterlin, J. Chem. Phys., *33*, 1799 (1960).

28. M. Fixman, J. Chem. Phys., *45*, 785; 793 (1966).

29. H.C. Öttinger, J. Chem. Phys., *83*, 6535 (1985); *84*, 4068 (1986).

30. H.C. Öttinger, J. Chem. Phys., *86*, 3731 (1987).

31. J. Magda, R.G. Larson, and M.E. Mackay, work in progress.

32. W. Kuhn and H. Kuhn, Helv. Chim. Acta, *28*, 1533 (1945).

33. (a) C.W. Manke and M.C. Williams, Macromolecules, *18*, 2045 (1985); (b) J. Non-Newtonian Fluid Mech., *19*, 43 (1985).

34. F.R. Cottrell, E.W. Merrill, and K.A. Smith, J. Polym. Sci. A-2, *7*, 1415 (1969).

35. S.P. Burow, A. Peterlin, and D.T. Turner, J. Polym. Sci. B, *2*, 67 (1964).

36. H. Giesekus, Rheol. Acta, *2*, 50 (1962); *5*, 29 (1966).

37. M. Keentok, A.G. Georgescu, A.A. Sherwood, and R.I. Tanner, J. Non-Newtonian Fluid Mech., *6*, 303 (1980).

38. P. Pincus, Macromolecules, *9*, 386 (1976).

39. H.R. Warner, Jr., I&EC Fund., *11*, 379 (1972).

40. A. Peterlin, Makromol. Chem. *44*, 338 (1961).

41. R.I. Tanner, *Engineering Rheology*, Oxford, New York (1985).

42. R. Cerf, J. Polym. Sci., *23*, 125 (1957); J. Phys. Radium, *19*, 122 (1958); J. Chim. Phys., *66*, 479 (1969).

43. H.C. Booij and P.H. van Wiechen, J. Chem. Phys., *52*, 5056 (1970).

44. G.G. Fuller and L.G. Leal, F. Non-Newtonian Fluid Mech., *8*, 271 (1981).

45. G.G. Fuller and L.G. Leal, Rheol. Acta, *19*, 580 (1980).

46. A. Keller and J.A. Odell, Colloid Polym. Sci., *263, 181 (1985).*

47. *G.K. Batchelor, J. Fluid Mech., 44*, 419 (1970).

48. P.R. Williams and R.W. Williams, J. Non-Newtonian Fluid Mech., *19*, 53 (1985).

49. P.G. de Gennes and R. Brochard, Macromolecules, *10*, 1157 (1977).

50. E.J. Hinch, Proc. Symp. Polym. Lubrification, Brest (1974).

51. P.G. de Gennes, J. Chem. Phys., *60*, 5030 (1974).

52. Y. Rabin, F.S. Henyey, and R.K. Pathria, Phys. Rev. Lett., *55*, 201 (1985).

53. X. Fan, R.B. Bird, and M. Renardy, J. Non-Newtonian Fluid Mech., *18*, 255 (1985).

54. E.J. Hinch and L.G. Leal, J. Fluid Mech., *76*, 187 (1976).

55. R.B. Bird, H.R. Warner, Jr., and D.C. Evans, Fortschr. Hochpolym. Forsch., *8*, 1 (1971).

56. N. Nemoto, J.L. Schrag, J.D. Ferry, and R.W. Fulton, Biopolymers, *14*, 407 (1975).

57. J.T. Yang, J. Am. Chem. Soc., *80*, 1783 (1958).

58. R.B. Bird and C.F. Curtiss, J. Non-Newtonian Fluid Mech., *14*, 85 (1984).

59. Y. Chikahisa, Kōbunshi, *15*, 228 (1966).

60. E.J. Hinch and L.G. Leal, J. Fluid Mech., *57*, 753 (1973).

61. E.J. Hinch, and L.G. Leal, *52*, 683, (1972).

62. C.J. Goh, N. Phan Thien, and J.D. Atkinson, J. Rheol., *30S*, S75 (1986); see, however, H. Brenner, J. Rheol., *31*, 423 (1987).

63. R.J. Gordon and W.R. Schowalter, Trans. Soc. Rheol., *16*, 79 (1972).

CHAPTER 9

Constitutive Equations for Special Flows

"Everything flows"

Heraclitus (c. 500 B.C.)

9.1 INTRODUCTION

As was emphasized in the preceding chapters, *the type of flow* with which one is concerned *influences the choice of constitutive equation.* The more restricted the flow type is, the less latitude there is for variation between different constitutive equations. For certain restricted flows, there are *universal constitutive relationships,* involving a few material parameters or functions, that relate the stress tensor to the deformation history.

9.2 FLOWS OF CONSTANT STRETCH HISTORY

9.2.1 Viscometric Flows

Among the most important of these special flows are flows that *from the point of view of a material element* have steady simple shearing histories. These are called *viscometric* flows.[1-3] More precisely, for a viscometric flow there is a frame that is possibly rotating and translating (with respect to the laboratory frame) in which the flow is a steady simple shear; i.e., in that frame the velocity gradient is

$$\underset{\sim}{\nabla}\underset{\sim}{\mathbf{v}} = \dot{\gamma}\begin{pmatrix} 0 & 0 & 0 \\ 1 & 0 & 0 \\ 0 & 0 & 0 \end{pmatrix} \tag{9-1}$$

where $\dot{\gamma}$ is a *constant* shear rate for each particle. Although $\dot{\gamma}$ may vary from particle to particle in a viscometric flow, *the deformation history of each fluid particle is constant.*

Viscometric flows include several different flows, different in the sense that fluid particles follow *pathlines of different geometry.* Figure 9-1, for example, illustrates that steady shearing can be produced both in a shear sandwich and in a cone and plate geometry. In the former geometry, pathlines are straight; in the latter, they are circular. Another viscometric flow occurs in the annulus between a stationary inner cylinder and a rotating outer cylinder; this is a *Couette flow.* These viscometric flows are easy to envision. Others are complex and for some there is no single global frame in which all particles are in steady motion.[4] In all

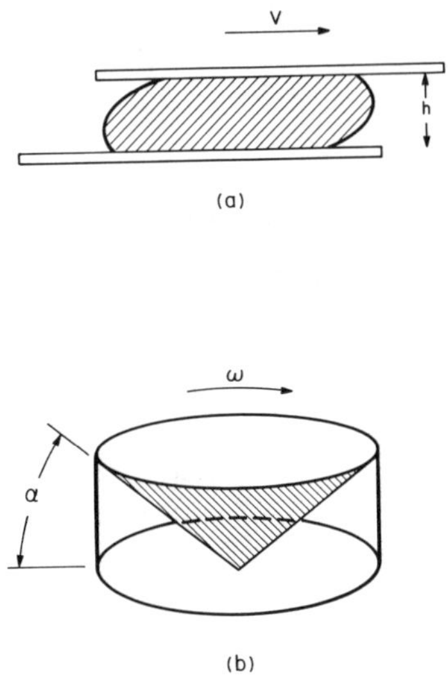

Fig. 9-1 Shearing in (a) the sandwich rheometer and (b) the cone and plate rheometer.

viscometric flows, however, there are ***shearing surfaces*** that are everywhere perpendicular to the velocity gradient. These are material surfaces that slip past each other; they may bend, but they do not stretch. Though they are not necessarily parallel to each other, the distance between any two material surfaces remains constant in time. Each shear surface is like a card in a possibly warped or splayed deck that is uniformly fanned or spread.

Since in all viscometric flows fluid particles see a steady shearing history, the stress tensor in a suitably rotating frame has the same components for any viscometric flow. In this frame, there are four components of the stress tensor, namely the shear stress, σ, plus the three normal stresses. For an incompressible fluid only the differences in the normal stresses can be measured; these differences are N_1, the first normal stress difference, and N_2, the second normal stress difference. At steady-state, σ, N_1, and N_2 are functions of the shear rate; they are the *three viscometric functions.* Because there are many different viscometric flows, several of which can easily be realized experimentally, the viscometric functions, especially $\sigma(\dot{\gamma})$ and $N_1(\dot{\gamma})$, are among the most often measured rheological functions.

9.2.1.1 Criminale-Ericksen-Filbey equation By rotating the frame of reference, one can obtain from the viscometric functions the stress tensor in the laboratory frame. In any frame, it can be shown that the stress tensor for viscometric flow is[5,6]

$$\underset{\approx}{\tau} = 2\eta \underset{\approx}{\mathbf{D}} - \Psi_1 \overset{0}{\underset{\approx}{\mathbf{D}}} + (2\Psi_1 + 4\Psi_2)\underset{\approx}{\mathbf{D}}^2 \tag{9-2}$$

where $\underset{\approx}{\mathbf{D}}$ is the symmetric part of the velocity gradient $\underset{\sim}{\nabla}\underset{\sim}{\mathbf{v}}$ and $\overset{0}{\underset{\approx}{\mathbf{D}}}$ is the corotational time derivative of $\underset{\approx}{\mathbf{D}}$. (See Section 3.4.4). $\eta = \sigma/\dot{\gamma}$ is the shear viscosity, $\Psi_1 \equiv N_1/\dot{\gamma}^2$ is the primary normal stress coefficient, and $\Psi_2 \equiv N_2/\dot{\gamma}^2$ is the secondary normal stress coefficient. These are functions of the second invariant of the rate-of-strain tensor, $\underset{\approx}{\mathbf{D}}$:

$$\mathrm{II_D} \equiv 2\underset{\approx}{\mathbf{D}} : \underset{\approx}{\mathbf{D}} \tag{9-3}$$

In a frame in which the flow field is a simple shear, $\mathrm{II_D} = \dot{\gamma}$, the shear rate. Equation 9-2 is called the *Criminale-Ericksen-Filbey (CEF) Equation.*

The CEF equation can be formulated with the upper-convected derivative, $\overset{\nabla}{\underset{\approx}{\mathbf{D}}}$, instead of the corotational derivative. Since

$$\overset{\nabla}{\underset{\approx}{\mathbf{D}}} = \overset{0}{\underset{\approx}{\mathbf{D}}} - 2\underset{\approx}{\mathbf{D}}^2 \tag{9-4}$$

Equation 9-2 can be rewritten as

$$\underset{\approx}{\tau} = 2\eta\underset{\approx}{\mathbf{D}} - \Psi_1 \overset{\nabla}{\underset{\approx}{\mathbf{D}}} + 4\Psi_2 \underset{\approx}{\mathbf{D}}^2 \tag{9-5}$$

If we assume that the stress in a fluid particle depends only on the deformation history of that particle, then Equation 9-2 applies to any fluid particle that sees a steady shear history, *even if the flow as a whole is not viscometric.*

9.2.1.2 Power-Law and Carreau Viscosities If one is primarily interested in the shearing component of the stress tensor, perhaps because the normal stresses do not affect the pressure gradient, then Equation 9-2 can be simplified to

$$\underset{\approx}{\tau} = 2\,\eta(\mathrm{II_D})\underset{\approx}{\mathbf{D}} \tag{9-6}$$

Equation 9-6 has the tensorial character of a Newtonian fluid, but can exhibit nonlinear viscous effects such as shear thinning. Since in shear it does not show any normal stress differences, which are the usual manifestations of elasticity, Equation 9-6 is regarded as a purely viscous constitutive equation. In a cylindrical† tube or channel of constant cross-sectional geometry, normal stresses do not affect the pressure gradient; thus Equation 9-6 can be used in the momentum balance equation to compute the pressure drop and viscous dissipation, if an accurate viscosity function, $\eta(\mathrm{II_D})$ is used. Equation 9-6 is also often used in complex polymeric flows for which it is not reliable, because realistic constitutive equations almost invariably lead to difficult or intractable numerical problems.

† If the tube cross section is not cylindrical, then a nonzero second normal stress difference can produce a secondary flow that is transverse to the tube axis[7]. Transverse flow is not predicted by Equation 9-6.

A popular form for $\eta(\mathrm{II_D})$ is the *power-law of Ostwald*[8] *and de Waele,*[9]

$$\eta = m\, \mathrm{II}_D^{(n-1)/2} \tag{9-7}$$

Here "n" is the power-law index; in shear $\sigma \sim \dot{\gamma}^n$. A second popular form is the *Carreau function,*[10]

$$\frac{\eta - \eta_\infty}{\eta_0 - \eta_\infty} = [1 + \lambda^2 \mathrm{II_D}]^{(n-1)/2} \tag{9-8}$$

which has Newtonian behavior at low *and* high shear rates, and power-law behavior in between.

9.2.2 Steady Extensional Flows

Consider a flow that is a steady-state extension from the viewpoint of a fluid element. In an extensional flow, each particle is stretched or compressed exponentially in time in each of the three principal directions. Because of incompressibility only two of these rates of stretch, $\dot{\epsilon}_1$ and $\dot{\epsilon}_2$, are independent. The general frame-invariant constitutive equation for such flows is

$$\underset{\approx}{\tau} = 2\eta_1(\mathrm{II_D}, \mathrm{III_D})\, \underset{\approx}{\mathbf{D}} + 2\eta_2(\mathrm{II_D}, \mathrm{III_D})\, \underset{\approx}{\mathbf{D}}^2 \tag{9-9}$$

The viscosity coefficients η_1 and η_2 depend on two invariants, $\mathrm{II_D}$, defined in Equation 9-3, and

$$\mathrm{III_D} \equiv \det \underset{\approx}{\mathbf{D}} \tag{9-10}$$

This is the *Reiner-Rivlin* equation.[11,12] In addition to being the general constitutive equation for steady extensional flows, the Reiner-Rivlin equation is the most general frame-invariant constitutive equation possible for isotropic incompressible fluids in which the stress depends only on the instantaneous velocity gradient, and not on any time derivatives of the velocity gradient.

9.2.3 General Flows of Constant Stretch History

Viscometric flows and steady extensional flows are special cases of flows that are steady in some rotating frame. These are called flows of ***constant stretch history*** (csh).[13–15] If a special frame —a frame that could be rotating with respect to the laboratory frame— can be found in which a fluid particle sees a constant velocity gradient, then that particle is said to experience a *motion* with constant stretch history. If all particles in the flow undergo motions of constant stretch history, then the flow is a csh flow.

In the laboratory frame, the motion may not be steady. But the particle is in a csh motion if there is a time-dependent rotation of frame, $\underset{\approx}{\mathbf{Q}}(t)$, that transforms a constant velocity gradient tensor, $\underset{\sim}{\nabla}\underset{\sim}{\mathbf{v}}^*$, into the laboratory velocity gradient, $\underset{\sim}{\nabla}\underset{\sim}{\mathbf{v}}$. From Equation 3-16, the transformation law for the velocity gradient tensor is

$$\underset{\sim}{\nabla}\underset{\sim}{\mathbf{v}} = \underset{\approx}{\mathbf{Q}}^{\mathrm{T}} \cdot \underset{\sim}{\nabla}\underset{\sim}{\mathbf{v}}^{*} \cdot \underset{\approx}{\mathbf{Q}} + \underset{\approx}{\mathbf{Q}}^{\mathrm{T}} \cdot \dot{\underset{\approx}{\mathbf{Q}}} \tag{9-11}$$

The dot above $\underset{\approx}{\mathbf{Q}}$ on the right side of Equation 9-11 denotes the *substantial* time derivative, $\partial/\partial t + \underset{\sim}{\mathbf{v}} \cdot \underset{\sim}{\nabla}$. In the special frame, the inverse of the deformation gradient is

$$(\underset{\approx}{\mathbf{F}}^{-1})^{*} = \exp[\underset{\sim}{\nabla}\underset{\sim}{\mathbf{v}}^{*}(t-t')] \tag{9-12}$$

In the laboratory frame,

$$\underset{\approx}{\mathbf{F}}^{-1} = \underset{\approx}{\mathbf{Q}}^{\mathrm{T}}(t') \cdot (\underset{\approx}{\mathbf{F}}^{-1})^{*} \cdot \underset{\approx}{\mathbf{Q}}(t) \tag{9-13}$$

Viscometric and steady extensional flows are special cases of csh flows for which

$$(\underset{\sim}{\nabla}\underset{\sim}{\mathbf{v}}^{*})^{2} = 0 \qquad \text{viscometric flows}$$

$$(\underset{\sim}{\nabla}\underset{\sim}{\mathbf{v}}^{*})^{\mathrm{T}} = \underset{\sim}{\nabla}\underset{\sim}{\mathbf{v}}^{*} \qquad \text{steady extensional flows}$$

The definition of csh flows is not in itself a useful criterion for determining whether a given velocity field or deformation history is of constant stretch history. To use the definition directly to determine whether a flow is of csh would require searching all time-dependent rotations, $\underset{\approx}{\mathbf{Q}}(t)$, to see if one of them renders the velocity gradient constant. Following Wang,[16] Huilgol[17] has established practical, though nontrivial, necessary and sufficient conditions for the velocity field to yield a motion of constant stretch history.

If it is known that the motion of a fluid particle is of constant stretch history, then the stress tensor in that particle is a function of a few frame-invariant kinematic tensors. We have seen that if the motion is a steady extension, then $\underset{\approx}{\mathbf{D}}$ (and powers of $\underset{\approx}{\mathbf{D}}$) determine the stress, Equation 9-9. If the flow is viscometric, both $\underset{\approx}{\mathbf{D}}$ and $\overset{0}{\underset{\approx}{\mathbf{D}}}$ are required. For other csh flows, at most $\underset{\approx}{\mathbf{D}}$, $\overset{0}{\underset{\approx}{\mathbf{D}}}$, and a third tensor, the corotational derivative of $\overset{0}{\underset{\approx}{\mathbf{D}}}$, are required.[16]

9.2.4 Planar Flows of Constant Stretch History

In three dimensions, the general constitutive equation for csh flows contains so many products of the three kinematic tensors that it is of little use. For *planar* csh flows, however, the general constitutive equation simplifies to[18]

$$\underset{\approx}{\tau} \equiv 2\eta(\mathrm{II_D}, \mathrm{S_f})\, \underset{\approx}{\mathbf{D}} - \Psi_1(\mathrm{II_D}, \mathrm{S_f})\, \overset{0}{\underset{\approx}{\mathbf{D}}} + [2\Psi_1(\mathrm{II_D}, \mathrm{S_f}) + 4\Psi_2(\mathrm{II_D}, \mathrm{S_f})]\, \underset{\approx}{\mathbf{D}}^{2} \tag{9-14}$$

This equation has the same *tensors* as the CEF equation for viscometric flows.

The term involving $\underset{\approx}{\mathbf{D}}^2$, however, does not couple to the planar momentum balance and so does not affect the flow. Equation 9-14 generalizes the CEF equation in that the coefficients depend not only on II_D but also on the *flow strength coefficient,*

$$S_f \equiv 2(\mathrm{tr}\underset{\approx}{\mathbf{D}}^2)^2/\overset{0}{\underset{\approx}{\mathbf{D}}}:\overset{0}{\underset{\approx}{\mathbf{D}}} \tag{9-15}$$

For strong planar csh flows, $S_f > 1$, and for weak planar csh flows, $S_f \leq 1$. In a strong flow, as defined in 1975 by Tanner and Huilgol,† at least one eigenvalue of $\underset{\sim}{\nabla}\underset{\sim}{\mathbf{v}}$ has a positive real part.[20] This means that some material lines are stretched exponentially in time. In a *strictly* weak flow, $S_f < 1$, all material lines oscillate sinusoidally. The marginal case is shear flow, $S_f = 1$; in it material lines grow monotonically but not exponentially in time. Almost all three-dimensional flows are, by this definition, strong.

The general constitutive relationship for *axisymmetric* csh flows is much more complex than Equation 9-14. Apparently frame invariance is a much less powerful simplifying principal in axisymmetric flows than it is in planar flows. For similar reasons, axisymmetric strains are more complicated than planar strains; incompressible axisymmetric strains are described by two invariants, I_1 and I_2, whereas incompressible planar strains have only one invariant, $I_1 = I_2$.

9.3 RETARDED MOTION EXPANSION

A flow that is not steady in any frame will still yield a special constitutive relationship if it is *slow* and *slowly varying.*

Consider, for example, the upper-convected Maxwell (UCM) equation

$$\lambda\overset{\nabla}{\underset{\approx}{\tau}} + \underset{\approx}{\tau} = 2G\lambda\underset{\approx}{\mathbf{D}} \tag{9-16}$$

with

$$\overset{\nabla}{\underset{\approx}{\tau}} \equiv \dot{\underset{\approx}{\tau}} - \underset{\sim}{\nabla}\underset{\sim}{\mathbf{v}}^{\mathrm{T}}\cdot\underset{\approx}{\tau} - \underset{\approx}{\tau}\cdot\underset{\sim}{\nabla}\underset{\sim}{\mathbf{v}} \tag{9-17}$$

The term $\lambda\overset{\nabla}{\underset{\approx}{\tau}}$ is of *higher order* than $\underset{\approx}{\tau}$ in that it becomes small relative to $\underset{\approx}{\tau}$ if either the relaxation time is small or the velocity gradient and stresses are small and change slowly in time. For then both $\dot{\underset{\approx}{\tau}}$ and the quadratic term, $\underset{\sim}{\nabla}\underset{\sim}{\mathbf{v}}^{\mathrm{T}}\cdot\underset{\approx}{\tau} + \underset{\approx}{\tau}\cdot\underset{\sim}{\nabla}\underset{\sim}{\mathbf{v}}$ can be neglected compared to the linear terms $\underset{\approx}{\tau}$ and $\underset{\approx}{\mathbf{D}}$. Neglecting $\lambda\overset{\nabla}{\underset{\approx}{\tau}}$, we get at first order

† In 1976 Tanner presented a somewhat different definition of strong and weak flows, one that is appropriate for the elastic dumbbell model.[19] Here we retain the original definition because it is independent of the molecular model.

$$\underset{\approx}{\tau}_{(1)} = 2G\lambda \underset{\approx}{\mathbf{D}} \tag{9-18}$$

the expected Newtonian result.† A *perturbation from the Newtonian result* is obtained if the lowest order approximation to the stress tensor, Equation 9-18, is used to approximate the "small" higher-order term $\lambda \overset{\nabla}{\underset{\approx}{\tau}}$. Thus $\overset{\nabla}{\underset{\approx}{\tau}}$ is approximated by the upper-convected derivative of $\underset{\approx}{\tau}_{(1)}$:

$$\overset{\nabla}{\underset{\approx}{\tau}} \approx \overset{\nabla}{\left(\underset{\approx}{\tau}_{(1)}\right)} = 2G\lambda \overset{\nabla}{\underset{\approx}{\mathbf{D}}} \tag{9-19}$$

Then the stress tensor can be obtained[21] to second order in λ by substituting Equation 9-19 into the constitutive equation, Equation 9-16,

$$\begin{aligned} \underset{\approx}{\tau}_{(2)} &= 2G\lambda \underset{\approx}{\mathbf{D}} - 2G\lambda^2 \overset{\nabla}{\underset{\approx}{\mathbf{D}}} \\ &= 2G\lambda \underset{\approx}{\mathbf{D}} - 2G\lambda^2 \overset{0}{\underset{\approx}{\mathbf{D}}} + 4G\lambda^2 \underset{\approx}{\mathbf{D}}^2 \end{aligned} \tag{9-20}$$

where we have used Equation 9-4 to relate the corotational derivative $\overset{0}{\underset{\approx}{\mathbf{D}}}$ to the upper-convected derivative $\overset{\nabla}{\underset{\approx}{\mathbf{D}}}$.

If other constitutive equations (with some exceptions discussed shortly) are expanded to second order, the same tensors appear with different coefficients. Thus the general constitutive equation of the *second-order fluid* (SOF) is

$$\underset{\approx}{\tau}_{(2)} = 2\eta \underset{\approx}{\mathbf{D}} - \Psi_1 \overset{0}{\underset{\approx}{\mathbf{D}}} + [2\Psi_1 + 4\Psi_2]\, \underset{\approx}{\mathbf{D}}^2 \tag{9-21}$$

The SOF equation is the same as the CEF equation, Equation 9-2, except that the SOF equation has constant coefficients. The CEF equation applies to any steady *shearing* flow; the SOF equation applies to *any* type of flow that is slow and slowly varying. The SOF equation implies that the coefficients of the CEF equation become constant at low shear rates.

Note that the coefficients of the equation of the second order fluid are combinations of the viscometric coefficients. Thus using only the stresses measured in a slow steady shear flow, *the stresses in any other steady flow, including extension, can be predicted up to second order in the velocity field.* The various extensional viscosities for the extensional flows defined in Section 6.2.3 are given by

† Perturbation expansions in which the small term is also the term with the highest order derivative, as is the case here, are known as *singular perturbations*. Singular perturbation expansions are frequently not uniformly convergent. Thus many flow problems have regions where the retarded motion expansion fails, even at very slow bulk flow rates.

$$\overline{\eta}_u = 3\eta\dot{\epsilon} + \frac{3}{2}\Psi_1(1 + 2\frac{\Psi_2}{\Psi_1})\dot{\epsilon}^2 \tag{9-22a}$$

$$\overline{\eta}_b = 6\eta\dot{\epsilon} - 6\Psi_1(1 + 2\frac{\Psi_2}{\Psi_1})\dot{\epsilon}^2 \tag{9-22b}$$

$$\overline{\eta}_{p,1} = 4\eta\dot{\epsilon} + 0\,\dot{\epsilon}^2 \tag{9-22c}$$

$$\overline{\eta}_{p,2} = 2\eta\dot{\epsilon} - 2\Psi_1(1 + 2\frac{\Psi_2}{\Psi_1})\dot{\epsilon}^2 \tag{9-22d}$$

In each case, the sign of the quadratic term is determined by $1 + 2\Psi_2/\Psi_1$. The ratio Ψ_2/Ψ_1 has been found experimentally to be negative and smaller in magnitude than 1/2.[22–24] Thus $1 + 2\Psi_2/\Psi_1$ is positive. This implies that as $\dot{\epsilon}$ is increased from zero, the first deviation from constancy in the uniaxial viscosity, $\overline{\eta}_u$, is an *increase* with $\dot{\epsilon}$. The biaxial viscosity, $\overline{\eta}_b$, should *decrease*, at least initially, as $\dot{\epsilon}$ is increased from zero. These predictions seem to be confirmed experimentally.[25]

9.3.1 Conditions for the Validity of the Retarded Motion Expansion

The above expansion, Equation 9-21, can be carried out to third order and higher. This expansion in powers of the relaxation time is called the *retarded motion expansion*. It was proposed by Rivlin and Ericksen.[26] Its convergence requires the following conditions:

1. The stress tensor must be a smooth frame-invariant function of the history of the velocity gradient.

2. The influence of velocity gradients in the remote past must die out sufficiently fast, for example exponentially in time.

3. The velocity field must be smooth.

"Smoothness" means differentiability. The higher the order to which the retarded motion expansion is carried, the higher the order of differentiability required of the constitutive equation and of the velocity field. Thus any lack of differentiability in the constitutive equation or the velocity field limits the order to which the retarded motion expansion can be carried. Even if the velocity field and the constitutive equation are smooth, we shall see that the retarded motion expansion may still fail to converge if the velocity gradient is high or rapidly varying in space or time.

9.3.1.1 Smoothness of the Constitutive Equation Most of the constitutive equations presented in this book contain only analytic functions of the deformation or stress and thus can be expanded to any order. The UCM equation, for example, can be expanded to all orders. By induction, the nth-order stress is found to be

$$\underset{\approx}{\tau}_{(n)} = G\sum_{i=1}^{n}(-1)^{i-1}\lambda^i \underset{\approx}{\mathbf{A}}_{(i)} \tag{9-23}$$

where $\underset{\approx}{\mathbf{A}}_{(i)}$ is the ith-order *Rivlin-Ericksen tensor.* $2\underset{\approx}{\mathbf{D}}$ is the first-order Rivlin-Ericksen tensor. Each subsequent Rivlin-Ericksen tensor is the upper-convected derivative of the tensor of next lower order; i.e., $\underset{\approx}{\mathbf{A}}_{(i)} \equiv \overset{\nabla}{\underset{\approx}{\mathbf{A}}}_{(i-1)}$. One hopes that as the order n increases, the expansion will converge to the exact result.

Consider, for example, a steady elongational flow; the upper-convected derivatives in the expansion are

$$\underset{\approx}{\mathbf{A}}_{(i)} = (-1)^{i-1}\begin{pmatrix} (2\dot{\epsilon})^i & 0 & 0 \\ 0 & (-\dot{\epsilon})^i & 0 \\ 0 & 0 & (-\dot{\epsilon})^i \end{pmatrix} \tag{9-24}$$

At infinite order, the normal stress difference from Equations 9-23 and 9-24 is

$$\begin{aligned} \tau_{(\infty),11} - \tau_{(\infty),22} &= G\sum_{i}^{\infty}[(2\lambda\dot{\epsilon})^i - (-\lambda\dot{\epsilon})^i] \\ &= G\left(\frac{1}{1-2\lambda\dot{\epsilon}} - \frac{1}{1+\lambda\dot{\epsilon}}\right) \end{aligned} \tag{9-25}$$

which is the answer obtained by solving the original UCM constitutive equation; see Section 1.6.3.2. Thus because the UCM equation is analytic (i.e., differentiable to all orders) and the velocity field, namely steady elongation, is analytic, the series exists to all orders and converges to the correct result. Figure 9-2 depicts the rate of convergence of the retarded motion expansion. The convergence of the series slows as the strain rate increases, and at $\dot{\epsilon}\lambda = 1/2$ the series *diverges.* This occurs because of the singularity or *pole* present in the UCM equation for elongation flow at this strain rate. Since the retarded motion expansion is an expansion about zero strain rate or zero relaxation time, its radius of convergence is limited by any occurrence of a stress singularity.

Each order in the retarded motion expansion, Equation 9-24, is a power of the product, $\dot{\epsilon}\lambda$, of a characteristic strain rate and a characteristic relaxation time. This dimensionless product is called the *Deborah number,* De. The retarded motion expansion is really a perturbation expansion in the Deborah number.

If the modulus or relaxation time of the UCM equation is made a nonanalytic function of $\underset{\approx}{\mathbf{D}}$ or $\underset{\approx}{\tau}$, as in Ide and White's version of the White-Metzner equation, Equations 5-58—5-59, then the retarded motion expansion fails at some order. Thus for these nonanalytic constitutive equations, a convergent expansion doesn't exist, no matter how small the Deborah number is.

The separable Kaye-BKZ equation

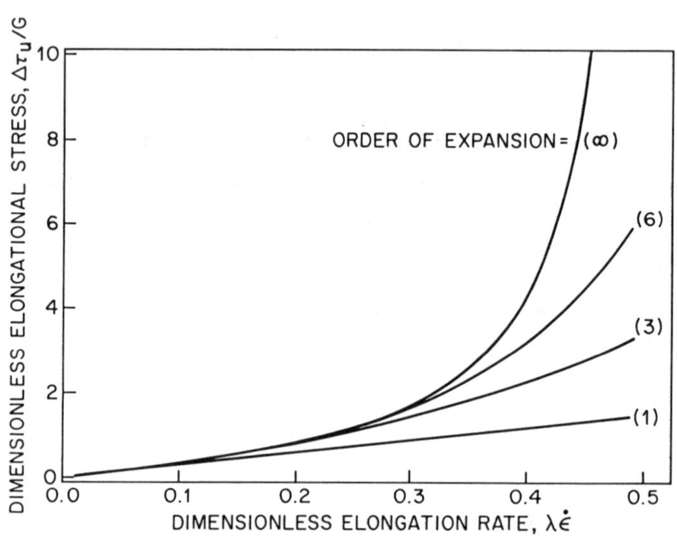

Fig. 9-2 Convergence of the retarded motion expansion of the upper-convected Maxwell equation in steady uniaxial extensional flow. The order of the expansion is in parentheses.

$$\underset{\approx}{\tau} = \int_{-\infty}^{t} m(t-t')\left(2\frac{\partial U}{\partial I_1}\underset{\approx}{C}^{-1} - 2\frac{\partial U}{\partial I_2}\underset{\approx}{C}\right) dt'$$

can also be expanded to all orders in Rivlin-Ericksen tensors if the potential function $U(I_1,I_2)$ is an analytic function of I_1 and I_2, and if the memory function $m(t-t')$ is a sum of decaying exponentials. Because the damping function of Wagner's equation,

$$\underset{\approx}{\tau} \equiv \int_{-\infty}^{t} m(t-t')\, h(I_1,I_2)\left(\underset{\approx}{C}^{-1} - \underset{\approx}{\delta}\right) dt' \tag{9-26}$$

$$h = \sum_i f_i \exp(-n\sqrt{I-3}); \qquad I \equiv \beta I_1 + (1-\beta) I_2 \tag{9-27}$$

has a nonanalytic square root, it fails to satisfy the retarded motion expansion[27] at the level of the second-order fluid. At small strain rates it does reduce to first-order Newtonian behavior, however, if m(s) decays exponentially at large s.

If the damping function is an analytic function of I_1 and I_2 but is made irreversible in the nonanalytic way proposed by Wagner (Section 5.5.1), then again the retarded motion expansion fails at some order.

9.3.1.2 Smoothness of the Velocity Field In addition to the requirement that the constitutive equation be a smooth functional of the velocity gradient, the validity of the retarded motion expansion also depends on the smoothness of the velocity gradient. Thus a boundary discontinuity, such as a sharp corner, invalidates the expansion, at least near the discontinuity.[28] Again, the higher the order of the nondifferentiability of the boundary shape, the higher the order at which the retarded motion expansion is expected to break down. Even if the boundary shape is analytic, nonanalytic stresses could appear because of stagnation points or shock surfaces where the flow problem *changes type* from elliptic to hyperbolic.[29] In these cases, too, the retarded motion expansion is expected to fail.

Because the stress is given explicitly in terms of derivatives of the *present* velocity field, the retarded motion expansion cannot predict stress relaxation after cessation of flow, no matter how many terms are included in the expansion. The stress is zero as soon as the velocities and their gradients vanish. Sudden cessation of flow is not a smooth history; hence the retarded motion expansion cannot describe it. Whereas in the integral constitutive equations the memory is stored in an integral over strain history, and in the implicit differential equations it is stored in the time derivative of the stress tensor, in the retarded motion expansion memory is stored in the *derivatives of the velocity field*. That memory is lost when discontinuities in space or time enter the velocity field.

9.3.1.3 Slow Flows If strain rates are high, the retarded motion expansion may either converge or diverge, depending on the constitutive equation and the type of flow. To illustrate, consider the corotational Maxwell equation (CRM) in steady elongational and shear flows. The CRM equation is

$$\lambda \overset{0}{\underset{\approx}{\tau}} + \underset{\approx}{\tau} = 2G\lambda \underset{\approx}{\mathbf{D}} \tag{9-28}$$

with

$$\overset{0}{\underset{\approx}{\tau}} \equiv \dot{\underset{\approx}{\tau}} - \underset{\approx}{\omega}^{\mathrm{T}} \cdot \underset{\approx}{\tau} - \underset{\approx}{\tau} \cdot \underset{\approx}{\omega} \tag{9-29}$$

$\underset{\approx}{\omega}$ is the vorticity tensor, $\underset{\approx}{\omega} \equiv \underset{\sim}{\nabla} \underset{\sim}{\mathbf{v}} - (\underset{\sim}{\nabla} \underset{\sim}{\mathbf{v}})^{\mathrm{T}}$.

In uniaxial extension, the first-order (or Newtonian) truncation of the retarded motion expansion for the CRM equation is

$$\Delta \tau_{(1)} = 3\eta_0 \dot{\epsilon} \tag{9-30}$$

where $\eta_0 = G\lambda$ is the zero-shear viscosity. This is also the *exact* normal stress difference in elongational flow of a CRM fluid; see Section 3.4.4. The higher-order terms of the retarded motion expansion are all zero in this case. Thus, for this flow, *the retarded motion expansion converges for any strain rate,* $\dot{\epsilon}$.

In steady shear the retarded motion expansion behaves differently, however. The exact shear stress for the CRM equation is (Section 3.4.4):

$$\tau = \frac{G\lambda\dot{\gamma}}{1 + \lambda^2\dot{\gamma}^2} \tag{9-31}$$

Expanding this expression in a Taylor series gives

$$\tau = G\lambda\dot{\gamma}\sum_{i=0}^{\infty}(-\lambda^2\dot{\gamma}^2)^i \tag{9-32}$$

which converges only if the Deborah number $\lambda\dot{\gamma}$ is less than unity. Thus, for the CRM model the retarded motion expansion converges in steady shear only at low Deborah numbers.

9.3.2 Numerical Calculations with the Retarded Motion Expansion

An attraction of the retarded motion expansion is that it is *independent of constitutive assumptions* (within limits to be discussed shortly). The same terms appear for all constitutive equations for which the expansion can be carried out. Thus one can treat the coefficients as material constants and solve flow problems *without choosing a particular constitutive equation.* One can then study the effect of the expansion coefficients, η, Ψ_1, and Ψ_2, for example, on the pressure and velocity field in much the same way as one studies the effect of the viscosity in Newtonian flows.

Another advantage of the retarded motion expansion is that *the stresses are given explicitly in terms of the instantaneous velocity field.* Thus the stress tensor can be eliminated from the momentum balance equation by substitution of the explicit equation for the stress tensor, and the number of primitive variables is limited to the velocity components and the pressure. This of course represents a big saving of computer storage and computation over that demanded by implicit differential equations, such as the UCM equation. For the UCM equation, stress components must be kept as primitive unknowns, and the constitutive equation must be solved simultaneously with the momentum and mass balance equations.

A major drawback to the retarded motion expansion is that it is limited to flows that are everywhere slow. Applicability to higher flow rates or Deborah numbers can only be had by increasing the order of the expansion. This exacts a heavy price, however, for the number of coefficients increases rapidly as the order increases. Thus at third order there are 6 terms and 14 at fourth order.[30] Furthermore, the differentiability required of the velocity field increases, and this also taxes the numerical scheme. Enthusiasm for going to higher order is further sapped by the uncertainty regarding the rate or even existence of convergence of the retarded motion series for complex flows.

9.3.2.1 Second-Order Fluid Thus terms beyond the second-order are rarely included in numerical calculations. The second-order truncation, however, has attracted considerable attention, though solving flow problems with even it has proven to be a challenge.

Several factors limit the range of usefulness of the equation of the second-order fluid. While the Newtonian description is valid for all velocity gradients below a certain threshold (which is high for many fluids), the second-order description is superfluous at strain rates low enough for the Newtonian description to be valid, and is erroneous at strain rates significantly higher than the Newtonian range. Thus the useful range of the second-order description is small; it is bounded at both the high and the low ends of the range of strain rates.

There are severe numerical problems encountered in attempting to solve flows of the second-order fluid; this may seem surprising when one considers that other nonlinear explicit constitutive equations, such as the equation of the power-law fluid, do not pose unusual difficulties. It is even more surprising in light of the fact that in *planar creeping flow* with velocity boundary conditions, *the Newtonian velocity field is always a solution for the SOF equation.* This result, proved by Giesekus[31] and later by Tanner,[32] implies that when elasticity begins to enter a two-dimensional flow problem, its first influence is felt by the stress field and not the velocity field. Numerical solutions for flows of the second-order fluid are sought, even when the velocity field is already known by the Giesekus-Tanner theorem to be Newtonian, in order to test and develop numerical schemes that could possibly be of use in solving viscoelastic constitutive equations that do not have known solutions for the velocity. The difficulties encountered in solving flows of second-order fluids suggest that numerical difficulties are to be expected whenever elastic effects of any kind are included in a constitutive equation. The SOF equation has viscoelasticity in the most limited sense. Though no relaxation time appears directly in the SOF equation, and the SOF equation cannot predict stress relaxation or elastic recovery, the ratio of coefficients Ψ_1/η is a relaxation time of sorts; its nonzero value produces normal stress differences that are a manifestation of elasticity.

If the constitutive equation of the second-order fluid is taken as a constitutive *law* that gives stresses for *any* flow history for some class of fluids, then bizarre conclusions follow. One strange result is that for physically reasonable values of the coefficients, the second law of thermodynamics can be shown to be violated in certain flow histories.[33] Another odd result is that steady-state flows, and even the rest state, are unstable to infinitesimal perturbations.[33–35] Thus a stagnant second-order fluid will spontaneously begin to flow, even without application of boundary stresses, boundary velocities, or body forces! The flows that are produced in this instability and the flows that violate the second law of thermodynamics, however, are not consistent with the assumptions of slow and slowly varying flow on which the second-order-fluid equation is based. Thus to avoid paradoxical results, *one should not regard the equation of the second-order fluid as a constitutive "law" describing some class of fluids, but as the lowest order of a perturbation expansion that describes the smallest deviations from Newtonian behavior.*[36]

9.3.2.2 Retarded Motion Expansion for Spinning Problem To illustrate the uses and limitations of the retarded motion expansion, we consider briefly the problem of *fiber spinning.*[37,38] Fiber spinning is an important commercial process in which liquid filaments are spun or drawn to small diameters, solidified, and then wound onto a spool, all in a continuous process, Figure 9-3. To show the influence of viscoelasticity, we limit consideration to a highly idealized version of the process wherein the temperature is taken as constant, and gravity, capillarity, and air drag —every force except viscoelasticity— is neglected. The thread of liquid is so thin that the flow field is nearly perfectly elongational. Although steady in an Eulerian sense (that is, in the laboratory frame), each fluid particle as it moves down the threadline sees a time-dependent velocity gradient and is therefore unsteady in the Lagrangian sense. The above simplifying assumptions imply that the tension in the threadline is uniform. The tension equals the product $a\Delta\tau$, where a is the local cross-sectional area of the filament and $\Delta\tau$ is the difference between the axial and the radial components of the stress tensor. Incompressibility implies that v, the local fluid velocity, is inversely proportional

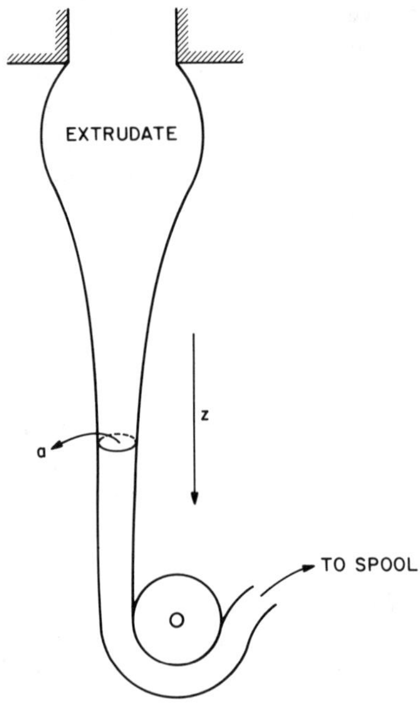

Fig. 9-3 Schematic diagram of the fiber spinning process.

to a. Thus in dimensionless units,

$$\Delta\tau = v \tag{9-33}$$

A constitutive equation is now required to relate $\Delta\tau$ to the velocity history. A choice that allows one to obtain an analytic solution of Equation 9-33 is the CRM equation. In elongational flow, the CRM equation gives

$$\Delta\dot{\tau} + \Delta\tau = 3\dot{\epsilon} \tag{9-34}$$

In light of Equation 9-33, we therefore have

$$\dot{v} + v = 3\dot{\epsilon} \tag{9-35}$$

We define $v' \equiv \partial v/\partial z$, where z is the distance along the spinline. Thus $\dot{\epsilon} = v'$ by definition. Also $\dot{v} \equiv \partial v/\partial t + v\partial v/\partial z = vv'$, since the flow is steady. Equation 9-35 therefore becomes

$$vv' + v = 3v' \tag{9-36}$$

This equation has the solution[39]

$$3 \log v - v = z + \text{const} \tag{9-37}$$

The constant must be determined by a boundary condition. Somewhat arbitrarily, we choose $v = 0.05$ at $z = 0$. Then const $= -9.04$. With this choice, the velocity gradients near $z=0$ are gentle enough that the fluid is nearly Newtonian, and the retarded motion expansion at low order gives nearly exact values for the stresses. Farther down the spinline the filament is thinner, the velocity and velocity gradients are higher, and low-order truncations of the retarded motion expansion are not expected to be accurate.

The analytic solution, Equation 9-37, allows us to assess the accuracy of the retarded motion expansion for the CRM equation with the spinning kinematics. The nth-order truncation of the retarded motion expansion for the dimensionless CRM equation is

$$\underset{\approx}{\tau}_{<n>} = 2 \sum_{i=1}^{n} \underset{\approx}{\mathbf{D}}_{<i>} \tag{9-38}$$

where $\underset{\approx}{\mathbf{D}}_{<i>}$ is the corotational derivative of $\underset{\approx}{\mathbf{D}}_{<i-1>}$; $\underset{\approx}{\mathbf{D}}_{<1>} = \underset{\approx}{\mathbf{D}}$, the rate-of-strain tensor. For an extensional (vorticity-free) flow that is steady in the Eulerian (but not in the Lagrangian) sense,

$$\underset{\approx}{\mathbf{D}}_{<i>} = \dot{\underset{\approx}{\mathbf{D}}}_{<i-1>} = v \frac{\partial}{\partial z} \underset{\approx}{\mathbf{D}}_{<i-1>} \tag{9-39}$$

Using the known exact kinematics, Equation 9-37, for spinning of a CRM fluid, we can compute the nth-order truncation of the retarded motion expansion for the stress, $\Delta\tau_{<n>}$, and compare it to the exact stress, $\Delta\tau$. Figure 9-4 shows the comparison. For $z < 7$ the flow is sufficiently slowly varying that the retarded motion expansion becomes ever more accurate as the order of the expansion increases. For $z > 7$, however, $\Delta\tau_{<n>}$ becomes catastrophically *worse* as n increases. Thus in the downstream region where v is rapidly varying, the first-order Newtonian estimate is better than any of the higher-order approximations shown in Figure 9-4.

This example illustrates an important limitation of the retarded motion expansion: *if the velocity gradient varies rapidly in the Lagrangian sense, then the retarded motion expansion might be slow to converge or even be divergent.*

9.4 FOUNDATIONS OF CONSTITUTIVE THEORY

The constitutive *relationships* presented in the first part of this chapter, such as the equation of the second-order fluid, transcend particular constitutive models. These relationships hold in general for viscoelastic liquids *if certain assumptions are met.* These assumptions are the postulates on which a mathematical theory of constitutive laws is built. Although some of the assumptions were stated explicitly earlier in this book, others are implied by the term "viscoelastic liquid," and still others are technical and can best be explained after the reader has become familiar with the many examples of constitutive equations presented in the earlier chapters.

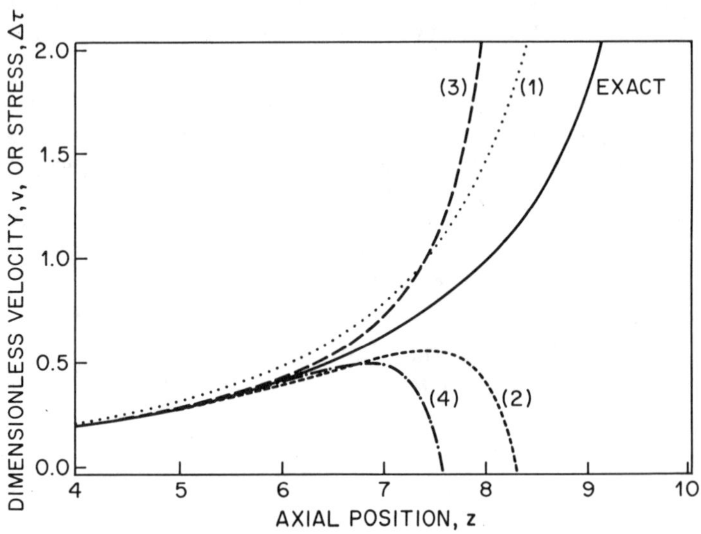

Fig. 9-4 Convergence for $z < 7$ and apparent divergence for $z > 7$ of the retarded motion expansion for the elongational stress in spinning of a corotational Maxwell fluid. The solid line is the exact stress or velocity. The other lines are the stresses given by the retarded motion expansion truncated at the *order* in parentheses.

The conclusions that can be drawn and theorems that can be proved depend of course on the assumptions one makes about the class of "ideal" viscoelastic fluids to be studied. The more specifically one defines the class, the more one can say about all constitutive equations of the class. Thus one would like to *include* as a postulate of the theory any property that seems to be reasonably descriptive of all or most materials that are thought of as viscoelastic, and *exclude* any postulate that overly restricts the theory. The postulates to be included are therefore a matter of judgment. We present first a very general set of assumptions, those of Oldroyd's[7] (1965) "general elastico-viscous liquid," and then present the additional postulates of Coleman and Noll's[40,41] (1958, 1961) viscoelastic "simple fluid." The following discussion avoids mathematical detail; for precise definitions and theorems, the reader is referred to the original papers.

9.4.1 Oldroyd's General 'Elastico-Viscous Liquid'

Oldroyd's general elastico-viscous liquid has the following properties:

9.4.1.1 Isotropy The flow properties of the material are independent of orientation. All anisotropy in the stress tensor is a consequence of anisotropy in the deformation history.

9.4.1.2 Incompressibility Each element of the liquid preserves its volume no matter how it is deformed. Incompressibility and isotropy imply that an extra isotropic stress can be uniformly superimposed on an element of the fluid in any flowing state with no effect on the flow.

9.4.1.3 Fluidity A fluid, according to Batchelor,[42] "cannot withstand any tendency by applied forces to deform it in a way which leaves the volume unchanged." No matter how small the force, a fluid will flow if given enough time. Combined with isotropy, fluidity implies that the material has no preferred shape or configuration. Given enough time, the material forgets both its initial state and old deformation history.

9.4.1.4 Local Action The stress in a small fluid element is determined by the rheological history of that fluid element. Thus if a deformation of a material body is not homogeneous, then the constitutive law can be applied independently to each small element of the body. The deformation and stress histories of the different elements are not coupled together by the constitutive equation, but by the laws of mass and momentum conservation together with the boundary conditions; in this way the deformation of the body as a whole is defined.

The above set of postulates constitute such a loose definition of a viscoelastic liquid that only a few results can be derived from it. Fluids that are not usually thought to be viscoelastic, such as thixotropic fluids, fall into the category of "elastico-viscous liquids," as defined by these postulates. Important general results, such as the retarded motion expansion, cannot be derived without further assumptions. Thus a more definitive, and hence more exclusive, definition of viscoelastic fluids is of value.

9.4.2 Coleman and Noll's Viscoelastic 'Simple Fluid'

9.4.2.1 Fading Memory The validity of the retarded motion expansion requires not only that the material forget its past configurations (in other words that it be an isotropic fluid as defined above) but that it forget *sufficiently rapidly*. This was first recognized by Coleman and Noll. As an illustration, consider the Lodge constitutive equation

$$\underset{\approx}{\tau} = \int_{-\infty}^{t} -\frac{dG}{dt}(t-t')\,\underset{\approx}{C}^{-1}(t,t')\,dt' \tag{9-40}$$

with the relaxation modulus

$$G(t) = \frac{G}{(1+t)^{p}} \tag{9-41}$$

instead of the usual sum of exponentials. This relaxation modulus is plotted in Figure 9-5. It has a *power-law* long-time tail. The power law is cut off at short times to keep the modulus and the stress bounded.

Consider a steady slow flow in which the velocity gradient is $\alpha\underset{\sim}{\nabla}\,\underset{\sim}{v}$, where $\underset{\sim}{\nabla}\,\underset{\sim}{v}$ is here a *dimensionless* velocity gradient and α is a small parameter. The magnitude of the velocity gradient can be made as small as desired by decreasing the scalar α. If the flow history is continuous, then the strain tensor can be expanded in a power series:

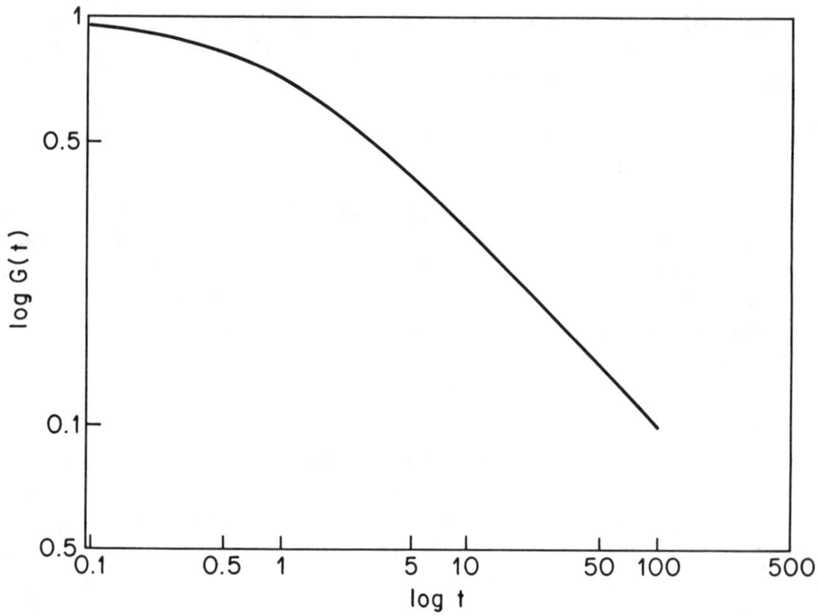

Fig. 9-5 Relaxation modulus with long-time power-law tail; $G(t)/G = (1+t)^{1/2}$.

$$\underset{\approx}{\mathbf{C}}^{-1}(t,t') = \underset{\approx}{\delta} + 2\alpha(t-t')\underset{\approx}{\mathbf{D}} - \alpha^2(t-t')^2\overset{\nabla}{\underset{\approx}{\mathbf{D}}} + \cdots \tag{9-42}$$

which is converges if the product of α and $(t-t')$ is small. Note that $-\overset{\nabla}{\underset{\approx}{\mathbf{D}}} = \underset{\sim}{\nabla}\underset{\sim}{\mathbf{v}}^T \cdot \underset{\approx}{\mathbf{D}} + \underset{\approx}{\mathbf{D}} \cdot \underset{\sim}{\nabla}\underset{\sim}{\mathbf{v}}$ for a steady flow.

If $\underset{\approx}{\tau}$ in Equation 9-40, with $G(t-t')$ given by Equation 9-41, is evaluated with this expansion, the term of zeroth-order in α converges as $t' \to -\infty$ if and only if $p > 0$. If it does converge, then the zeroth order term is

$$\underset{\approx}{\tau}_{(0)} = G\underset{\approx}{\delta} \int_{-\infty}^{t} \frac{p}{(1+t-t')^{p+1}}\, dt' = G\underset{\approx}{\delta} \qquad (p>0) \tag{9-43}$$

The first-order term converges as t' approaches $-\infty$ if $p > 1$; the first-order term then is

$$\underset{\approx}{\tau}_{(1)} = 2G\alpha\underset{\approx}{\mathbf{D}} \int_{-\infty}^{t} \frac{p(t-t')}{(1+t-t')^{p+1}}\, dt' = \frac{2G}{p-1}\alpha\underset{\approx}{\mathbf{D}} \qquad (p>1) \tag{9-44}$$

The second-order term, which converges if $p > 2$, is

$$\underset{\approx}{\tau}_{(2)} = -\alpha^2 \overset{\nabla}{\underset{\approx}{\mathrm{D}}} \int_{-\infty}^{t} \frac{p(t-t')^2}{(1+t-t')^{p+1}}\, dt'$$

$$= -G\left(\frac{p}{p-2} - \frac{2p}{p-1} + 1\right)\alpha^2 \overset{\nabla}{\underset{\approx}{\mathrm{D}}} \qquad (p>2) \tag{9-45}$$

Thus the higher the order one wishes to reach in the retarded motion expansion, the larger p must be, and therefore *the faster the fluid memory must fade as* $t' \rightarrow -\infty$.

Note that if $p < 1$, even the first-order Newtonian description fails. If, however, $G(t-t')$ is the usual sum of exponentials, then as $t' \rightarrow -\infty$, $G(t-t')$ decays faster then Equation 9-41 does for any p, and the retarded motion expansion is valid to all orders.

Though they do not satisfy the retarded motion expansion to all orders, constitutive equations with power-law relaxation moduli can be of value in describing some polymeric liquids that have an especially broad distribution of relaxation times.[43,44] Thus, consider fluids whose relaxation moduli obey the power law

$$G(t) = ct^{-p} \tag{9-46}$$

with $0 < p < 1$. The materials that obey Equation 9-46 most closely are cross-linking polymer liquids very near the gel point; these have been shown to follow Equation 9-46 over some five decades of frequency.[45] Such fluids have no Newtonian regime; i.e., the zero shear viscosity is infinite. These materials can be considered borderline solids. In steady shearing, the shear stress predicted by the Lodge equation, Equation 9-40 with $G(t) = ct^{-p}$, is unbounded, because the Finger tensor $\underset{\approx}{\mathrm{C}}^{-1}$ has no strain softening character. Suppose, however, the Finger tensor is replaced by a "softer" elastic strain measure, for example,

$$\underset{\approx}{\tau} = \int_{-\infty}^{t} -\frac{dG}{dt}(t-t')\frac{1}{1+(\xi'/3)(I_1-3)}\underset{\approx}{\mathrm{C}}^{-1}(t,t')\, dt' \tag{9-47}$$

This constitutive equation always gives a bounded steady-state stress even if G(t) is a power law, Equation 9-46, with $0 < p < 1$. Equation 9-46 and 9-47 with $0<p<1$ violate the retarded motion expansion at all orders beyond the zeroth. However, for a *planar flow of constant stretch history*, the stress tensor for this constitutive equation can be shown to reduce to[18]

$$\underset{\approx}{\tau} = 2\alpha_1(S_f)\, II_D^{(p-1)/2}\, \underset{\approx}{\mathrm{D}} - \alpha_2(S_f)\, II_D^{(p-2)/2}\, \overset{\nabla}{\underset{\approx}{\mathrm{D}}} \tag{9-48}$$

α_1 and α_2 are functions of the *flow strength*, S_f, defined in Equation 9-15. Equation 9-48 is a generalization of the power-law fluid, Equation 9-7. Like the power-law fluid, all stress components in Equation 9-48 have power-law dependencies on strain rate, but unlike the power-law fluid, Equation 9-48 has a first normal stress difference in simple shear.

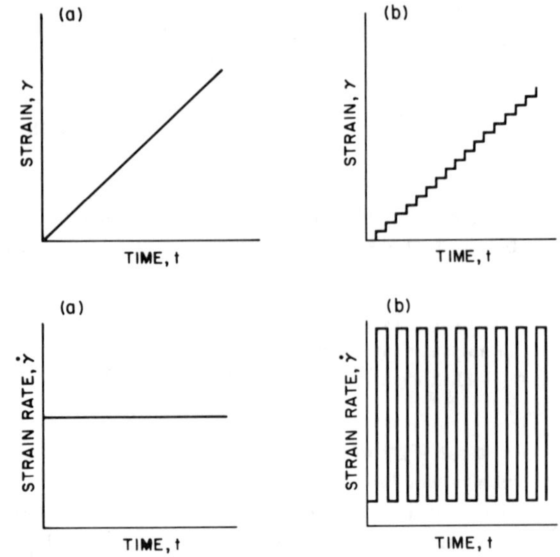

Fig. 9-6 (a) and (b) have similar ***strain*** histories, but dissimilar strain-***rate*** histories. For simple fluids, the stress is not sensitive to the differences between histories (a) and (b).

9.4.2.2 Continuity in Strain History For the viscoelastic simple fluid of Coleman and Noll, the stress is a ***continuous*** functional of the Finger ***strain*** history and is therefore insensitive to small changes in ***strain*** history. This concept can be understood by considering the two shear deformation histories depicted in Figure 9-6. The ***strain*** histories (top) are similar: one is a smooth ramp; the other approximates the ramp by many small steps. The strain-***rate*** histories, however, are obviously different.

For a fluid satisfying the continuity postulate of a viscoelastic simple fluid, the stress is ***not*** sensitive to the differences in deformation history depicted in Figure 9-6, if the steps in Figure 9-6(b) are spaced closely enough in time. A simple fluid is only sensitive to large differences in ***strain*** history, not large differences in strain-***rate*** history.

9.4.2.2.1 Step strains The continuity postulate makes sensible the ***step*** strain history, wherein strain rates are made very large for a brief period of time, just long enough to apply a given strain. Simple-fluid theory implies that the stress resulting from the step (shear, say) is not sensitive to the magnitude of the strain rate ($\dot{\gamma}$) or the duration of the step (Δt), as long as Δt is small enough and the total strain ($\gamma = \dot{\gamma}\Delta t$), is held fixed. If this implication of simple-fluid theory were false, one would be obliged to speak not of step strains but rather of ***high strain rates of short duration***, where the rate and duration would both need to be specified. In experiments with polymer melts, it seems that only the total strain influences the stress, if the measurement of stress is delayed until a time lag after the step that is long compared to the duration of the step. Experience with step

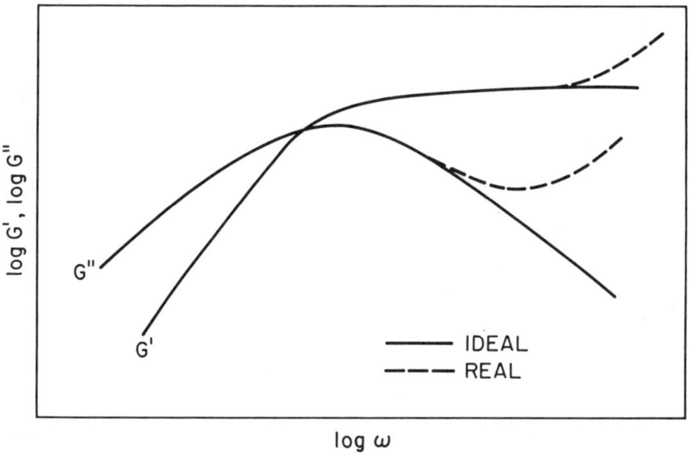

Fig. 9-7 According to ideal simple-fluid theory, G′ is constant and G″ approaches zero at high frequency. Real viscoelastic fluids deviate from this behavior.

strains thus supports the continuity postulate of simple-fluid theory.

9.4.2.2.2 Lodge-Meissner relationship Coleman and Noll's definition of continuity together with frame invariance and isotropy imply that the Lodge-Meissner relationship[46] holds between the first normal stress difference, N_1, and the shear stress, σ, after a step shear strain, γ:

$$N_1 = \sigma\gamma \tag{9-49}$$

The observed validity of this relationship for melts (Section 3.5.5), affirms simple-fluid theory.

9.4.2.2.3 Transition from linear to nonlinear viscoelasticity For viscoelastic simple fluids the transition from linear to nonlinear behavior occurs at a finite *strain*, no matter how high the strain rate becomes. This result is supported by experiment; see Section 6.3.7.

9.4.2.2.4 Elastic limit Continuity in *strain* history implies that if a deformation is imposed rapidly, then the stress is determined only by the cumulative strain and not by the *strain rate* or *strain path*. Thus in a rapid deformation, a viscoelastic simple fluid behaves like an elastic rubber.

Strictly speaking, this implies that in a small-amplitude oscillatory deformation, G′ reaches a plateau and G″ approaches zero as the frequency, ω, approaches infinity, Figure 9-7. Nearly monodisperse polymer melts *do* have a high-frequency plateau region in G′; in this region G″ decreases as ω increases. But if ω is increased enough, G″ and G′ invariably turn upward again. The upturn is produced by *glassy relaxation modes* that are excited at high frequency.[47]

These glassy modes are often not important in polymer flows, however; to include them in a viscoelastic description would be a needless complication. The

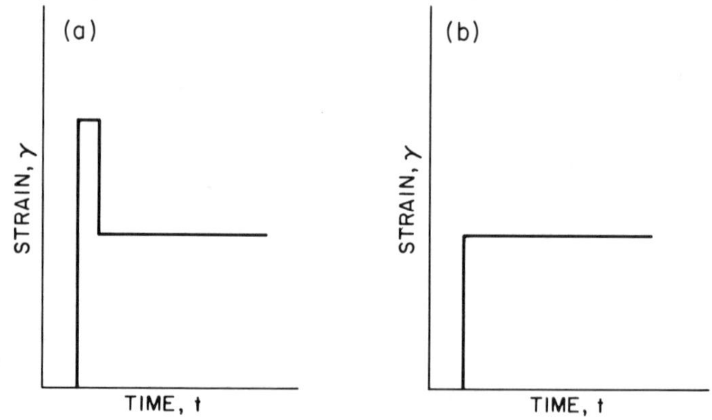

Fig. 9-8 Viscoelastic simple fluids are not sensitive to the difference between deformations (a) and (b); strains that are reversed quickly enough have little effect on the stress.

elastic limit assumed in the simple fluid theory is therefore not so much a description of real viscoelastic fluid behavior as it is a means of *cutting off* the high-frequency end of the relaxation spectrum.

The elastic high-frequency cutoff postulated by the simple fluid theory is not the only cutoff possible. A *dissipative cutoff* is another alternative. The high-frequency cutoff is dissipative, for example, if the stress tensor contains a Newtonian contribution, such as $2\eta_s \underset{\approx}{D}$, where η_s is a "solvent" viscosity. A term of this kind occurs in the Oldroyd-B model (Section 3.4.3). When this term is added to the stress tensor, G'' turns up at high frequency; $G'' \rightarrow \eta_s\omega$ as $\omega \rightarrow \infty$. $2\eta_s\underset{\approx}{D}$ makes no contribution to G'. Dilute solutions of rigid particles or flexible molecules with internal viscosity have a dissipative contribution from a term proportional to $\langle \underset{\sim\sim\sim\sim}{RRuu} \rangle : \underset{\approx}{D}$; see Section 8.7. This term gives a high-frequency contribution,[48] $2\omega/5$, to G''.

A high-frequency cutoff in the relaxation spectrum can be avoided. The relaxation modulus

$$G(t) = ct^{-p} e^{-t/\lambda_1} \tag{9-50}$$

has a longest relaxation time, λ_1, but no shortest relaxation time. Constitutive equations with this relaxation modulus or with a dissipative high-frequency cutoff are *not* in the simple-fluid class of constitutive equations.

9.4.2.2.5 Reversibility The existence of an elastic limit implies that strains, even large strains, if removed quickly enough, have no effect on the stress. Thus the two strain histories depicted in Figure 9-8 produce nearly identical stresses if the fluid is a simple fluid. Because of this characteristic, simple-fluid equations often produce unrealistic predictions for strain histories with strain reversal, such as reversing double-step shearing and elongational elastic recovery; see Section 5.5.1. Two equations that have been found to make reasonably accurate

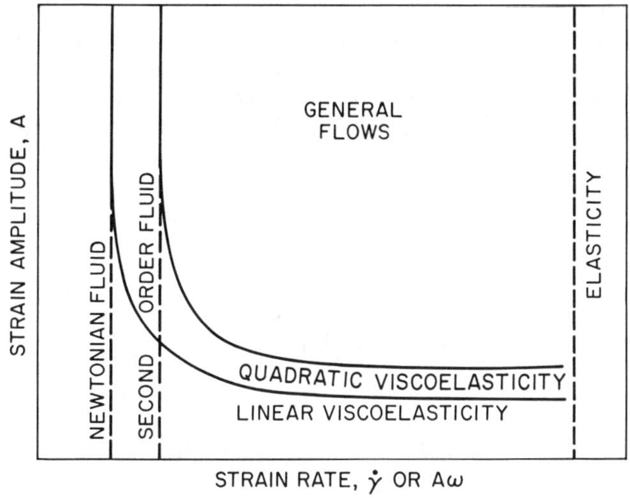

Fig. 9-9 Modified Pipkin diagram showing the behavior of all viscoelastic simple fluids in the limits of low strain rates, low amplitude deformations, and high strain rates.

predictions of reversing strain histories, namely the Wagner II equation and the Doi-Edwards equation without the independent alignment approximation, are *not* simple-fluid equations.

9.4.2.3 Reasonableness of the Simple Fluid Postulates In sum, simple-fluid theory is realistic for melts in that it predicts that the stress response after a step strain is a function only of the step size, gives the Lodge-Meissner relationship in step shears and predicts that the transition from linear to nonlinear viscoelasticity occurs at a finite strain no matter how high the strain rate becomes. Unrealistic are the predictions that viscoelastic fluids are perfectly elastic at high strain rates and the corollary that if a strain is both imposed and then removed rapidly enough the stress is not affected and no energy is dissipated. These difficulties in simple-fluid theory can be avoided by including relaxation processes that are faster than any allowed strain rate. For some purposes, however, it may be more convenient to include purely dissipative terms in the constitutive equation.

The behavior of viscoelastic simple fluids in the limits of high and low strain rates and of high and low strain amplitudes is illustrated in the modified Pipkin diagram[36] shown in Figure 9-9. All constitutive equations belonging to the simple-fluid class, which includes many of the equations presented in this book, are governed by these limits. The Newtonian, second-order fluid, and linear viscoelastic limits are achievable in experiments; the elastic limit is not.

9.4.3 Perturbation Expansions for Small Strain Amplitudes

From the postulates of simple-fluid theory, the limiting behavior at low strain rates and low amplitudes in the Pipkin diagram can be derived. In addition, under the simple-fluid postulates, one can move inward by perturbation expansions from the left side or from the bottom of the diagram toward general viscoelastic flows. The expansion from the slow-flow limit (left side) is the retarded motion expansion. The expansion from the small-amplitude limit (bottom of the diagram) is a *multiple integral expansion*, valid for deformations of arbitrary rate but limited amplitude. The first-order term of this expansion is the single integral†

$$\underset{\approx}{\tau}_1 = \int_{-\infty}^{t} m(t-t')\,\underset{\approx}{C}^{-1}(t,t')\,dt' \tag{9-51}$$

which for small strains is equivalent to the *general equation of linear viscoelasticity* (Section 3.2). All simple fluids obey Equation 9-51 at low strain amplitudes.

The second-order term is a double integral[15]

$$\underset{\approx}{\tau}_2 = \int_{-\infty}^{t}\int_{-\infty}^{t} \alpha(t-t',t-t'')\,\underset{\approx}{C}^{-1}(t,t')\cdot\underset{\approx}{C}^{-1}(t,t'')\,dt'\,dt'' + \int_{-\infty}^{t}\int_{-\infty}^{t} \beta(t-t',t-t'')\,[\mathrm{tr}\underset{\approx}{C}^{-1}(t,t'')]\underset{\approx}{C}^{-1}(t,t')\,dt'\,dt'' \tag{9-52}$$

which contains bivariate functions $\alpha(t-t',t-t'')$ and $\beta(t-t',t-t'')$. At second order in strain, $\underset{\approx}{\tau}_1 + \underset{\approx}{\tau}_2$ is equivalent to the *general equation of quadratic viscoelasticity*. The nth-order term contains n nested integrals over kernel functions of n time variables. The kernel function, $m(t-t')$, of the first-order equation is determined by linear viscoelastic measurements (G' and G'', for example), and its measurement is commonplace. The second-order functions, $\alpha(t-t',t-t'')$ and $\beta(t-t',t-t'')$ are much harder to measure. Thus, like the retarded motion expansion, the multiple integral expansion requires unmanageably many measurements even at low order.

The multiple integral expansion can, however, be *collapsed into a single integral* under the assumption that *the influence of each strain increment on the stress is independent of the other strain increments.*[15] For then $\alpha(t-t',t-t'')$ and $\beta(t-t',t-t'')$ are zero unless $t' = t''$. This assumption is exactly fulfilled for single-step strains. For other strain histories such as double-step strains or steady straining, it is an approximation. Under this approximation, the multiple integral expansion at second order reduces to a single integral

† The choice of $\underset{\approx}{C}^{-1}$ in Equation 9-51 is arbitrary; in the linear limit all the strain tensors give the same predictions.

$$\underset{\approx}{\tau} = \underset{\approx}{\tau}_1 + \underset{\approx}{\tau}_2 =$$

$$\int_{-\infty}^{t} \left\{ m(t-t')\underset{\approx}{C}^{-1} + \alpha(t-t',t-t')\underset{\approx}{C}^{-2} + \beta(t-t',t-t')[\mathrm{tr}(\underset{\approx}{C}^{-1})]\underset{\approx}{C}^{-1} \right\} dt' \quad (9\text{-}53)$$

Third- and higher-order expansions contain higher powers of $\underset{\approx}{C}^{-1}$. By the Cayley-Hamilton theorem (Section 3.4.5), $\underset{\approx}{C}^{-n}$ with $n \geq 2$ can be expressed in terms of $\underset{\approx}{C}^{-1}$, $\underset{\approx}{C}$, and $\underset{\approx}{\delta}$, and the invariants $I_1 \equiv \mathrm{tr}\underset{\approx}{C}^{-1}$ and $I_2 \equiv \mathrm{tr}\underset{\approx}{C}$. Equation 9-53, and the expansion *out to any order* can therefore be reduced to

$$\underset{\approx}{\tau} = \sum_{i=1}^{\infty} \underset{\approx}{\tau}_i = \int_{-\infty}^{t} [\,\phi_1(t-t',I_1,I_2)\,\underset{\approx}{C}^{-1} - \phi_2(t-t',I_1,I_2)\,\underset{\approx}{C}\,]\, dt' \qquad (9\text{-}54)$$

to within an isotropic term. This is the *superposition integral equation* mentioned in Section 3.6.2. The functions $\phi(t-t',I_1,I_2)$ and $\phi(t-t',I_1,I_2)$ can be obtained by stress measurements in single step strains. If ϕ_1 and ϕ_2 are given by a potential function, $u(t-t',I_1,I_2)$, such that $\phi_i = \partial u / \partial I_i$, then Equation 9-54 is the *Kaye-BKZ equation*.

9.5 SUMMARY

There are universal constitutive equations for special flows such as flows of constant stretch history (csh), slow and slowly varying flows, and small-amplitude deformations. The general constitutive equation for flows of constant stretch history is very complex, but there are simple forms for viscometric, steady extensional, and planar csh flows. For flows that are slow and slowly varying, the general constitutive equation is a truncation of the retarded motion expansion. The higher the order at which the expansion is truncated, the more accurate the constitutive equation becomes for slow flows. The convergence of the retarded motion expansion is, in general, not uniform over all strain rates; if the velocity gradient is large or very unsteady, the expansion may diverge. The multiple integral expansion is valid for deformations of any rate but of limited amplitude. The first-order term of this expansion is equivalent at small strains to the equation of linear viscoelasticity. Higher-order terms contain multiple integrals.

These limiting constitutive equations can be derived from Coleman and Noll's theory of viscoelastic "simple fluids." Simple fluids are isotropic, incompressible fluids in which the stress in a particle is determined by the deformation history of that particle only. They possess a fading memory and are insensitive to small changes in strain history. Some of the most popular constitutive equations belong to the simple-fluid class, but there are many important exceptions. Simple-fluid equations include the upper-convected Maxwell equation, the Giesekus equation, and the Kaye-BKZ equation. Exceptions include the Johnson-Segalman equation, the White-Metzner equation, the Doi-Edwards equation without the independent alignment approximation, Wagner's first and second equations, equations with a retardation term, and equations with a power-law spectrum of relaxation times.

References

1. B.D. Coleman, H. Markovitz, and W. Noll, *Viscometric Flows of Non-Newtonian Fluids,* Springer-Verlag, New York (1966).
2. A.C. Pipkin and R.I. Tanner, Mechanics Today, *1,* 262 (1972).
3. R.I. Tanner, *Engineering Rheology,* Oxford, New York (1985).
4. W.L. Yin and A.C. Pipkin, Arch. Ratl. Mech. Anal., *37,* 111 (1970).
5. W.O. Criminale, Jr., J.L. Ericksen, and G.L. Filbey, Jr., Arch. Ratl. Mech. Anal., *1,* 410 (1958).
6. R.B. Bird, R.C. Armstrong, and O. Hassager, *Dynamics of Polymeric Liquids,* Vol. 1, John Wiley & Sons, New York (1977).
7. J.G. Oldroyd, Proc. Roy. Soc., *A283,* 115 (1965).
8. W. Ostwald, Kolloid-Z., *36,* 99 (1925).
9. A. de Waele, Oil and Color Chem. Assoc. J., *6,* 33 (1923).
10. P.J. Carreau, Ph. D. Thesis, Univ. of Wisc., Madison (1968).
11. M. Reiner, Am. J. Math., *67,* 350 (1945).
12. R.S. Rivlin, Proc. Roy. Soc. (London), *A193,* 260 (1948); Proc. Cambridge Phil. Soc., *45,* 88 (1949).
13. B.D. Coleman, Trans. Soc. Rheol., *6,* 293 (1962); Arch. Ratl. Mech. Anal., *9,* 273 (1962).
14. W. Noll, Arch. Ratl. Mech. Anal., *11,* 97 (1962).
15. G. Astarita and G. Marrucci, *Principles of Non-Newtonian Fluid Mechanics,* McGraw-Hill, London (1974).
16. C.-C. Wang, Arch. Ratl. Mech. Anal., *20,* 329 (1965).
17. R.R. Huilgol, Rheol. Acta, *15,* 120 (1976).
18. R.G. Larson, Rheol. Acta, *24,* 443 (1985).
19. R.I. Tanner and R.R. Huilgol, Rheol. Acta, *14,* 959 (1975).
20. R.I. Tanner, AIChE J., *22,* 910 (1976).
21. M.A. Mendelsohn, P.-W. Yeh, R.A. Brown, and R.C. Armstrong, J. Non-Newtonian Fluid Mech., *10,* 31 (1982).
22. M. Keentok, A.G. Georgescu, A.A. Sherwood, and R.I. Tanner, J. Non-Newtonian Fluid Mech., *6,* 303 (1980).
23. G.A. Alvarez, A.S. Lodge, and H.-J. Cantow, Rheol. Acta, *24,* 368 (1985).
24. S. Ramachandran, H.W. Gao, and E.B. Christiansen, Macromolecules, *18,* 695 (1985).
25. J. Meissner, S.E. Stephenson, A. Demarmels, and P. Portmann, J. Non-Newtonian Fluid Mech., *11,* 221 (1982).

26. R.S. Rivlin and J.L. Ericksen, J. Ratl. Mech. Anal., *4*, 323, (1955).

27. K. Osaki, S. Ohta, M. Fukuda, and M. Kurata, J. Polym. Sci., Polym. Phys. Ed., *14*, 1701 (1976).

28. G.G. Lipscomb, R. Keunings, and M.M. Denn, J. Non-Newtonian Fluid Mech., submitted 1987.

29. D.D. Joseph, J. Non-Newtonian Fluid Mech., *20*, 117 (1986).

30. R.B. Bird, R.C. Armstrong, and O. Hassager, *Dynamics of Polymeric Liquids*, Vol. 1, John Wiley & Sons, New York (1977), p. 577.

31. H. Giesekus, Rheol. Acta, *3*, 59 (1963).

32. R.I. Tanner, Phys. Fluids, *9*, 1246 (1968).

33. J.E. Dunn and R.L. Fosdick, Arch. Ratl. Mech. Anal., *56*, 191 (1974).

34. B.D. Coleman, R.J. Duffin, and V.J. Mizal, Arch. Ratl. Mech. Anal., *19*, 100 (1965).

35. A.D.D. Craik, J. Fluid Mech., *33*, 33 (1968).

36. A.C. Pipkin, *Lectures on Viscoelasticity Theory,* Springer-Verlag (1972).

37. M.A. Matovich and J.R.A. Pearson, I. & E.C. Fund., *8*, 512 (1969).

38. M.M. Denn, Ann. Rev. Fluid Mech., *12*, 365 (1980).

39. R.G. Larson, J. Rheol., *27*, 475 (1983).

40. W. Noll, Arch. Ratl. Mech. Anal., *2*, 197 (1958).

41. B.D. Coleman and W. Noll, Rev. Mod. Phys., *33*, 239 (1961).

42. G.K. Batchelor, *An Introduction to Fluid Dynamics,*, Cambridge Univ. Press, London (1967), p. 1.

43. W. Heindl and H. Giesekus, Rheol. Acta, *24*, 327 (1985).

44. R.G. Larson, Rheol. Acta, *24*, 327 (1985).

45. H.H. Winter and F. Chambon, J. Rheol., *30*, 367 (1986); F. Chambon and H.H. Winter, J. Rheol., to appear in 1987.

46. A.S. Lodge and J. Meissner, Rheol. Acta, *11*, 351 (1972).

47. J.D. Ferry, *Viscoelastic Properties of Polymers*, 3rd ed., John Wiley & Sons, New York (1980).

48. R.B. Bird, O. Hassager, R.C. Armstrong, and C.F. Curtiss, *Dynamics of Polymer Liquids*, Vol. 2, John Wiley & Sons, New York (1977).

CHAPTER 10

Theories for Nondilute Solutions of Rodlike Molecules

"and if a tree fall toward the south
or toward the north,
in the place where the tree falleth,
there it shall be."

Ecclesiastes 11:3

10.1 INTRODUCTION

Chapter 8 contains a discussion of dilute solutions of flexible polymers and to a lesser extent dilute solutions of rigid polymers. "Dilute" here means that the concentration is low enough that interactions between different molecules are negligible, Figure 10-1a. We have seen in Chapter 4 that when the concentration of long *flexible* polymers is increased, and chains begin to overlap, the dilute region gives way to a concentrated regime wherein the dynamics are affected by topological constraints or *entanglements*. Solutions of rodlike polymers also show a transition to a regime wherein the dynamics are dominated by topological constraints.[1] For rodlike polymers, the transition occurs at a concentration, ν_1, in numbers of rods per unit volume, just high enough that rotation of a rod is blocked by surrounding rods, Figure 10-1b. A crude estimate of ν_1 is the concentration at which there is overlap of imaginary spheres of rotation each defined as the locus of the positions swept by the ends of the rod as it rotates about its center. This estimate gives $\nu_1 \sim 1/L^3$, where L is the length of the rods. Experimentally,[2] the transition occurs at more than 30 times this estimate of ν_1, apparently because a rod can "dodge" several other rods that invade its sphere of rotation.

If the rods are long and slender enough, then at concentrations of order $\nu_1 \sim 1/L^3$, a new rod can be randomly inserted into the solution with negligible probability of it intersecting other rods, Figure 10-1b. Thus although the *dynamic* properties of the solution are strongly affected by the interactions of neighboring rods, the static or equilibrium properties are negligibly affected. There is little correlation between orientations of neighboring rods, and the rods can then be thought of as lines with zero volume. At higher concentrations, $\nu \geq \nu_2 = 1/bL^2$, where b is the rod diameter, randomly oriented rods have a significant probability

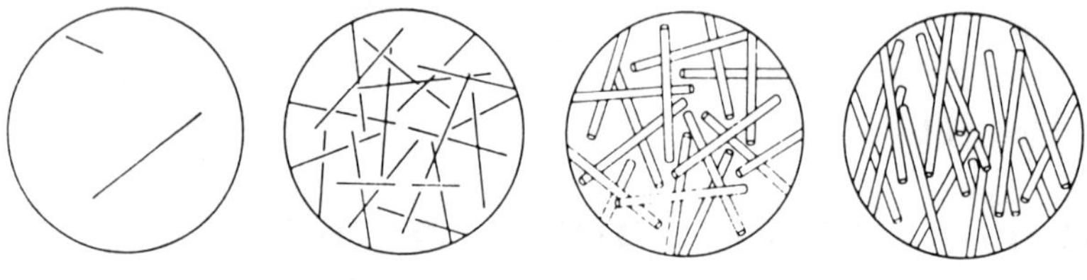

Fig. 10-1 Concentration regimes for rodlike molecules. (a) dilute; (b) semidilute; (c) isotropic concentrated; and (d) nematic liquid crystalline (from Doi and Edwards,[1] reprinted with permission from *The Theory of Polymer Dynamics*, Oxford University Press, 1986).

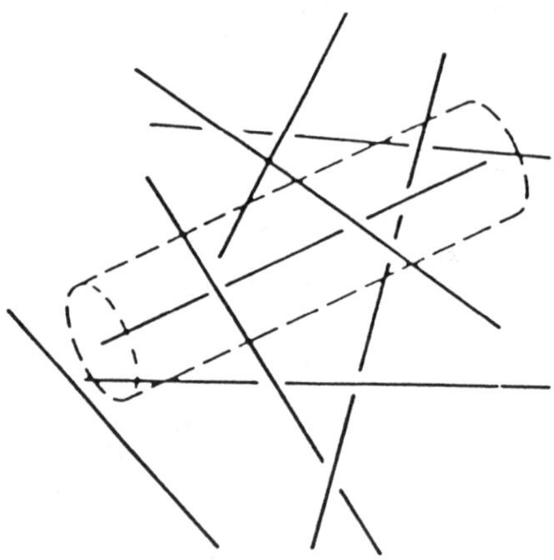

Fig. 10-2 In the Doi and Edwards model for semidilute solutions of rods, a given rod is confined by surrounding rods to a tubelike region (from Doi and Edwards,[1] reprinted with permission from *The Theory of Polymer Dynamics*, Oxford University Press, 1986).

of intersecting each other; see Figure 10-1c. In this concentration regime, the excluded volume of the rods becomes important, and the orientation of one rod is no longer independent of the orientation of its neighbors. Thus in this regime the *static* properties of the solution of rods are affected by the concentration. Note that if the rod aspect ratio, b/L, is very small, then the volume fraction of rods at which static interference occurs, namely $\phi_2 = \nu_2 \pi b^2 L/4 = \pi b/4L$, is small.

Finally, if the rod concentration is increased above a critical value, ν^*, the rods spontaneously align. This is the transition to the nematic liquid crystalline state. Doi and Edwards[1] call the concentrations between ν_1 and ν_2 *semidilute*; those between ν_2 and ν^* are in the *isotropic concentrated* regime; and those with concentrations greater than ν^* are *nematic*. Again if $b/L << 1$, ν^* can be much smaller than unity. Though this chapter is mainly concerned with the liquid crystalline regime, we first discuss briefly the much better understood semidilute regime.

10.2 SEMIDILUTE REGIME

10.2.1 Diffusion Coefficient

10.2.1.1 Isotropic Solutions In the semidilute regime, the rod diameter is much smaller than the distance between rods, and therefore a rod can translate along its own axis with negligible interference. Lateral motion of the rod is blocked by other rods, however. Doi and Edwards[1,3] modeled this by assuming that each rod is confined by surrounding rods to a tube-like region, Figure 10-2. This model is analogous to Doi and Edward's tube model for concentrated solutions of flexible molecules, discussed in Chapter 4. There is a crucial difference, however. The rod must remain almost completely oriented as long as any part of it lies in the tube, whereas any portion of the flexible molecule that escapes the tube becomes disoriented. When the flexible molecule escapes its tube, its conformation is completely relaxed or uncorrelated with the orientation it had in the tube. When a rod leaves its tube, it can only change its orientation by a small angle, $\epsilon \approx a/L$, where a is the tube radius and L is the length of the rod and of the tube. The tube is thought of as a cylinder surrounding the rod, into which other rods do not penetrate. By dimensional arguments, the radius of this tube must scale as* $a \sim 1/\nu L^2$. Thus $\epsilon \approx 1/\nu L^3$. Each time the rod escapes its tube, it therefore changes its orientation by a small random angle, ϵ. If τ_D is the time required to escape the tube, the time required for the rod to become totally disoriented by this mechanism is approximately τ_D/ϵ^2. The reciprocal of this is the rotational diffusion coefficient

$$D_r \approx \epsilon^2/\tau_D \tag{10-1}$$

The time required by the rod to escape its tube, τ_D, is the time it takes to diffuse along its own axis. Since the other rods do not impede this motion, τ_D is a property of dilute solutions of rods and is approximately equal to the inverse of D_{ro}, the rotary diffusion constant in dilute solution. Thus from Equation 10-1,

$$D_r \approx \epsilon^2 D_{ro} = D_{ro}(1/\nu L^3)^2 \tag{10-2}$$

By definition, $\nu L^3 >> 1$ in the semidilute regime. Hence the diffusion coefficient in the semidilute regime is greatly suppressed compared to its value in the dilute regime.

Now $\nu \propto \rho/M$, and $L \propto M$, where ρ is the mass of rods per unit volume and M is the molecular weight of a rod. Furthermore, $D_{ro} \propto \log(M)/M^3$. Therefore

$$D_r \propto \frac{\log M}{M^3}\left(\frac{1}{\rho M^2}\right)^2 = \rho^{-2}M^{-7}\log M \tag{10-3}$$

* See Doi and Edwards[1,3] for a more detailed calculation of the tube radius.

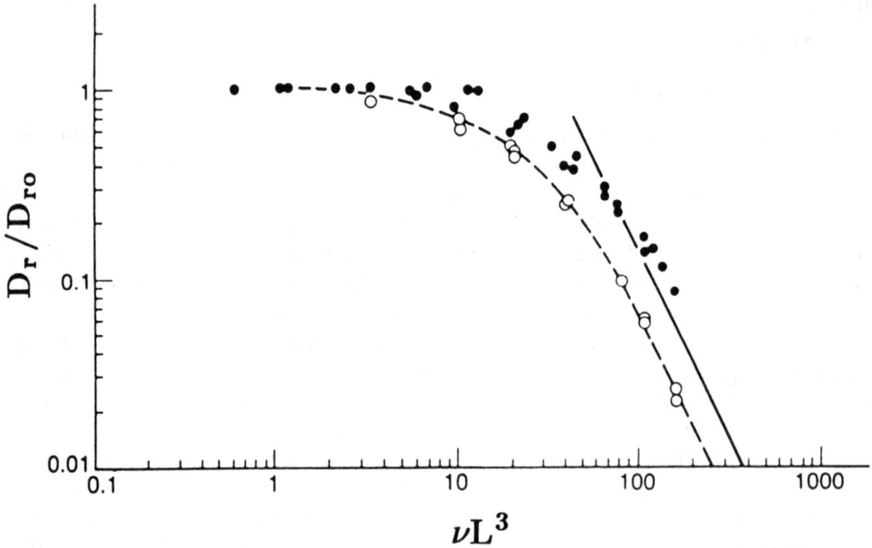

Fig. 10-3 Reduced rotary diffusivity, Dr/Dro, versus dimensionless concentration νL^3. The filled circled are experimental data for rodlike poly(γ-benzyl-L-glutamate), and the open circles are from computer simulation. The solid line has the slope of −2 predicted by Doi and Edwards (from Doi and Edwards,[1] reprinted with permission from *The Theory of Polymer Dynamics*, Oxford University Press, 1986).

Thus for a given mass density of rods in solution, the diffusivity is predicted to decrease with increasing molecular weight as the seventh power of the molecular weight! Measurements[2] of dynamic electric birefringence in solutions of rodlike poly(γ-benzyl-L-glutamate) (PBLG) show that the dependence of D_r on M is indeed steep; see Figure 10-3. Figure 10-3 shows that the transition from the dilute to the semidilute regime occurs at $\nu L^3 \sim 30$.

10.2.1.2 Oriented Solutions In the above estimate of the diffusion coefficient, the rods were assumed to be randomly oriented. If the orientation distribution of the rods becomes *anisotropic*, due to the application of an external magnetic or electric field, or due to imposition of a flow field, then the average diffusivity increases. This is illustrated in Figure 10-4, where the effective tube diameter is seen to increase as the orientations of surrounding rods become more nearly parallel to the tube. Suppose the matrix rods are, on the average, oriented at an angle, β, with respect to the test rod. Then the tube diameter, a, is inversely proportional to sin β. Since the orientation-dependent diffusion coefficient, $\hat{D}_r$, is proportional to $\epsilon^2 \propto a^2$, $\hat{D}_r \propto \sin^{-2}\beta$. Thus in a matrix of oriented rods, the diffusion coefficient of a rod is expected to be orientation dependent.

More precisely, suppose $\underset{\sim}{u}'$ is a unit vector oriented in the direction of a matrix rod, and let $\psi(\underset{\sim}{u}')$ be the distribution of orientations $\underset{\sim}{u}'$ of the matrix rods. Let $\underset{\sim}{u}$ be the orientation of the test rod. Doi and Edwards[1] showed that $\hat{D}_r \sim \langle\sin\beta\rangle^{-2}$, where $\langle\sin\beta\rangle$ is the average of $\sin\beta$ over all orientations $\underset{\sim}{u}'$ of the

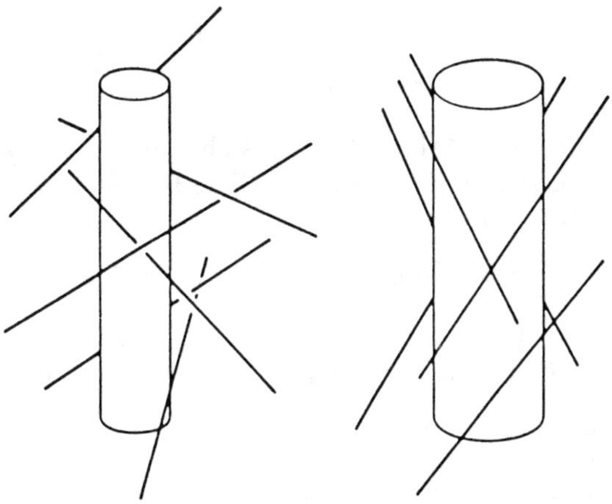

Fig. 10-4 The effective tube diameter for a rod increases when the surrounding rods are oriented in approximately the same direction (from Doi and Edwards,[1] reprinted with permission from *The Theory of Polymer Dynamics*, Oxford University Press, 1986).

matrix rods:

$$\langle \sin\beta \rangle = \frac{4}{\pi} \int |\underset{\sim}{u} \times \underset{\sim}{u}'| \, \psi(\underset{\sim}{u}') \, du^2 \tag{10-4}$$

Here $|\underset{\sim}{u} \times \underset{\sim}{u}'|$, the cross product of $\underset{\sim}{u}$ and $\underset{\sim}{u}'$, is the sin of the angle between $\underset{\sim}{u}$ and $\underset{\sim}{u}'$. Then

$$\hat{D}_r(\underset{\sim}{u}) = D_r \left[\frac{4}{\pi} \int |\underset{\sim}{u} \times \underset{\sim}{u}'| \, \psi(\underset{\sim}{u}') \, du'^2\right]^{-2} \tag{10-5}$$

As the rods become more oriented, $\hat{D}_r(\underset{\sim}{u})$ becomes more direction dependent, and its average value increases.

10.2.2 Doi-Edwards Constitutive Equation for Semidilute Solutions of Rods

Except for the dependence of $\hat{D}_r$ on $\underset{\sim}{u}$, the Smoluchowski Equation for rigid rods in a flow field is the same as that for rigid dumbbells given by Equation 8-85:*

* $\frac{\partial}{\partial \underset{\sim}{u}} = (\frac{\partial}{\partial \theta}, \frac{1}{\sin\theta}\frac{\partial}{\partial \phi})$, is the gradient operator on the unit sphere. θ and ϕ are the polar and azimuthal angles, respectively,

$$\frac{\partial \psi}{\partial t} + \frac{\partial}{\partial \underset{\sim}{u}} \cdot [(\underset{\sim}{u} \cdot \nabla \underset{\sim}{v} - \underset{\sim}{uuu} : \underset{\approx}{D}) \psi - \hat{D}_r \frac{\partial \psi}{\partial \underset{\sim}{u}}] = 0 \qquad (10\text{-}6)$$

An important simplification of this Equation is obtained by replacing $\hat{D}_r$ with an orientation-independent average diffusivity, $\overline{D}_r$. Doi and Edwards suggested using an average of Equation 10-5:

$$\overline{D}_r \equiv D_r [\frac{4}{\pi} \int\int \psi(\underset{\sim}{u}) \psi(\underset{\sim}{u}') \, |\underset{\sim}{u} \times \underset{\sim}{u}'| \, du^2 du'^2]^{-2} \qquad (10\text{-}7)$$

Replacing $\hat{D}_r(\underset{\sim}{u})$ with $\overline{D}_r$ in Equation 10-6, one can multiply the Equation by $\underset{\sim\sim}{uu}$ and integrate over all orientations to obtain

$$\lambda \overset{\nabla}{<\underset{\sim\sim}{uu}>} + 2\lambda <\underset{\sim\sim\sim\sim}{uuuu}> : \underset{\approx}{D} + <\underset{\sim\sim}{uu}> - \frac{1}{3} \underset{\approx}{\delta} = \underset{\approx}{0} \qquad (10\text{-}8)$$

where $\lambda = 1/6\overline{D}_r$ is the relaxation time. The symbol ∇ denotes the upper-convected derivative. Acting on an arbitrary tensor $\underset{\approx}{S}$, it is defined as

$$\overset{\nabla}{\underset{\approx}{S}} \equiv \dot{\underset{\approx}{S}} - \nabla \underset{\sim\sim}{v}^T \cdot \underset{\approx}{S} - \underset{\approx}{S} \cdot \nabla \underset{\sim\sim}{v}$$

where $\dot{\underset{\approx}{S}}$ is the *substantial* time derivative, $\dot{\underset{\approx}{S}} \equiv \partial \underset{\approx}{S} / \partial t + \underset{\sim}{v} \cdot \nabla \underset{\approx}{S}$.

10.2.2.1 The Stress Tensor The stress tensor is given by the analog of Equation 8-88,

$$\underset{\approx}{\tau} = \nu\varsigma <\underset{\sim\sim\sim\sim}{uuuu}> : \underset{\approx}{D} + 3\nu kT \{<\underset{\sim\sim}{uu}> - \frac{1}{3} \underset{\approx}{\delta}\}$$

$$\text{(viscous)} \qquad\qquad \text{(entropic)} \qquad (10\text{-}9)$$

For the rod, the friction coefficient, ς, is given by[1]

$$\varsigma = \frac{\pi \eta_s L^3}{6 \log(L/b)} \qquad \text{(with hydrodynamic interaction)} \qquad (10\text{-}10)$$

This formula accounts for the hydrodynamic interaction between different parts of the rod. In the semidilute regime this hydrodynamic interaction is probably partly screened out by the surrounding rods. If hydrodynamic interaction were completely screened out, the relevant friction coefficient would be

$$\varsigma = \frac{\pi \eta_s L^3}{4} \qquad \text{(no hydrodynamic interaction)} \qquad (10\text{-}11)$$

Thus ς should lie between Equations 10-10 and 10-11. Fortunately these two estimates of ς are not vastly different. For either value, the friction coefficient is approximately related to the dilute solution rotary diffusion constant* by[4]

$$\varsigma = \frac{kT}{D_{ro}} \tag{10-12}$$

In dilute solutions, where $D_r = D_{ro}$, $\varsigma = 6\lambda kT$; Equation 10-9 can then be written as

$$\underset{\approx}{\tau} = \underbrace{6\nu kT <\underset{\sim\sim\sim\sim}{uuuu}> : \lambda\underset{\approx}{D}}_{\text{(viscous)}} + \underbrace{3\nu kT\{<\underset{\sim\sim}{uu}> - \frac{1}{3}\underset{\approx}{\delta}\}}_{\text{(entropic)}} \quad \text{(dilute)} \tag{10-13}$$

Since the coefficients of the viscous and the entropic contributions are similar, in the linear or moderately nonlinear flow regime, where $\lambda\underset{\approx}{D}$ is of order unity, these two contributions to the stress tensor are therefore of comparable magnitude.

If the concentration of rods is increased into the semidilute regime, then ς remains of the same order of magnitude, $\varsigma \approx kT/D_{ro}$, but $\overline{D}_r$ decreases dramatically, $\overline{D}_r \sim D_{ro}(\nu L^3)^{-2}$, where $(\nu L^3)^{-2} << 1$ in the semidilute regime. Now $\lambda = 1/6\overline{D}_r$. Hence $\varsigma \approx kT(\nu L^3)^{-2}/\overline{D}_r = 6kT\lambda(\nu L^3)^{-2}$. Equation 10-9 can therefore be rewritten as

$$\underset{\approx}{\tau} = \underbrace{6\nu kT(\nu L^3)^{-2} <\underset{\sim\sim\sim\sim}{uuuu}> : \lambda\underset{\approx}{D}}_{\text{(viscous)}} + \underbrace{3\nu kT\{<\underset{\sim\sim}{uu}> - \frac{1}{3}\underset{\approx}{\delta}\}}_{\text{(entropic)}} \quad \text{(semidilute)} \tag{10-14}$$

Thus if $\underset{\approx}{D}\lambda$ is not vastly greater than unity, then in the semidilute regime the viscous contribution to the stress tensor becomes smaller than the entropic contribution by the factor $(\nu L^3)^{-2}$. We conclude that the viscous term in Equation 10-9 can be dropped for most purposes in the semidilute regime. Thus

$$\underset{\approx}{\tau} = 3\nu kT[<\underset{\sim\sim}{uu}> - \frac{1}{3}\underset{\approx}{\delta}] = 3\nu kT\underset{\approx}{S} \tag{10-15}$$

where

$$\underset{\approx}{S} \equiv <\underset{\sim\sim}{uu}> - \frac{1}{3}\underset{\approx}{\delta} \tag{10-16}$$

Since the optical anisotropy tensor is also proportional to $\underset{\approx}{S}$, the stress and optical anisotropy tensors are proportional to each other. *The stress-optical law therefore applies to semidilute solutions of rods.*[1]

* In the absence of hydrodynamic interaction, Equation 10-12 is exact; with hydrodynamic interaction, $\varsigma = kT/2D_{ro}$.

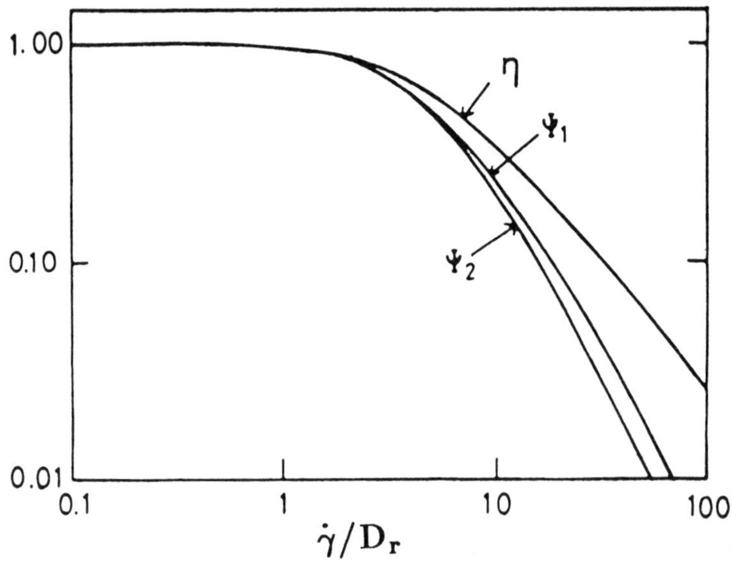

Fig. 10-5 Normalized viscosity, first normal stress coefficient, and second normal stress coefficient as functions of shear rate for the Doi-Edwards model for semidilute solutions of rods (from Kuzuu and Doi,[5] reprinted with permission from Polymer J.).

10.2.2.2 Predictions of the Doi-Edwards Theory Because Equation 10-8, the equation of change for the second moment, $<\underset{\sim\sim}{uu}>$, of the distribution function, depends on the fourth moment, $<\underset{\sim\sim\sim\sim}{uuuu}>$, and because $\overline{D}_r$ depends on $\psi(\underset{\sim}{u})$, accurate solutions for $<\underset{\sim\sim}{uu}>$ are obtained by first solving the Smoluchowski Equation, Equation 10-6, for ψ and then averaging $\underset{\sim\sim}{uu}$ over ψ to obtain $<\underset{\sim\sim}{uu}>$. Doi and Edwards[3] and Kuzuu and Doi[5] have carried out these calculations for simple shearing and elongational flows. In general, the rheological behavior is similar to that of the Doi-Edwards reptation theory for concentrated flexible polymers discussed in Chapter 4. Figure 10-5, for example, shows the shear rate dependence of the shear viscosity and first and second normal stress coefficients predicted by the Doi-Edwards theory for semidilute rods; these results are similar to those predicted by the Doi-Edwards theory for entangled flexible molecules; see Figure 4-12.

At low shear rates, the viscosity and first and second normal stress coefficients are[5]

$$\eta_0 = \frac{\nu k T}{10 D_r}$$

$$\Psi_{1,0} = \frac{\nu k T}{30 D_r}$$

$$\frac{\Psi_{2,0}}{\Psi_{1,0}} = -\frac{1}{7} \tag{10-17}$$

The value for $\Psi_{2,0}/\Psi_{1,0}$ is exactly that of the Doi-Edwards theory for flexible chains. The theory for flexible molecules is similar to the theory for rods because of the fast *retraction process* that prevents flexible strands from stretching in the flow field; flexible strands therefore respond to the flow as rods do.

10.2.2.3 Premises of the Doi-Edwards Theory for Semidilute Rods The following summarizes the assumptions from which the Doi-Edwards theory is derived:

1. Surrounding rods create an effective tube out of which a test rod can escape only by longitudinal diffusion. The tube is defined as a region just small enough that the surrounding rods scarcely penetrate it; as the rods become oriented, the tube diameter grows.

2. After escaping the tube by longitudinal diffusion, the test rod can rotate through a small angle approximately equal to the tube diameter. Further rotation of the rod requires that it diffuse out of the new tube in which it finds itself. The repeated processes of tube escape by longitudinal diffusion, and rotation through the allowed small random angle, define a rotary diffusion process and a rotary diffusion coefficient for the test rod that depends on the orientation of the test rod with respect to the surrounding rods.

In addition, the following mathematical approximation makes the theory tractable:

3. The orientation-dependent rotary diffusion coefficient is replaced by an average value that depends on the average orientation of the rods but not on the orientation of the test rod with respect to the surrounding rods.

10.2.2.4 Marrucci-Grizzuti Theory for Polydisperse Rods Because the rotary diffusivity of a rod has such a strong dependence on the molecular weight of the rod, scaling as $\rho^{-2}M^{-7}$ in semidilute solution, the effects of even small levels of polydispersity are large. It is therefore difficult to obtain material of sufficient monodispersity to test the Doi-Edwards theory. Marrucci and Grizzuti,[6] however, have extended the theory to account for a discrete distribution of rod lengths. They assumed that a test rod, i, can rotate through an elementary angle, ϵ_i, if rod i itself diffuses longitudinally out its tube, *or if a constraining rod, j, diffuses out of the way of the test rod.* If the constraining rod is shorter than the test rod, i.e., $L_j < L_i$, then the latter process is assumed to occur first. If the test rod is shorter, then it must diffuse out of its tube. From this and other assumptions and approximations, Marrucci and Grizzuti derived a formula for $\overline{D}_{ri}$

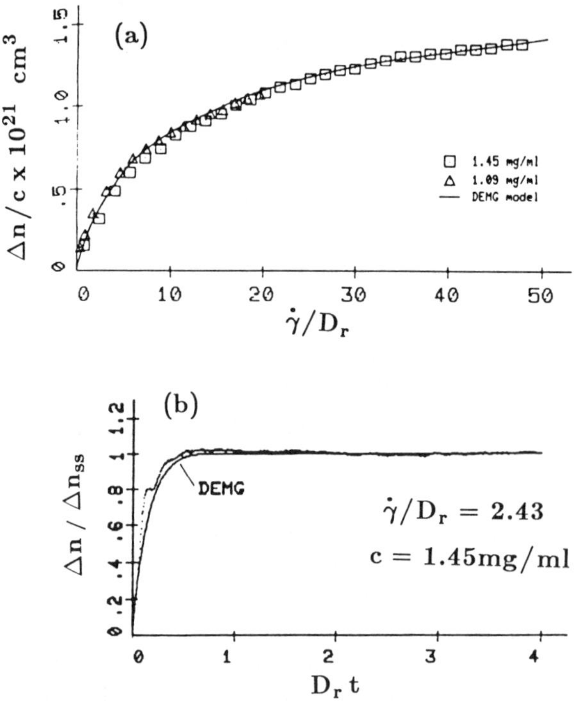

Fig. 10-6 (a) Steady-state birefringence divided by rod concentration versus dimensionless shear rate for two collagen solutions. (b) Growth of reduced birefringence with reduced time for a collagen solution. In both (a) and (b) comparison is made to Marrucci and Grizzuti's modification (DEMG) of the Doi-Edwards theory for semidilute rods (from Chow et al.,[7] reprinted with permission from Macromolecules ©1985 American Chemical Society).

for each rod of length L_i in a polydisperse mixture.

Chow et al.[7] compared the predictions of Marrucci and Grizzuti's generalization of the Doi-Edwards theory to birefringence measurements on semidilute solutions of slightly flexible collagen molecules for start-up of steady shearing. Figure 10-6 shows that the steady-state birefringence and extinction angle in shear as well as the growth of birefringence after start-up of steady shearing for polydisperse solutions of rigid collagen molecules agree almost quantitatively with the Doi-Edwards theory as modified by Marrucci and Grizzuti. The predicted scaling, $\lambda \propto \nu^2$, of relaxation time with concentration is also observed. Note that for nearly rigid molecules, the birefringence overshoot is small, even though $\lambda\dot{\gamma} \approx 6\dot{\gamma}/D_r$ is large. For entangled flexible polymer molecules, large birefringence overshoots occur when $\lambda\dot{\gamma} >> 1$. This indicates that when flexible molecules are suddenly subjected to high rates of shear, the molecules are stretched significantly before the retraction process can reduce the stretch and, consequently, the

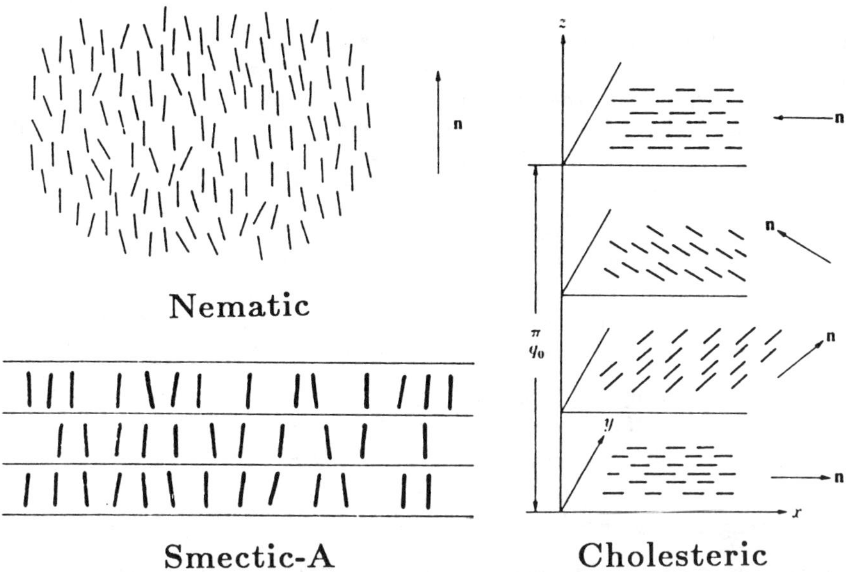

Fig. 10-7 Molecular arrangements in nematic, cholesteric, and smectic-A phases (from de Gennes,[8] reprinted with permission from *The Physics of Liquid Crystals*, Oxford University Press, 1974).

birefringence; see Sections 4.6.3.4 and 6.3.5.2. Such stretching, of course, does not occur in solutions of rods.

10.3 THE ISOTROPIC TO NEMATIC TRANSITION

When the rod concentration is so increased that the rods can no longer orient randomly without significant interference with each other, the rods spontaneously orient to form a *nematic liquid crystalline* phase. Liquid crystals, or *mesophases*, are intermediate between simple liquids, which have no long-range order, and crystalline solids, which possess long-range three-dimensional positional order.[8] The simplest liquid crystals have long-range *orientational order*, but no long-range positional order. There are two types of such liquid crystals: *nematics* and *cholesterics*. In a nematic phase, molecules tend to orient in a preferred direction, Figure 10-7. This preferred orientation is described by a vector, $\underset{\sim}{n}$, called the *director*. The molecules in a nematic either possess fore-aft symmetry or point forward as frequently as backward so that $\underset{\sim}{n}$ and $-\underset{\sim}{n}$ describe the same state of orientation. In cholesterics, on the other hand, the preferred direction is constant only in special planes; the director rotates helically as one moves in a direction perpendicular to these planes, Figure 10-7.

There is, in addition, a class of liquid crystals with long-range positional order in one dimension. These are the *smectics*. Smectics have a layered structure with long-range order in the direction perpendicular to the layers. The simplest is the smectic-A, depicted in Figure 10-7, in which each layer is a two-dimensional liquid with no long-range positional order. The director in a smectic-A phase is perpendicular to the layers. There are other more complex smectic phases (B,C,D,F, and G), in which the director may not be perpendicular to the layers or in which some positional order may exist within each layer. For more details regarding the various liquid crystalline phases, the reader is referred to ref. 8. Most theoretical work to date has focused on the simple nematics.

The transition from an isotropic to a liquid crystalline phase is induced in some materials by increasing the concentration of rodlike molecules in solution at a given temperature. These are called *lyotropic* liquid crystals. Other liquid crystalline materials contain no solvent; these, the *thermotropics*, are brought into a liquid crystalline state by suitably varying the temperature.

In lyotropics, a transition from an isotropic to a nematic phase occurs when rodlike molecules become concentrated enough to frustrate packing in random orientations. The first theoretical treatment of this transition is that of Onsager[9] (1949).

10.3.1 Onsager Theory

Onsager expressed the free energy for a concentrated solution of rods as[8]

$$A = \text{const} + \underbrace{\nu kT \int \psi(\underset{\sim}{u}) \log[\psi(\underset{\sim}{u})]\, du^2}_{\text{(orientation)}} + \underbrace{\tfrac{1}{2}\nu^2 kT \int\int \psi(\underset{\sim}{u})\psi(\underset{\sim}{u}')\,\beta(\underset{\sim}{u},\underset{\sim}{u}')\, du^2 du'^2}_{\text{(excluded volume)}} \tag{10-18}$$

The "orientation" term

$$A_1 \equiv \nu kT \int \psi \log\psi \, du^2 \tag{10-19}$$

is the usual entropic contribution that also appears in dilute solution theories; see Section 8.7.1.2. The additional term

$$A_2 \equiv \tfrac{1}{2}\nu^2 kT \int\int \psi(\underset{\sim}{u})\psi(\underset{\sim}{u}')\,\beta(\underset{\sim}{u},\underset{\sim}{u}')\, du^2 du'^2 \tag{10-20}$$

comes from pairwise *excluded volume* effects. Expression 10-18 is valid if there are no interactions between rods, except the requirement that the rods not interpenetrate, and if three-body and higher order interactions are negligible. Both of these assumptions become more reasonable as the rod volume fraction, $\phi = \pi L b^2 \nu/4$, decreases. As noted earlier, a nematic transition occurs at small values of ϕ when the aspect ratio, L/b, is large. $\beta(\underset{\sim}{u},\underset{\sim}{u}')$ is the excluded volume per rod; it depends on the angle between rod orientations $\underset{\sim}{u}$ and $\underset{\sim}{u}'$. As depicted in Figure 10-8, the center of mass of rod O′ is excluded from the shaded region around rod O. This shaded region has area $L^2|\underset{\sim}{u} \times \underset{\sim}{u}'|$. The thickness of the excluded region is twice the rod diameter. Thus the volume excluded to O′ is

$$\beta(\underset{\sim}{u},\underset{\sim}{u}') = 2L^2 b\, |\underset{\sim}{u} \times \underset{\sim}{u}'| \tag{10-21}$$

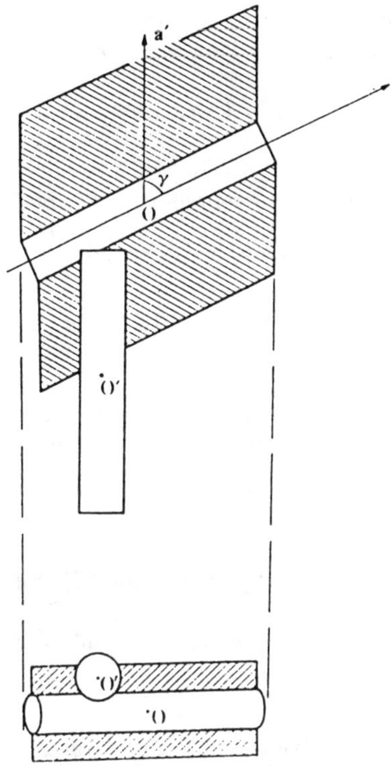

Fig. 10-8 Rod O excludes the center of mass of any other rod, such as O′, from the shaded region. The excluded volume decreases as the angle, γ, between the rods decreases (from de Gennes,[8] reprinted with permission from *The Physics of Liquid Crystals*, Oxford University Press, 1974).

At equilibrium, the rod orientation distribution function adjusts itself to minimize the free energy, A. To find this minimization condition, we vary $\psi(\underset{\sim}{u})$ subject to the constraint of conservation of probability,

$$\int \psi \, du^2 = 1 \tag{10-22}$$

This constraint is imposed by the method of Langrange multipliers. Thus we minimize

$$A[\psi(\underset{\sim}{u})] - \xi \int \psi(\underset{\sim}{u}) \, du^2 \tag{10-23}$$

with respect to changes in $\psi(\underset{\sim}{u})$, where ξ is the Langrange multiplier. To find this local minimum by the calculus of variations, we add a small perturbation, $\delta\psi(\underset{\sim}{u})$, to $\psi(\underset{\sim}{u})$ and find the function $\psi(\underset{\sim}{u})$ for which $A[\psi] - \xi \int \psi du^2$ is independent of this perturbation. This is equivalent to finding ψ such that the *functional derivative* of A with respect to ψ is zero. Thus we seek $\psi(\underset{\sim}{u})$ such that

$$A[\psi + \delta\psi] - A[\psi] - \xi\left(\int (\psi + \delta\psi) du^2 - \int \psi du^2\right) = 0 \qquad (10\text{-}24)$$

for small $\delta\psi$. For the orientational contribution to the free energy, A_1,

$$A_1[\psi + \delta\psi] - A_1[\psi] = \nu kT \int (\psi + \delta\psi)\log(\psi + \delta\psi) du^2 - \nu kT \int \psi \log\psi du^2$$

$$= \nu kT \int \delta\psi \log\psi du^2 + \nu kT \int \psi \log\left(\frac{\psi + \delta\psi}{\psi}\right) du^2 = \nu kT \int \delta\psi \log\psi du^2 + \nu kT \int \delta\psi du^2$$

where only terms first order in $\delta\psi$ have been kept. Performing similar manipulations on the other terms of Equation 10-24, one obtains

$$\nu kT \int \delta\psi \{ \log\psi - 1 + \nu \int \beta(\underset{\sim}{u}, \underset{\sim}{u}')\psi(\underset{\sim}{u}') du'^2 - \xi \} du^2 = 0 \qquad (10\text{-}25)$$

Equation 10-25 holds for any small perturbation $\delta\psi(\underset{\sim}{u})$ if and only if

$$\log\psi - 1 + \nu \int \beta(\underset{\sim}{u}, \underset{\sim}{u}')\psi(\underset{\sim}{u}') du'^2 - \xi = 0 \qquad (10\text{-}26)$$

It remains, therefore, to solve the above integral Equation for $\psi(\underset{\sim}{u})$. Although it cannot be solved analytically, Onsager obtained an approximate solution by assuming a form for the solution, namely

$$\psi(\underset{\sim}{u}) = \frac{\alpha}{4\pi \sinh\alpha} \cosh(\alpha \underset{\sim}{u} \cdot \underset{\sim}{n}) \qquad (10\text{-}27)$$

where $\underset{\sim}{n}$ is an arbitrary unit vector. The coefficient in front of $\cosh(\alpha \underset{\sim}{u} \cdot \underset{\sim}{n})$ is required so that $\int \psi du^2 = 1$. α is a constant determined by inserting the above expression into Equation 10-18 and minimizing the free energy, A, with respect to variations in α.

The free energy, A, as a function of α and of concentration, ν, is plotted[1] in Figure 10-9. At small concentrations, A is a monotonically increasing function of α. Thus the minimum free energy is at $\alpha = 0$, the isotropic state. As ν increases, A develops a second local minimum, corresponding to a metastable state that eventually decreases until it becomes the global minimum, and therefore the stable state. Thus at higher ν, a nematic state with $\alpha \neq 0$ becomes the stable phase, and the isotropic state becomes first metastable and then at still higher ν it becomes unstable relative to the nematic.

Analysis of the dependence of A on α and ν reveals that there is a range of overall rod concentration in which there is *coexistence* of an isotropic and a nematic phase. In the coexistence region, the volume fraction of rods in the nematic phase is $\phi_n^c = 4.5b/L$, and in the coexisting isotropic phase it is $\phi_i^c = 3.3b/L$. The higher the aspect ratio, L/b, the lower the concentration, ϕ_i^c, at

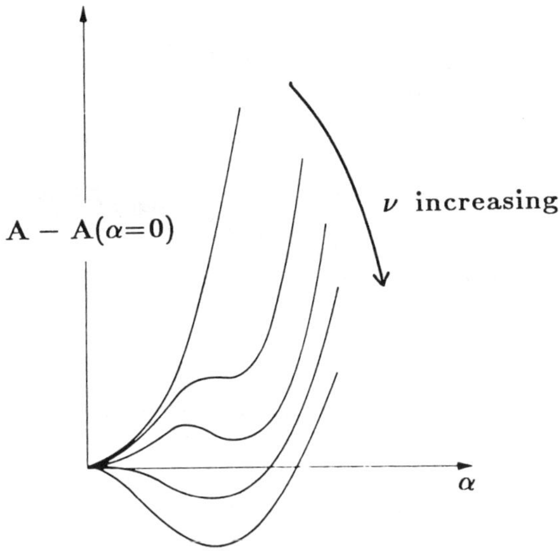

Fig. 10-9 Onsager free energy versus parameter α for various concentrations ν (from Doi and Edwards,[1] reprinted with permission from *The Theory of Polymer Dynamics,* Oxford University Press, 1986).

which a nematic phase begins to appear.

In the nematic phase, α in Equation 10-27 turns out to be large, and ψ is therefore a strongly peaked function of the angle between $\underset{\sim}{u}$ and $\underset{\sim}{n}$. Thus the degree of orientation in the nematic phase is high. A measure of this degree of orientation is provided by the *order parameter,* S, defined as

$$S \equiv \frac{3}{2}\int \psi\,[(\underset{\sim}{u}\cdot\underset{\sim}{n})^2 - \frac{1}{3}]\,du^2 \qquad (10\text{-}28)$$

The order parameter varies from zero for a randomly oriented system to unity for complete orientation. Nematic phases have uniaxial symmetry, which means that the distribution function is isotropic in the plane perpendicular to $\underset{\sim}{n}$. $\psi(\underset{\sim}{u})$ is therefore a function only of the angle, θ, between $\underset{\sim}{u}$ and $\underset{\sim}{n}$. Hence

$$S = \frac{\pi}{2}\int_0^{\pi} \psi(\theta)\sin\theta\,[3\cos^2\theta - 1]\,d\theta \qquad (10\text{-}29)$$

For the Onsager theory, the order parameter, S, in the nematic phase is 0.84 at the transition concentration. S approaches unity as the concentration is further increased.

There is also a *tensor order parameter,* $\underset{\approx}{S}$, defined as

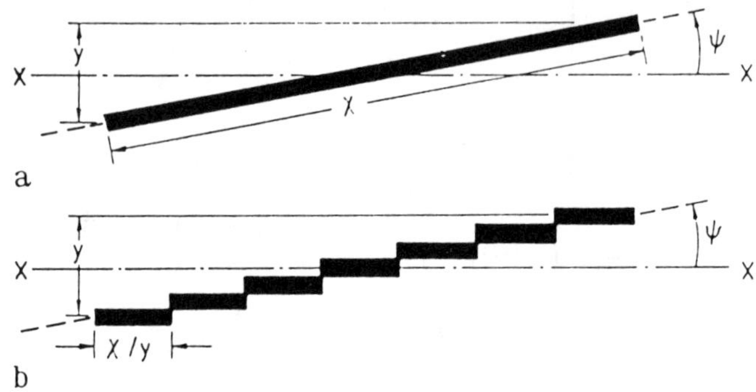

Fig. 10-10 (a) Rod oriented at an angle ψ with respect to the preferred direction; (b) its representation on a cubic lattice by subdivision of the X units of rod length into y segments each of length X/y (from Flory,[11] reprinted with permission from Advances in Polymer Science, ©1984 Springer Verlag).

$$\underset{\approx}{S} \equiv \langle \underset{\sim\sim}{uu} \rangle - \frac{1}{3}\underset{\approx}{\delta} = S(\underset{\sim\sim}{nn} - \frac{1}{3}\underset{\approx}{\delta}) \tag{10-30}$$

The scalar order parameter, S, which can be obtained from $\underset{\approx}{S}$, is

$$S = \frac{3}{2}\underset{\approx}{S} : \underset{\sim\sim}{nn} \tag{10-31}$$

10.3.2 Flory Lattice Theory

A more accurate estimate of the critical volume fraction, ϕ_i^c, at which a nematic phase appears is provided by the Flory lattice theory.[10,11] Flory (1956) fit imperfectly oriented rods on a cubic lattice by segmenting the rods as depicted in Figure 10-10. Using only excluded volume interactions, Flory computed the numbers of ways of packing rods on the lattice in this way, and thereby obtained the partition function and thus the thermodynamic properties for rods in a solvent. For very high aspect ratio, $L/b \rightarrow \infty$, Flory's theory predicts that the threshold concentration for the appearance of a nematic phase is

$$\phi_n^c \approx 8\, b/L \tag{10-32}$$

This value seems to be in better agreement with experimental data than the value given by the Onsager theory. Table 10-1 compares the predictions of the two theories.

Table 10-1 Coexistent Volume Fractions at the Nematic Transition

	$\phi_i^c \cdot L/b$	ϕ_n^c/ϕ_i^c
Onsager, approximate	3.34	1.34
Onsager, exact	3.29	1.28
Flory, approximate $L/b \to \infty$	8.28	1.59
Flory, exact $L/b \to \infty$	7.89	1.47
Flory exact $L/b = 20$	7.28	1.37

For a *neat fluid*, that is, a single component fluid with $\phi = 1$, the exact Flory theory[11] predicts that the nematic phase forms at a minimum aspect ratio of $L/b = 6.42$. A small rigid molecule rodlike molecule that forms a neat phase is p-azoxyanisole (PAA) with the structure

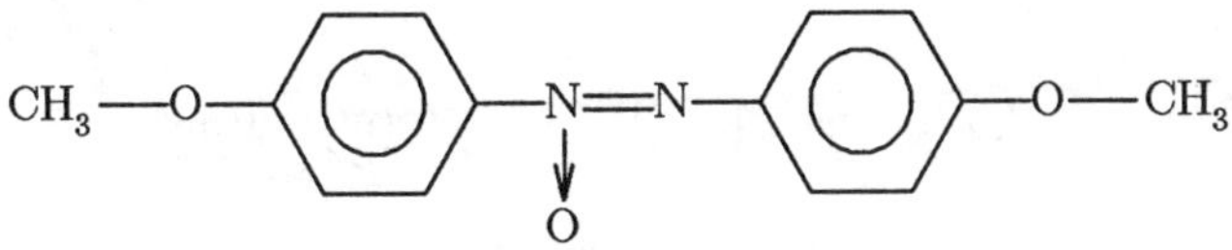

10.3.3 Maier-Saupe Theory

Maier and Saupe[12] proposed a phenomenological theory for nematic phases. In the Maier-Saupe theory, the free energy is given by

$$A = \text{const} + \underset{\text{(orientation)}}{\nu kT \int \psi \log\psi \, du^2} + \underset{\text{(molecular interactions)}}{A_2} \qquad (10\text{-}33)$$

A_2 is regarded as a material function that depends on the order parameter and can also depend on temperature and pressure. Thus in the Maier-Saupe theory, the simple entropic excluded volume interaction assumed by the Onsager and Flory theories can be augmented by energetic interactions such as orientation-dependent van der Waals forces between the molecules. The temperature and pressure dependence of A_2 make the Maier-Saupe theory especially suitable for thermotropic liquid crystals.

It is usually assumed that the dependence of A_2 on S is

$$A_2 = \text{const} - \frac{1}{2}\nu U_{MS} S^2 \quad (10\text{-}34)$$

where U_{MS} is a parameter that can depend on temperature and pressure.

The Maier-Saupe theory is simpler than the Onsager theory in that A_2 depends on the distribution function $\psi(\underset{\sim}{u})$ only through the order parameter $S = \frac{3}{2}\int \psi[(\underset{\sim}{u}\cdot\underset{\sim}{n})^2 - \frac{1}{3}]\,du^2$. Thus in the Maier-Saupe theory, A_2 is a *mean-field* quantity. A relationship between the Onsager and Maier-Saupe theories is illuminated by Doi and Edwards, who write the Onsager excluded volume free energy as[1]

$$A_2 = \frac{\nu}{2}\int \psi(\underset{\sim}{u})\,U_{scf}(\underset{\sim}{u})\,du^2 \quad (10\text{-}35)$$

where U_{scf} is the average potential acting on a rod and is given by

$$U_{scf} = \nu kT \int \psi(\underset{\sim}{u}')\,\beta(\underset{\sim}{u},\underset{\sim}{u}')\,du'^2 \quad (10\text{-}36)$$

If we replace Onsager's choice for $\beta(\underset{\sim}{u},\underset{\sim}{u}')$ by

$$\beta(\underset{\sim}{u},\underset{\sim}{u}') = 2L^2 b\,[1 - (\underset{\sim}{u}\cdot\underset{\sim}{u}')^2] \quad (10\text{-}37)$$

then

$$U_{scf} = 2\nu kTL^2 b\,[1 - \underset{\sim}{u}\underset{\sim}{u} : \int \psi(\underset{\sim}{u}')\,\underset{\sim}{u}'\underset{\sim}{u}'\,du'^2]$$

$$= \text{const} - 2\nu kTL^2 b\,\underset{\sim}{u}\underset{\sim}{u} : \underset{\approx}{S} \quad (10\text{-}38)$$

A_2 then follows from Equation 10-35:

$$A_2 = \text{const} - \nu^2 kTL^2 b\underset{\approx}{S} : \int \psi(\underset{\sim}{u})\underset{\sim}{u}\underset{\sim}{u}\,du^2 = \text{const} - \nu^2 kTL^2 b\underset{\approx}{S} : \underset{\approx}{S}$$

$$= \text{const} - \frac{4}{9}\nu^2 kTL^2 bS^2 \quad (10\text{-}39)$$

which has precisely the S-dependence proposed by Maier and Saupe. Thus, except for possible T and p dependence of U_{MS} in the Maier-Saupe theory, the Maier-Saupe form for A_2 can be obtained from that of Onsager by replacing the excluded volume term $|\underset{\sim}{u} \times \underset{\sim}{u}'| = \sin\theta$ with $1 - (\underset{\sim}{u}\cdot\underset{\sim}{u}')^2 = 1 - \cos^2\theta = \sin^2\theta$.

Figure 10-11 compares the Onsager interaction, $\beta(\underset{\sim}{u},\underset{\sim}{u}') \propto \sin\theta$ with that of Maier and Saupe, $\beta(\underset{\sim}{u},\underset{\sim}{u}') \propto \sin^2\theta$. The Maier-Saupe form is "softer"; the penalty, $\beta(\underset{\sim}{u},\underset{\sim}{u}')$, for a modest angular deviation, θ, of a test rod's orientation from that of surrounding rods is much less for the Maier-Saupe theory than for the Onsager theory. In the Maier-Saupe theory, the rods in the nematic are

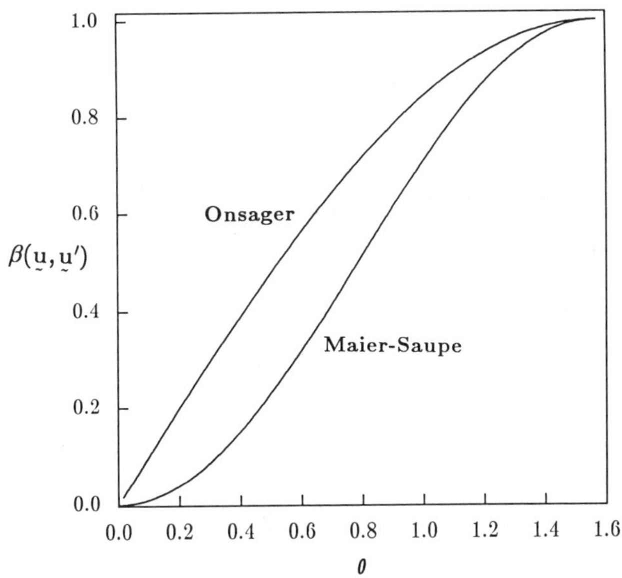

Fig. 10-11 Excluded volume factor, β, as a function of θ, the angle at which one rod is oriented with respect to the other, according to the Onsager theory and a Maier-Saupe theories.

therefore less aligned, and the order parameter S is consequently less than in the Onsager or Flory theories. In the nematic phase at the transition, $S = 0.44$ for the Maier-Saupe theory, compared to 0.84 for the Onsager theory. Experimentally, typical values for S above the transition are 0.4 to 0.7 for low molecular weight liquid crystals.[13] The smaller values predicted for the order parameter are therefore a welcome feature of the Maier-Saupe theory.

If the free energy, from Equations 10-33 and 10-34, of the Maier-Saupe theory is minimized using the variational principle, as was done in Section 10.3.1 for the Onsager theory, an analytic expression for $\psi(\underset{\sim}{u})$ emerges, namely

$$\psi(\underset{\sim}{u}) = \exp[m(\underset{\sim}{u} \cdot \underset{\sim}{n})^2] \,/\, 4\pi Z \qquad (10\text{-}40)$$

with

$$m \equiv \frac{3}{2}(U_{MS}S/kT)$$

and a normalization constant

$$Z \equiv \frac{1}{4\pi}\int_0^1 e^{mx^2}\,dx$$

A first-order transition between isotropic and nematic phases occurs at a temperature given by

$$\frac{kT}{U_{MS}(T)} = \frac{1}{4.55} \tag{10-41}$$

10.4 DOI CONSTITUTIVE EQUATION FOR NEMATIC POLYMERS

Doi[14,1] extended the theory for semidilute rods discussed in Section 10-2 to nematics by incorporating into it a Maier-Saupe potential. The Smoluchowski Equation for rotary motion of rods in the presence of a potential, U_{scf}, is

$$\frac{\partial \psi}{\partial t} + \frac{\partial}{\partial \underset{\sim}{u}} \cdot [(\underset{\sim}{u} \cdot \nabla \underset{\sim}{v} - \underset{\sim}{u}\underset{\sim}{u}\underset{\sim}{u} : \underset{\approx}{D})\psi]$$

$$- \bar{D}_r \frac{\partial}{\partial \underset{\sim}{u}} \cdot [\frac{\partial \psi}{\partial \underset{\sim}{u}} + \psi \frac{\partial}{\partial \underset{\sim}{u}}(\frac{U_{scf}}{kT})] = 0 \tag{10-42}$$

When $U_{scf} = 0$, Equation 10-42 reduces to Equation 10-6 for the semidilute regime.

For the Maier-Saupe theory,

$$U_{scf} = \text{const} - \frac{3}{2} UkT \, \underset{\sim}{u}\underset{\sim}{u} : \underset{\approx}{S} \tag{10-43}$$

where U, according to Equation 10-38, is a constant proportional to $\nu L^2 b$. L and b are the rod length and diameter, respectively, and ν is the rod number concentration per unit volume.

Equation 10-42 uses a preaveraged form for the rotary diffusivity, $\bar{D}_r$, which depends on rod orientation only in an average sense through its dependence on the order parameter, $\underset{\approx}{S}$. Although this dependence can be calculated from Equation 10-7 for special cases, for convenience Doi approximated it by*

$$\bar{D}_r = D_r [1 - \frac{3}{2} \underset{\approx}{S} : \underset{\approx}{S}]^{-2} \tag{10-44}$$

where D_r is the rotary diffusivity in an isotropic solution; see Equation 10-3.

10.4.1 Dynamic Equation for the Order Parameter

Multiplying Equation 10-42 by $\underset{\sim}{u}\underset{\sim}{u} - \frac{1}{3}\underset{\approx}{\delta}$ and integrating over ψ gives, after integrating by parts,[1]

---

* For highly oriented nematics, a more accurate expression is $\bar{D}_r = D_r[1 - \frac{3}{2}\underset{\approx}{S} : \underset{\approx}{S}]^{-1}$.

$$\frac{\partial}{\partial t}\underset{\approx}{S} = \underset{\approx}{F}[\underset{\approx}{S}] + \underset{\approx}{G}[\underset{\approx}{S}] \tag{10-45}$$

Here $\underset{\approx}{F}[\underset{\approx}{S}]$ is the diffusion term

$$\underset{\approx}{F} \equiv -6\bar{D}_r\underset{\approx}{S} + 6\bar{D}_r U[\underset{\approx}{S}\cdot<\underset{\sim\sim}{uu}> - \underset{\approx}{S}:<\underset{\sim\sim\sim\sim}{uuuu}>] \tag{10-46}$$

and $\underset{\approx}{G}[\underset{\approx}{S}]$ is the convection term

$$\underset{\approx}{G} \equiv \frac{2}{3}\underset{\approx}{D} + \underset{\sim\sim}{\nabla v}^T\cdot\underset{\approx}{S} + \underset{\approx}{S}\cdot\underset{\sim\sim}{\nabla v} - 2\underset{\approx}{D}:<\underset{\sim\sim\sim\sim}{uuuu}> \tag{10-47}$$

As always in rod dynamics, a fourth moment of $\underset{\sim}{u}$ appears in the evolution Equation. If the nematic is highly oriented, this average can be approximated by a product of second moments*

$$<\underset{\sim\sim\sim\sim}{uuuu}> \approx <\underset{\sim\sim}{uu}><\underset{\sim\sim}{uu}> \tag{10-48}$$

Equations 10-46 and 10-47 then become

$$\underset{\approx}{F} \approx -6\bar{D}_r[(1-\frac{U}{3})\underset{\approx}{S} - U(\underset{\approx}{S}\cdot\underset{\approx}{S} - \frac{1}{3}\underset{\approx}{S}:\underset{\approx\approx}{S\delta}) + U\underset{\approx\approx}{SS}:\underset{\approx}{S}] \tag{10-49}$$

$$\underset{\approx}{G} \approx \frac{2}{3}\underset{\approx}{D} + \underset{\sim\sim}{\nabla v}^T\cdot\underset{\approx}{S} + \underset{\approx}{S}\cdot\underset{\sim\sim}{\nabla v} - \frac{2}{3}\underset{\approx}{D}:\underset{\approx\approx}{S\delta} - 2\underset{\approx}{D}:\underset{\approx\approx}{SS} \tag{10-50}$$

At equilibrium in the absence of flow, $d\underset{\approx}{S}/dt = \underset{\approx}{0}$, and $\underset{\approx}{G} \equiv \underset{\approx}{0}$; thus $\underset{\approx}{F} = \underset{\approx}{0}$. Since at equilibrium there is uniaxial symmetry, $\underset{\approx}{S} = S(\underset{\sim\sim}{nn} - \frac{1}{3}\underset{\approx}{\delta})$, and Equation 10-49 with $\underset{\approx}{F} = \underset{\approx}{0}$ implies that

$$(1-\frac{U}{3})S - \frac{U}{3}S^2 + \frac{2}{3}S^3 = 0 \tag{10-51}$$

Equation 10-51 always has a solution corresponding to the isotropic state, that is $S = 0$. This is the only solution when $U < U_1^* = 8/3$. A second solution exists for $U > 8/3$, namely

---

* Preaveraging is usually dangerous, but especially so when applied to $<\underset{\sim\sim\sim\sim}{uuuu}>$. In a dilute or semidilute solution of rods only weakly oriented by flow, preaveraging produces large errors. Since nematics are oriented even at equilibrium, the errors are likely to be less; nevertheless we shall point out some important qualitative predictions of the Doi theory that are qualitatively altered by preaveraging.

$$S_{eq} = \frac{1}{4} + \frac{3}{4}(1 - \frac{8}{3U})^{1/2} \tag{10-52}$$

This solution is the only stable† one when $U > U_2^* = 3$. Thus according to the Doi theory, a stable nematic phase exists when $U > U_2^*$; in the nematic phase the order parameter takes a value given by

$$S_{eq} \geq S_{eq}(U_2^*) = 0.50 \tag{10-53}$$

As U or, equivalently, ν rises, S_{eq} increases, reaching 0.95 when $U = 20$.

10.4.2 The Stress Tensor

In the nematic or in the isotropic phase close to the nematic transition, the rod interaction potential U_{scf} contributes directly to the stress tensor. Doi obtained this contribution to the stress tensor from the *principle of virtual work*, discussed in Section 2.3.2. From Equations 10-35 and 10-43, the contribution of the rod interaction potential to the free energy is

$$A_2 = -\frac{3}{4} U\nu kT \int\int \psi(\underset{\sim}{u})\psi(\underset{\sim}{u}') \underset{\sim}{u}\underset{\sim}{u} : \underset{\sim}{u}'\underset{\sim}{u}' \, du'^2 du^2 \tag{10-54}$$

The power or rate of doing work required to deform the nematic is $\underset{\sim}{\nabla} \underset{\sim}{v}^T : \underset{\approx}{\tau}$. In the absence of dissipation, the contribution of U_{scf} to this power is

$$\begin{aligned} \underset{\sim}{\nabla} \underset{\sim}{v}^T : \underset{\approx}{\tau}_{scf} &= \frac{\partial A_2}{\partial t} \\ &= -\frac{3}{4} U\nu kT \int\int \psi(\underset{\sim}{u})\psi(\underset{\sim}{u}') \frac{\partial}{\partial t} (\underset{\sim}{u} \cdot \underset{\sim}{u}')^2 \, du'^2 du^2 \end{aligned} \tag{10-55}$$

where $\underset{\approx}{\tau}_{scf}$ is the contribution of the interaction potential to the stress tensor. Equation 10-55 applies to a hypothetical deformation that is so fast that no Brownian motion occurs. For rods convected in a velocity gradient, $\underset{\sim}{\nabla} \underset{\sim}{v}$, with no Brownian motion,

$$\frac{\partial}{\partial t} \underset{\sim}{u} = \underset{\sim}{u} \cdot \underset{\sim}{\nabla} \underset{\sim}{v} - \underset{\sim}{u}\underset{\sim}{u}\underset{\sim}{u}\underset{\sim}{u} : \underset{\sim}{\nabla} \underset{\sim}{v} \tag{10-56}$$

With this expression, Equation 10-55 can be converted to

$$\underset{\sim}{\nabla} \underset{\sim}{v}^T : \underset{\approx}{\tau}_{scf} = -3U\nu kT \int\int \psi(\underset{\sim}{u})\psi(\underset{\sim}{u}') \underset{\sim}{\nabla} \underset{\sim}{v}^T : (\underset{\sim}{u}'\underset{\sim}{u}' \cdot \underset{\sim}{u}\underset{\sim}{u} - \underset{\sim}{u}'\underset{\sim}{u}' : \underset{\sim}{u}\underset{\sim}{u}\underset{\sim}{u}\underset{\sim}{u})$$

† Equilibrium coexistence of the nematic and isotropic phases occurs at a value of U slightly less than U_2^*. For practical purposes, however, it can be assumed that U_2^* is the lowest value of U at which the nematic phase is more stable than the isotropic phase.

$$= \; -\underset{\sim}{\nabla}\underset{\sim}{v}^T : 3U\nu kT(\underset{\approx}{S}\cdot<\underset{\sim\sim}{uu}> - \underset{\approx}{S} : <\underset{\sim\sim\sim\sim}{uuuu}>) \qquad (10\text{-}57)$$

Since Equation 10-57 holds for any large velocity gradient, $\underset{\sim}{\nabla}\underset{\sim}{v}$, it follows that

$$\underset{\approx}{\tau}_{scf} \; = \; -3U\nu kT(\underset{\approx}{S}\cdot\underset{\approx}{S} - \underset{\approx}{S} : <\underset{\sim\sim\sim\sim}{uuuu}>) \qquad (10\text{-}58)$$

Using the preaveraging approximation, $<\underset{\sim\sim\sim\sim}{uuuu}> \approx <\underset{\sim\sim}{uu}><\underset{\sim\sim}{uu}>$, and $<\underset{\sim\sim}{uu}> = \underset{\approx}{S} + \frac{1}{3}\underset{\approx}{\delta}$, we obtain

$$\underset{\approx}{\tau}_{scf} \; = \; -3U\nu kT(\underset{\approx}{S}^2 - \underset{\approx}{S} : \underset{\approx\approx}{SS} + \frac{1}{3}\underset{\approx}{S} - \frac{1}{3}\underset{\approx}{S} : \underset{\approx\approx}{S\delta}) \qquad (10\text{-}59)$$

The total (elastic) stress tensor is therefore

$$\underset{\approx}{\tau} \; = \; 3\nu kT\underset{\approx}{S} + \underset{\approx}{\tau}_{scf}$$

$$= \; 3\nu kT[\underset{\approx}{S}(1 - \frac{U}{3}) - U(\underset{\approx}{S}\cdot\underset{\approx}{S} - \frac{1}{3}\underset{\approx}{S} : \underset{\approx\approx}{S\delta}) + U\underset{\approx}{S} : \underset{\approx\approx}{SS}] \; = \; \frac{3\nu kT}{-6\overline{D}_r}\underset{\approx}{F} \qquad (10\text{-}60)$$

The formula for the stress tensor, Equation 10-60, together with the evolution Equation for $\underset{\approx}{S}$, Equations 10-44, 10-45, 10-49 and 10-50, constitute the *Doi constitutive Equation for nematics.* The only unknown parameter in the Doi theory is $\overline{D}_r$, and even its dependence on concentration, ν, is known. Since the isotropic rotary diffusivity scales as $D_r \propto \nu^2$, from Equation 10-44 we have

$$\overline{D}_r \; = \; D_r^* \left(\frac{\nu^*}{\nu}\right)^2 (1 - S^2)^{-2} \qquad (10\text{-}61)$$

where D_r^* is the isotropic rotary diffusivity at $U = U_2^*$, and $\nu^*/\nu \equiv U_2^*/U$.

10.4.3 Shear Viscosity Predictions

Thus the Doi theory predicts the entire range of concentration-dependent viscoelastic phenomena with only two adjustable parameters, D_r^* and ν^*. Figure 10-12 shows the predicted dependence of the shear viscosity on concentration and on shear rate, $\dot{\gamma}$. The zero-shear viscosity is predicted by the Doi theory to rise with concentration as ν^3 in the isotropic region, and then drop suddenly and continue dropping as the nematic transition is crossed. Shear thinning, which occurs as the shear rate is increased, is predicted to blunt the viscosity maximum. Experimental data shown in Figure 10-13 for polybenzylglutamate rodlike molecules are qualitatively consistent with these predictions.

Other viscoelastic predictions of the Doi theory, especially predictions of unsteady phenomena, are in poor accord with experiment, as will be discussed in Section 10.7. The major theoretical limitation of the Doi theory, which is likely responsible to a large extent for its failure to conform to experimental behavior, is

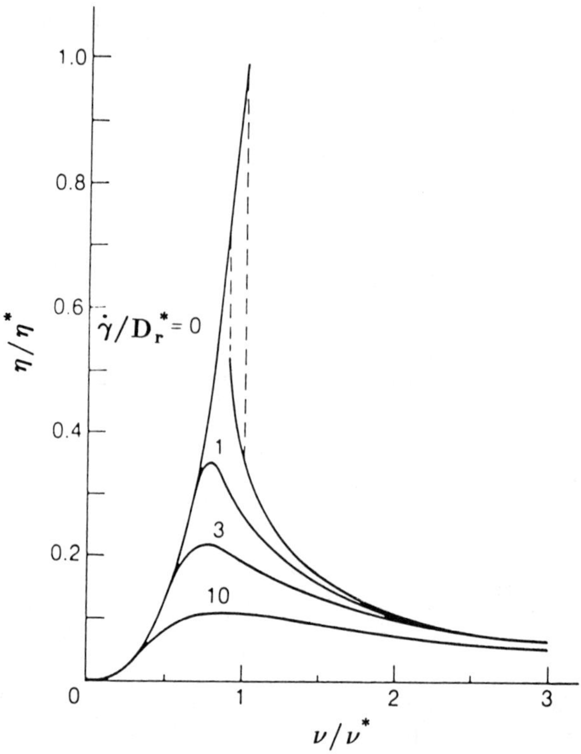

Fig. 10-12 Reduced shear viscosity as a function of reduced concentration at various reduced shear rates, $\dot{\gamma}/Dr^*$, according to the Doi theory for concentrated rods (from Doi and Edwards,[1] reprinted with permission from *The Theory of Polymer Dynamics*, Oxford University Press, 1986).

that it neglects the effect of *defects* in orientational order. Defects are ubiquitous in polymeric liquid crystals, and certainly affect rheological properties, probably in a profound way.

The influence of defects on the rheology of liquid crystal polymers is poorly understood, though the area is under active investigation. In small-molecule liquid crystals, defects are less prevalent and their influence is much better understood. Furthermore, small-molecule liquid crystals can be freed entirely from defects; the rheology of defect-free small molecule nematics is well understood, at least at a phenomenological level. Since much of what is currently understood or theorized about the rheology and the role of defects in liquid crystal polymers has a basis in studies of small-molecule liquid crystals, the next two sections are primarily, though not exclusively, directed toward small-molecule nematics.

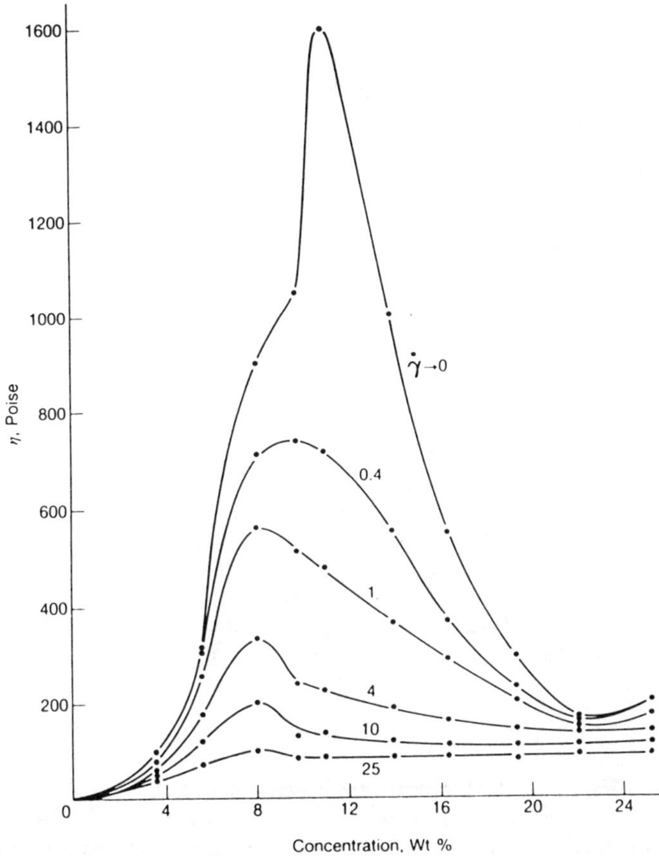

Fig. 10-13 Steady-state viscosity of poly(γ-benzyl-glutamate) solutions as a function of polymer concentration at various shear rates, $\dot{\gamma}$. (from Kiss and Porter,[34] reprinted from J. Polym. Sci., Polym. Symp., ©1978 by permission of John Wiley & Sons, Inc.).

10.5 STATICS OF LIQUID CRYSTALS

10.5.1 Theory of Spatially Varying Orientation in Nematics

10.5.1.1 Frank Constants In the nematic phase, molecules tend to orient, with some dispersion, about an orientation defined by a director, $\underset{\sim}{n}$. Intermolecular interactions are responsible for the existence and intensity of the anisotropy, but do not by themselves control the specific direction, $\underset{\sim}{n}$, toward which the molecules tend to orient. This direction is determined not by the material itself, but by the interaction of the material with body forces (electric or magnetic), or with boundaries, or with flow fields.

In the absence of body forces or flow fields, the influence of the surface interactions between the nematic fluid and the bounding material extends into

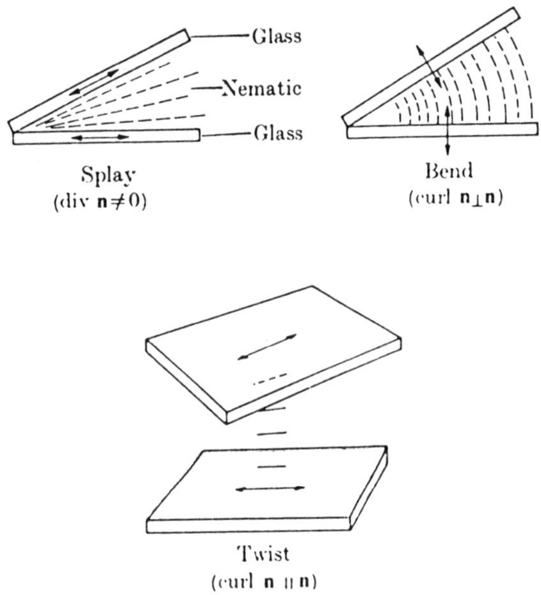

Fig. 10-14 Three types of distortion occurring in nematics (from de Gennes,[8] reprinted with permission from *The Physics of Liquid Crystals*, Oxford University Press, 1974).

the bulk of the nematic. With special treatment of the surface of the bounding material, one can control the *easy direction*, i.e., the orientation of the nematic at the surface. Buffing, rubbing, or ruling a glass surface in a single direction can force the nematic at the surface to align in this direction. An easy direction that is perpendicular to a solid surface can be achieved by suitable chemical treatment of the solid surface. If the molecules are *strongly anchored* at the surface, then the easy direction remains fixed even if a different orientation is forced on the bulk by a magnetic field, by velocity gradients, or by incompatible orientations at other boundaries.

Suppose, for example, that strong anchoring exists at the boundaries with an easy direction acting parallel or perpendicular to the boundaries as depicted in Figure 10-14. If the boundaries are not parallel, then the orientation of the director must vary spatially from one boundary to the other.

Spatial variations in the director field incur a free-energy penalty; the minimization of this penalty determines the equilibrium or static dependence of the director $\underset{\sim}{n}$ on position $\underset{\sim}{x}$. If the gradient of the director is everywhere small enough that the change in $\underset{\sim}{n}$ over the length of a single molecule is insignificant, then the excess free energy produced by that gradient can be obtained from a continuum theory. The most general continuum theory for nematics was developed by Oseen,[15] Zochar,[16] Frank,[17] and Ericksen.[18] The bulk free energy, W_d, resulting from a small gradient, $\underset{\sim}{\nabla}\,\underset{\sim}{n}$, in the director is

$$2W_d = K_1(\nabla \cdot \underset{\sim}{n})^2 + K_2(\underset{\sim}{n} \cdot \nabla \times \underset{\sim}{n})^2 + K_3(\underset{\sim}{n} \times \nabla \times \underset{\sim}{n})^2$$

$$\text{(splay)} \qquad \text{(twist)} \qquad \text{(bend)} \tag{10-62}$$

The three contributions to W_d are associated with the three independent modes of distortion: splay, twist, and bend, depicted in Figure 10-14.

Terms linear in $\underset{\sim}{n}$, such as $\underset{\sim}{\nabla} \cdot \underset{\sim}{n}$ and $\nabla^2 \underset{\sim}{n}$, do not appear in Equation 10-62 because nematics must be invariant to the transformation, $\underset{\sim}{n} \rightarrow -\underset{\sim}{n}$. The term $\underset{\sim}{n} \cdot \underset{\sim}{\nabla} \times \underset{\sim}{n}$ does not possess chiral symmetry; i.e., it is not invariant with respect to the transformation $\underset{\sim}{x} \rightarrow -\underset{\sim}{x}$, and thus must be dropped for nematics (although it is present for cholesterics). The omission of $\underset{\sim}{\nabla} \cdot \underset{\sim}{n}$ and $\underset{\sim}{n} \cdot \underset{\sim}{\nabla} \times \underset{\sim}{n}$ leave only terms quadratic in $\underset{\sim}{\nabla} \underset{\sim}{n}$. Terms of higher order than quadratic in $\underset{\sim}{\nabla} \underset{\sim}{n}$ are only required if spatial distortions become severe. The *Frank constants* K_1, K_2, and K_3 are of the order of magnitude 10^{-6} dynes for small-molecule nematics. For p-azoxyanisole (PAA), for example,[8]

$$K_1 = 0.7 \times 10^{-6} \text{ dyn} \; ; \; K_2 = 0.43 \times 10^{-6} \text{ dyn} \; ; \text{ and } K_3 = 1.7 \times 10^{-6} \text{ dyn}$$

Analyses are simplified by taking all three constants to be equal; $K \equiv K_1 = K_2 = K_3$. It can then be shown that

$$2W_d = K \underset{\sim}{\nabla} \underset{\sim}{n} : (\underset{\sim}{\nabla} \underset{\sim}{n})^T = K n_{j,i} n_{j,i} \tag{10-63}$$

Here T denotes transpose. The Equation is expressed both in Gibb's and Einstein's notation with the conventional summation over repeated indices. The subscript ",i" means partial differentiation with respect to x_i.

10.5.1.2 Minimization of Free Energy Now at equilibrium the spatial variation in the bulk of the director must be such that the free energy $W_d[\underset{\sim}{\nabla} \underset{\sim}{n}]$ is minimized subject to the constraint that $\underset{\sim}{n} \cdot \underset{\sim}{n} = 1$. The minimizing director field, $\underset{\sim}{n}(\underset{\sim}{x})$, can be found from the calculus of variations technique discussed in Section 10.3.1; one finds that $\underset{\sim}{n}(\underset{\sim}{x})$ is a solution of the Euler-Lagrange Equation

$$\underset{\sim}{\nabla} \cdot \left(\frac{\partial W_d}{\partial \underset{\sim}{\nabla} \underset{\sim}{n}} \right) - \frac{\partial W_d}{\partial \underset{\sim}{n}} = -\gamma \underset{\sim}{n} \tag{10-64a}$$

In Einstein's notation,

$$\frac{\partial}{\partial x_j} \left(\frac{\partial W_d}{\partial n_{i,j}} \right) - \frac{\partial W_d}{\partial n_i} = -\gamma n_i \tag{10-64b}$$

Here γ is a Lagrange multiplier introduced to satisfy the constraint $\underset{\sim}{n} \cdot \underset{\sim}{n} = 1$. de

Gennes calls the left side of Equation 10-64 the *molecular field,* $\underset{\sim}{h}$:

$$h_i \equiv \frac{\partial}{\partial x_j}\left(\frac{\partial W_d}{\partial n_{i,j}}\right) - \frac{\partial W_d}{\partial n_i}$$

At equilibrium, it follows from Equation 10-64 that

$$\underset{\sim}{n} \times \underset{\sim}{h} = \underset{\sim}{0} \qquad \text{(equilibrium)} \tag{10-65}$$

Thus *the director* $\underset{\sim}{n}$ *is parallel to the molecular field* $\underset{\sim}{h}$ *at equilibrium.*

10.5.1.3 Torque Balance Equation 10-65 can be interpreted as a *balance of torques*:

$$\underset{\sim}{n} \times \underset{\sim}{\nabla} \cdot \frac{\partial W_d}{\partial \underset{\sim}{\nabla}\underset{\sim}{n}} - \underset{\sim}{n} \times \frac{\partial W_d}{\partial \underset{\sim}{n}} = \underset{\sim}{0} \tag{10-66}$$

(surface torque) (body torque)

Here $\underset{\sim}{n} \times \underset{\sim}{\nabla} \cdot (\partial W_d / \partial \underset{\sim}{\nabla}\underset{\sim}{n})$ is the net torque transmitted to a small fluid element across its surfaces from neighboring fluid elements and $\underset{\sim}{n} \times (\partial W_d / \partial \underset{\sim}{n})$ is the torque exerted on the fluid element by molecules within the element.[19] Thus splayed, twisted, or bent nematics can transmit torque from one surface to another. For the single constant approximation $W_d = ½K \underset{\sim}{\nabla}\underset{\sim}{n} : (\underset{\sim}{\nabla}\underset{\sim}{n})^T$, we obtain

$$\frac{\partial W_d}{\partial \underset{\sim}{\nabla}\underset{\sim}{n}} = K \underset{\sim}{\nabla}\underset{\sim}{n} \tag{10-67}$$

$$\frac{\partial W_d}{\partial \underset{\sim}{n}} = \underset{\sim}{0} \tag{10-68}$$

10.5.1.4 Hydrostatics Another picture of the equilibrium state can be derived by multiplying Equation 10-64b by $n_{i,k}$:

$$\frac{\partial}{\partial x_j}\left(\frac{\partial W_d}{\partial n_{i,j}}\right) n_{i,k} - \frac{\partial W_d}{\partial n_i} n_{i,k} = \gamma n_i n_{i,k} = 0 \tag{10-69}$$

The last equality follows because $\partial(\underset{\sim}{n} \cdot \underset{\sim}{n})/\partial x_k$ is zero. Then using the product rule for $\partial/\partial x_j$, we rearrange Equation 10-69 as

$$\frac{\partial}{\partial x_j}\left\{\left(\frac{\partial W_d}{\partial n_{i,j}}\right) n_{i,k}\right\} - \left(\frac{\partial W_d}{\partial n_{i,j}}\right) n_{i,jk} - \frac{\partial W_d}{\partial n_i} n_{i,k} = 0 \tag{10-70}$$

Now

$$\tau_{ij}^{d} \equiv \left(\frac{\partial W_d}{\partial n_{i,j}}\right) n_{i,k} = \frac{\partial W_d}{\partial \nabla n} \cdot (\nabla n)^T$$

$$= K \nabla n \cdot (\nabla n)^T \qquad (\text{if } K_1 = K_2 = K_3) \tag{10-71}$$

is the *distortional stress* tending to cause an element of the nematic to change its shape in response to gradients in the director field. In the single constant approximation, τ^d is symmetric, but when $K_1 \neq K_2 \neq K_3$, it is *asymmetric*.[8]

From the definitions of $\partial W_d / \partial \nabla n$ and $\partial W_d / \partial n$, the other two terms in Equation 10-70 can be combined:

$$\frac{\partial W_d}{\partial n_{i,j}} n_{i,jk} + \frac{\partial W_d}{\partial n_i} n_{i,k} = \frac{\partial W_d}{\partial x_k} \tag{10-72}$$

Thus Equation 10-69 becomes

$$- \nabla \cdot \tau^d - \nabla W_d = 0 \qquad (\text{balance of forces}) \tag{10-73}$$

Equation 10-73 is a *force balance*, or a balance of linear momentum at hydrostatic equilibrium. Noting that $-W_d = p$, the hydrostatic pressure produced by the distorted nematic, Equation 10-73, is identical to Cauchy's momentum Equation in the absence of body forces and inertia.

The torque balance, Equation 10-66, and the force balance, Equation 10-73, were both derived from a minimization of free energy and are thus redundant under equilibrium conditions. This redundancy, of course, disappears when viscous flow occurs.

10.5.2 Magnetic Fields

Although most organic molecules are only slightly diamagnetic, the large number of such molecules and the high degree of correlation in their orientation make nematics responsive to external magnetic fields. If a magnetic field H is applied to a diamagnetic material, the resulting magnetization is[8]

$$M = \chi_{perp} H + (\chi_{paral} - \chi_{perp}) H \cdot n\, n \tag{10-74}$$

where χ_{perp} and χ_{paral} are the diamagnetic susceptibilities respectively perpendicular and parallel to the director n. We need only be concerned with the anisotropic contribution, so $\chi_{perp} H$ is hereafter dropped from Equation 10-74. The magnetic free energy is

$$W_M = -\int_0^{\underset{\sim}{H}} \underset{\sim}{M} \cdot d\underset{\sim}{H} = -\frac{1}{2}\chi_a(\underset{\sim}{n} \cdot \underset{\sim}{H})^2 \tag{10-75}$$

where

$$\chi_a \equiv \chi_{paral} - \chi_{perp} \tag{10-76}$$

is the anisotropy in the magnetic susceptibility; in nematics it is usually positive and of order of magnitude[8] $\chi_a \sim 10^{-7}$ (c.g.s. units). In the presence of both a magnetic field and boundary-induced gradients in the director, the total free energy is

$$W = W_M + W_d \tag{10-77}$$

The molecular field $\underset{\sim}{h}$ is influenced by the magnetic field:

$$\underset{\sim}{h} = \underset{\sim}{h}_d + \underset{\sim}{h}_M = \underset{\sim}{\nabla} \cdot \left(\frac{\partial W_d}{\partial \underset{\sim}{\nabla} \underset{\sim}{n}}\right) - \frac{\partial W_d}{\partial \underset{\sim}{n}} + \chi_a(\underset{\sim}{n} \cdot \underset{\sim}{H})\underset{\sim}{H} \tag{10-78}$$

At equilibrium, $\underset{\sim}{h}$ is again parallel to $\underset{\sim}{n}$:

$$\underset{\sim}{n} \times \underset{\sim}{h} = \underset{\sim}{0} \qquad \text{(torque balance)} \tag{10-79}$$

This torque balance Equation can again be arranged into a form that expresses the balance of linear momentum:

$$\underset{\sim}{\nabla} \cdot \underset{\approx}{\tau}^d + \underset{\sim}{\nabla}(W_d + W_M) = \underset{\sim}{0} \qquad \text{(force balance)} \tag{10-80}$$

Note from Equation 10-78 that if $\underset{\sim}{h}_M$ becomes much larger than $\underset{\sim}{h}_d$, then the tendency of bulk nematic to align in a direction compatible with the boundary alignment is overwhelmed by the magnetic field, which tends to orient the nematic parallel to $\underset{\sim}{H}$. If the anchoring at the surface is strong, then even though the bulk nematic is not parallel to the easy direction of the surface, there remains near the boundary a layer of thickness $\xi = (K/\chi_a)^{1/2}/H$ that remains parallel to the easy direction.[8] In a field of 1000 gauss, the wall boundary layer in a liquid crystal with $K = 10^{-6}$ dyn and $\chi_a = 10^{-7}$ (in c.g.s. units) is only about 30 microns thick.

10.5.3 Textures of Nematics

Liquid crystals have nonuniform director fields, not only because of wall effects, but also because of persistent metastable *defects*. Defects are points, lines, or planes where the director field, $\underset{\sim}{n}(\underset{\sim}{x})$, changes discontinuously. Figure 10-15 illustrates some line defects or *disclinations* in which the defect line is directed out of the plane of the page.[8,19] The director is continuous except at the

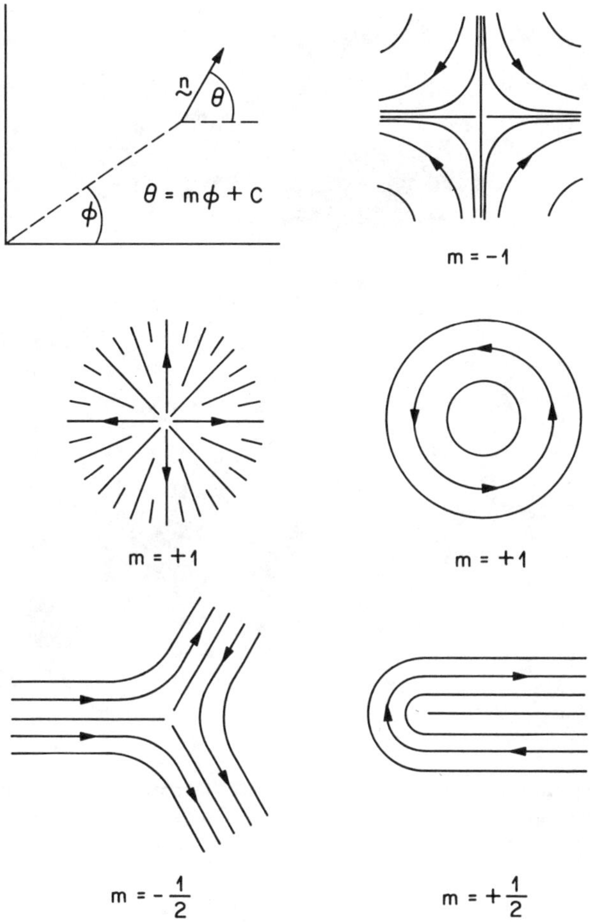

Fig. 10-15 Classes of defects in a nematic. As one revolves around the defect, the director rotates in the same direction as one is revolving for a "+" defect and in the opposite direction for a "-" defect; For the half-integral defects, an arbitrary line must be chosen along which the director reverses direction discontinuously.

defect. Thus starting at a point on the x axis in Figure 10-15, as one moves in a circle around the defect, the angle that the director makes with the x axis must vary continuously, until one returns to the starting point, where the director must return to its original orientation, or to an orientation opposite to the original, since the states $\underset{\sim}{n}$ and $-\underset{\sim}{n}$ are indistinguishable in a nematic. Thus as one moves through an angle of $\phi = 2\pi$ with respect to the x axis, the director rotates through an angle of $\theta = 2\pi m$, where m is a positive or negative multiple of 1/2. Different values of m define distinct classes of defects. Figure 10-15 illustrates the four important classes, for which

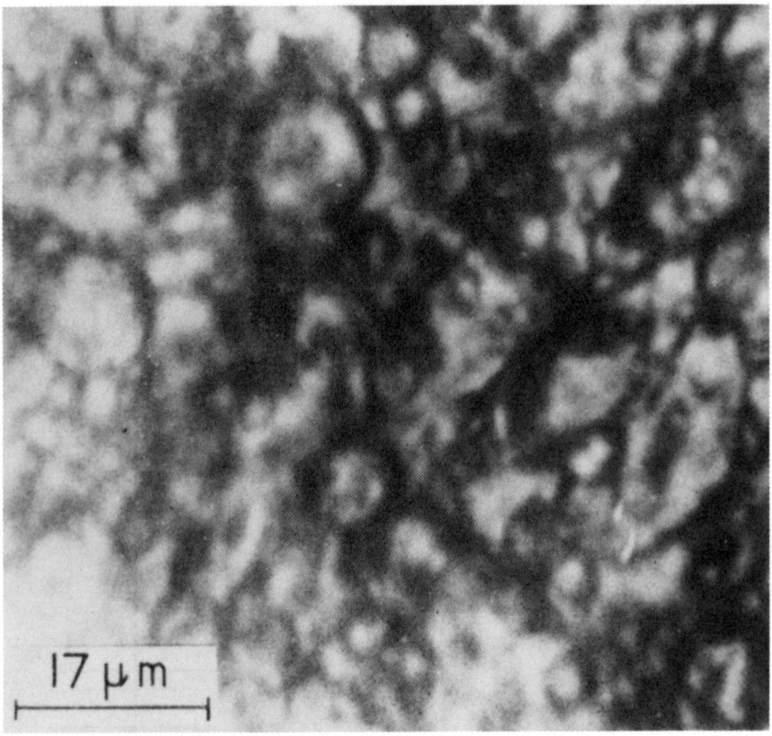

Fig. 10-16 Texture of a thermotropic copolyester nematic in a polarizing microscope at 350° C (from Mackley et al.,[21] reprinted from Polymer, © 1981 Butterworths).

$$m = \pm 1 \quad \text{and} \quad \pm 1/2 \tag{10-81}$$

The curvature energy density per unit length of disclination diverges because of the ever-growing curvatures encountered as the singularity is approached. Disclinations with integral m ($\pm$ 1) can avoid this singularity by "escaping to the third dimension." The director near the singularity simply rotates out of the plane and, at the disclination, becomes parallel to the disclination line. Steep director gradients near the escape region can then smooth out by diffusion and the near-singularity can gradually vanish. Half-integral disclinations have no such escape route and are therefore more stable, at least in small-molecule nematics.

Near the singularity in a half-integral disclination, curvatures lead to a breakdown in the continuum theory and produce an isotropic "core region" around the defect believed to be ~ 50-100 Angstroms in radius.[20]

In addition to line defects, there are integral *point defects,* analogous to the integral defects depicted in Figure 10-15. *Sheet defects* are mathematically possible, but can be shown to be unstable to a broadening process that would dissipate any sheet discontinuity.[8]

Polymeric nematics differ from small-molecule nematics in that the former typically have many more defects. Whereas small-molecule nematics can be freed of defects by careful preparation technique, especially with respect to surfaces, polymeric nematics are rarely defect free and are usually riddled with extremely persistent disclination lines;[21] see Figure 10-16. So thick with defects are most polymeric nematics that light is strongly attenuated by scattering when it passes through bulk samples. In thermotropics, the scattering is so intense that samples thicker than 15-20 microns are all but opaque.[21]

At a high rate of shear, disclination loops multiply in polymeric liquid crystals.[22] These disclination loops diminish in size with increased rates of shear until they vanish from the view of an optical microscope.

The density of defects can be quantified as the combined length of all disclinations in a unit volume of nematic; this is roughly $1/a^2$, where a is the average distance between neighboring disclinations.[23] Thus a represents the size of a "domain" — a region of approximately constant orientation that shows up in a polarizing microscope as a light region between extinction bands, and constitutes the smallest unit of the observed texture.[24]

10.6 VISCOUS FLOW OF NEMATICS

For small-molecule nematics, molecular relaxation times are short, and flow-producing mechanical stresses are usually small or modest. As a result, though viscous stresses affect the director orientation, $\underset{\sim}{n}$, they do not significantly perturb the scalar order parameter, S. Thus under flow the small-molecule nematic retains its equilibrium structure. Most small-molecule *isotropic* fluids also retain their equilibrium structure under flow; for these fluids, the stress is linear in the velocity gradient, and Newton's law of viscosity applies. For *anisotropic* small-molecule fluids, the stress tensor is also linear in the velocity gradient, but not all symmetry relations that simplify the constitutive equation for isotropic fluids apply to anisotropic fluids.

10.6.1 Ericksen's Transversely Isotropic Fluid

10.6.1.1 Constitutive Equation The simplest liquid crystalline materials are nematics; these are uniaxial fluids in which the anisotropy is described by a director $\underset{\sim}{n}$. The material is isotropic in the plane perpendicular to $\underset{\sim}{n}$. In 1960 Ericksen derived a general form for the stress tensor for these incompressible *transversely isotropic fluids*.[25] When $\underset{\approx}{\tau}$ is symmetric and linear in the velocity gradient, Ericksen's Equation* is

$$\underset{\approx}{\tau} = 2\mu \underset{\approx}{D} + 2\mu_1 \underset{\approx}{D} : \underset{\sim}{n}\underset{\sim}{n}\underset{\sim}{n}\underset{\sim}{n} + \mu_2 (\underset{\sim}{n}\underset{\sim}{n} \cdot \underset{\approx}{D} + \underset{\approx}{D} \cdot \underset{\sim}{n}\underset{\sim}{n}) \tag{10-82}$$

---

* An additional term proportional to $\mu_1 \underset{\sim}{n}\underset{\sim}{n}$, present in Ericksen's original theory, is here omitted because it produces a *yield stress,* that is, a stress that persists indefinitely in the absence of flow.

where μ, μ_1, and μ_2 are constants. Three analogous terms appear in the expression for the stress tensor for suspensions of rigid ellipsoids; see Equation 8-93.

Thus Equation 10-82 is for nematics the analog of Newton's law of viscous flow. Equation 10-82 contains explicitly the director $\underset{\sim}{n}$. Since the director can be influenced by the flow, an additional dynamic equation must be provided to couple $\underset{\sim}{n}$ to the velocity gradient. The most general relationship that gives $\partial \underset{\sim}{n}/\partial t$ in terms of $\underset{\sim}{n}$ and $\underset{\sim}{\nabla}\underset{\sim}{n}$ that satisfies frame invariance, the constraint $\underset{\sim}{n}\cdot\underset{\sim}{n}$, and that is linear in $\underset{\sim}{\nabla}\underset{\sim}{v}$ is

$$\frac{\partial}{\partial t}\underset{\sim}{n} - \underset{\sim}{n}\cdot\underset{\approx}{\omega} - \lambda(\underset{\sim}{n}\cdot\underset{\approx}{D} - \underset{\approx}{D}:\underset{\sim\sim\sim}{nnn}) \tag{10-83}$$

The term $-\underset{\approx}{D}:\underset{\sim\sim\sim}{nnn}$ is included to satisfy the constraint $\underset{\sim}{n}\cdot\underset{\sim}{n} = 1$. Alternatively, this constraint is satisfied if Equation 10-83 is replaced by

$$\underset{\sim}{n} \times (\frac{\partial}{\partial t}\underset{\sim}{n} - \underset{\sim}{n}\cdot\underset{\approx}{\omega} - \lambda\underset{\sim}{n}\cdot\underset{\approx}{D}) = \underset{\sim}{0} \tag{10-84}$$

The phenomenological constant λ should not be confused with a relaxation time. Equation 10-84 also appears in Section 8.7.2, where it describes the motion of axisymmetric ellipsoids in a flow field. For ellipsoids, $\underset{\sim}{n}$ is the orientation of the axis of symmetry and $\lambda = (r^2-1)/(r^2+1)$, where r is the aspect ratio of the ellipsoids, L_1/L_2, with L_1 being the length of the symmetry axis, and L_2 the length of each of the other two axes. For an ellipsoid, $\lambda \leq 1$, with $\lambda \rightarrow 1$ for a long slender rod. In a shear field without Brownian motion, the ellipsoids tumble in Jeffery orbits with a period $2\pi(r + r^{-1})/\dot{\gamma}$, where $\dot{\gamma}$ is the shear rate.[26]

For nematics, λ need not be less than or equal to unity. If $\lambda \geq 1$, straining dominates over vorticity in Equation 10-83, and the director does not tumble but tends toward a fixed angle, ϕ, with respect to the shearing direction, where

$$\tan\phi = [\frac{\lambda-1}{\lambda+1}]^{1/2} \tag{10-85}$$

As $\lambda \rightarrow \infty$, ϕ approaches an angle of 45° with respect to the flow direction; as $\lambda \rightarrow 1$, ϕ approaches 0°.

Since all terms in Equation 10-83 except the time derivative are proportional to the velocity gradient, the director orientation in a shear field is a function only of the initial orientation and of the total shear strain, $\dot{\gamma}t$. It is not a function of time, t, and strain rate, $\dot{\gamma}$, separately.

10.6.1.2 Viscosity Coefficients In incompressible linear *isotropic* fluids, the viscosity in any type of flow, say uniaxial extension, can be obtained directly from the shear viscosity, η; thus $\overline{\eta}_u = 3\eta$, where $\overline{\eta}_u$ is the Trouton elongational viscosity. For *anisotropic* Ericksen fluids, there are four independent viscosities obtained from the parameters μ, μ_1, μ_2, and λ. In simple shear, there are three stress components: the shear stress τ_{12}; the first normal stress difference $N_1 \equiv \tau_{11} - \tau_{22}$; and the second normal stress difference $N_2 \equiv \tau_{22} - \tau_{33}$. These are related to the Ericksen coefficients by

$$\frac{\tau_{12}}{\dot{\gamma}} = \eta = \mu + \frac{\mu_2}{2} + \frac{\lambda^2 - 1}{2\lambda^2}\mu_1$$

$$\frac{N_1}{\dot{\gamma}} = \frac{\sqrt{\lambda^2 - 1}}{\lambda^2}\mu_1$$

$$\frac{N_2}{\dot{\gamma}} = \frac{\sqrt{\lambda^2 - 1}}{2\lambda^2}\left[\lambda\mu_2 + (\lambda - 1)\mu_1\right] \tag{10-86}$$

For isotropic viscoelastic fluids, $N_1 \propto N_2 \propto \dot{\gamma}^2$ at low shear rates; in contrast for anisotropic fluids, $N_1 \propto N_2 \propto \tau_{12} \propto \dot{\gamma}$ at low shear rates. Thus in liquid crystals, normal stresses are as important at low shear rates as they are at high shear rates.

The uniaxial, planar, and biaxial extensional viscosities (defined in Section 6.2.3.2) for an Ericksen fluid are given by

$$\bar{\eta}_u = 3\mu + 2\mu_2 + 2\mu_1 \qquad \text{(uniaxial viscosity, with } n_x = 1 \text{)}$$

$$\left.\begin{aligned} \bar{\eta}_{p,1} &= 4\mu + 2\mu_2 + 2\mu_1 \\ \bar{\eta}_{p,2} &= 2\mu \end{aligned}\right. \qquad \text{(planar viscosities, with } n_x = 1 \text{)}$$

$$\bar{\eta}_b = 6\mu + 2\mu_2 + 2\mu_1 \qquad \text{(biaxial viscosity, with}^* \ n_x = 1) \tag{10-87}$$

10.6.2 Leslie-Ericksen Theory

Flow coefficients are most readily measured if the director $\underset{\sim}{n}$ is kept at a uniform in space, at some specified orientation. In small-molecule liquid crystals in the absence of body torques, the director field is determined by the competition between torques generated at the wall and those generated by viscous flow, and is generally nonuniform. The director can be made nearly uniform by imposing a magnetic field of sufficient strength to overwhelm the viscous torque, provided that the magnetically induced orientation is also either compatible with the wall orientation or is strong enough to compress the wall boundary layer to a small fraction of the bulk sample. A balance of viscous, magnetic, and wall-induced distortional torques gives[8]

$$\underset{\sim}{n} \times \underset{\sim}{h}_d + \underset{\sim}{n} \times \underset{\sim}{h}_M - \underset{\approx}{\underset{\sim}{\epsilon}} : \underset{\approx}{\tau} = \underset{\sim}{0} \tag{10-88}$$

* In biaxial extension, all director orientations in the x-y plane are equally stable; we arbitrarily choose the x axis so that it is parallel to the director.

The symbol $\underset{\approx}{\epsilon}$ is the *alternator*; it is defined as

$$\epsilon_{ijk} \equiv \begin{cases} +1 \text{ if } ijk = 123,\ 231,\ \text{or } 312 \\ -1 \text{ if } ijk = 213,\ 321,\ \text{or } 132 \\ 0 \quad \text{if any two indices are the same} \end{cases} \tag{10-89}$$

Acting on the stress tensor, $\underset{\approx}{\tau}$, the alternator produces a vector

$$\underset{\approx}{\epsilon} : \underset{\approx}{\tau} = (\tau_{32} - \tau_{23},\ \tau_{13} - \tau_{31},\ \tau_{21} - \tau_{12})$$

In the absence of magnetic and wall-induced distortional torques, the balance of torques, Equation 10-88, therefore leads to the condition

$$\tau_{ij} = \tau_{ji}, \quad j \neq i \qquad \text{(viscous torques only)} \tag{10-90}$$

that the stress tensor be symmetric. In general, however, *magnetic and wall-induced torques break the symmetry of the stress tensor.* Thus the Ericksen theory is not suitable when magnetic fields or wall effects are significant. An extension of Ericksen's theory, developed by Leslie, incorporates these effects.

10.6.2.1 Constitutive and Conservation Equations In Leslie's formulation,[27] the most general form is written for the asymmetric stress tensor, $\underset{\approx}{\tau}$, as a function of $\underset{\sim}{n}$, $\dot{\underset{\sim}{n}}$, and as a linear function of $\underset{\approx}{D}$:

$$\underset{\approx}{\tau} = \alpha_1 \underset{\sim}{n}\underset{\sim}{n}\underset{\sim}{n}\underset{\sim}{n} : \underset{\approx}{D} + \alpha_4 \underset{\approx}{D} + \alpha_5 \underset{\sim}{n}\underset{\sim}{n} \cdot \underset{\approx}{D} + \alpha_6 \underset{\approx}{D} \cdot \underset{\sim}{n}\underset{\sim}{n} + \alpha_2 \underset{\sim}{n}\underset{\sim}{N} + \alpha_3 \underset{\sim}{N}\underset{\sim}{n} \tag{10-91}$$

Here $\underset{\sim}{N}$ is defined as

$$\underset{\sim}{N} \equiv \dot{\underset{\sim}{n}} - \underset{\sim}{n} \cdot \underset{\approx}{\omega} \tag{10-92}$$

Thus $\underset{\sim}{N}$ is the rate of change of $\underset{\sim}{n}$ relative to the background fluid. The antisymmetric part of the viscous stress is

$$\underset{\approx}{\tau} - \underset{\approx}{\tau}^T = (\alpha_5 - \alpha_6)(\underset{\sim}{n}\underset{\sim}{n} \cdot \underset{\approx}{D} - \underset{\approx}{D} \cdot \underset{\sim}{n}\underset{\sim}{n}) + (\alpha_2 - \alpha_3)(\underset{\sim}{n}\underset{\sim}{N} - \underset{\sim}{N}\underset{\sim}{n}) \tag{10-93}$$

It can readily be shown that

$$\underset{\approx}{\epsilon} : \underset{\approx}{\tau} = \underset{\sim}{n} \times [(\alpha_6 - \alpha_5)\underset{\sim}{n} \cdot \underset{\approx}{D} + (\alpha_3 - \alpha_2)\underset{\sim}{N}] = \underset{\sim}{n} \times [\gamma_2 \underset{\sim}{n} \cdot \underset{\approx}{D} + \gamma_1 \underset{\sim}{N}] \tag{10-94}$$

where

$$\gamma_1 \equiv \alpha_3 - \alpha_2, \quad \gamma_2 \equiv \alpha_6 - \alpha_5 \tag{10-95}$$

The angular momentum balance Equation, 10-88, thus becomes

$$\underset{\sim}{n} \times (\underset{\sim}{h} - \gamma_2 \underset{\sim}{n} \cdot \underset{\approx}{D} - \gamma_1 \underset{\sim}{N}) = \underset{\sim}{0} \tag{10-96}$$

where the molecular field $\underset{\sim}{h}$ is given by Equation 10-78. The balance of linear momentum is

$$\underset{\sim}{\nabla} \cdot \underset{\approx}{\tau}^{d} + \underset{\sim}{\nabla} \cdot \underset{\approx}{\tau} - \underset{\sim}{\nabla} p = 0 \tag{10-97}$$

Here body forces, such as gravity, have also been neglected. Inertial terms have been neglected in Equations 10-96 and 10-97.

The conservation Equations, 10-96 and 10-97, together with the constitutive Equation, 10-91, the definition of the molecular field, Equation 10-80, and the continuity equation for an incompressible fluid,

$$\underset{\sim}{\nabla} \cdot \underset{\sim}{v} = 0 \tag{10-98}$$

define a complete set of equations for the unknowns $\underset{\sim}{n}$, $\underset{\sim}{v}$, $\underset{\approx}{\tau}$, and p.

In the absence of magnetic fields and director distortional stresses, the Ericksen Equation for transversely isotropic liquids, Equation 10-82, is recovered. The parameters in Ericksen's theory are then obtained from Leslie's parameters:

$$\lambda = -\frac{\gamma_2}{\gamma_1} = \frac{\alpha_5 - \alpha_6}{\alpha_3 - \alpha_2}$$

$$\mu_1 = \frac{\alpha_1}{2} - \frac{(\alpha_2 + \alpha_3)(\alpha_6 - \alpha_5)}{2(\alpha_2 - \alpha_3)}$$

$$\mu_2 = \frac{\alpha_2\alpha_6 - \alpha_3\alpha_5}{\alpha_2 - \alpha_3}$$

$$\mu = \frac{\alpha_4}{2} \tag{10-99}$$

10.6.2.2 Values of the Leslie Coefficients Of the six Leslie coefficients, only five are independent, according to a relationship derived by Parodi,[28]

$$\alpha_6 = \alpha_2 + \alpha_3 + \alpha_5 \qquad \text{(Parodi's relationship)} \tag{10-100}$$

Parodi derived Equation 10-100 from an Onsager relationship, but Currie[29] showed that it is a necessary condition if the state of uniform orientation is to be stable under small disturbances.

Complete sets of values for the Leslie coefficients are rarely measured. The following values, in centipoise, were obtained for N-(p-methoxybenzylidene)-p-butylaniline (MBBA), a small-molecule liquid crystal, at room temperature:[30]

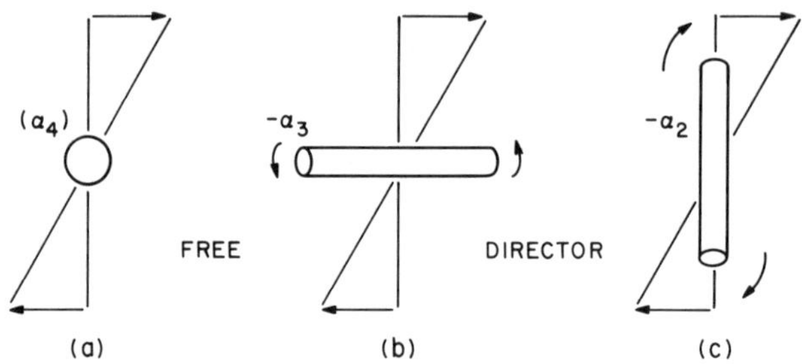

Fig. 10-17 The Leslie coefficients α_3 and α_2 determine the rate and direction of free rotation of the director in a shear field. (adapted from Skarp, et al.,[31]).

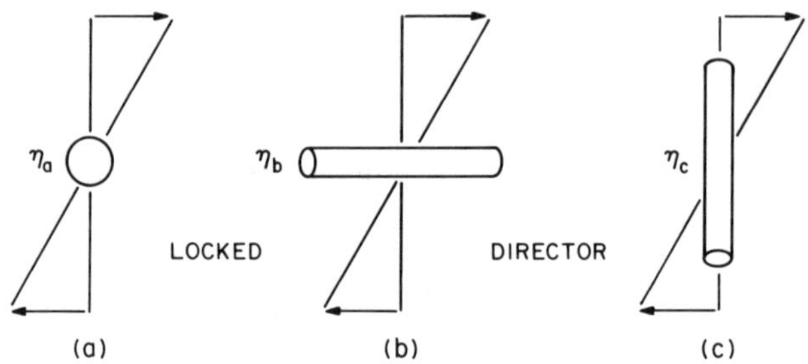

Fig. 10-18 The Miesowicz viscosities η_a, η_b, and η_c are viscosities measured with the director locked in the positions shown (adapted from Skarp, et al.,[31]).

$$\alpha_1 = 6.5 \pm 4; \quad \alpha_2 = -77.5 \pm 1.6; \quad \alpha_3 = -1.2 \pm 0.1$$

$$\alpha_4 = 83.2 \pm 1.4; \quad \alpha_5 = 46.3 \pm 4.5; \quad \alpha_6 = -32.4 \pm 2.2$$

The most important of these values are α_3, α_2, and α_4. α_4 is the coefficient of the Newtonian term $\underset{\approx}{D}$. It produces viscous drag even when the rods lie perpendicular to both the velocity and velocity-gradient directions, as in Figure 10-17a. α_3 determines the torque transmitted from the background fluid to a rod oriented in the direction of flow as in Figure 10-17b. Unless α_3 is negative, a free rod tumbles through the shearing plane and does not reach a steady-state orientation. α_2 determines the torque transmitted to a rod aligned in the direction of the velocity gradient. It is always found to be negative, corresponding to clockwise rotation in Figure 10-17c.

10.6.2.3 Miesowicz Viscosities Since the viscous properties of the liquid crystal depend on the orientation of the director, it is convenient to measure the viscosity under conditions in which the director is held fixed by a magnetic field. When the director is held fixed in each of the three orientations shown in Figure 10-18a–c, the three *Miesowicz shear viscosities,*[32] η_a, η_b, and η_c, are measured. The Miesowicz viscosities are related to the Leslie coefficients by

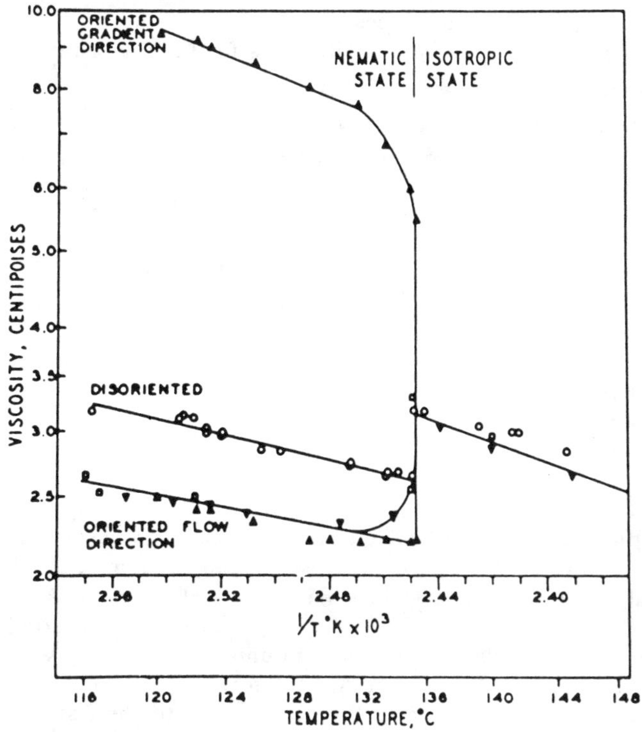

Fig. 10-19 Dependence of Miesowicz viscosities η_b (director oriented in the flow direction) and η_c (director oriented in the gradient direction) and the viscosity in the absence of a magnetic field on temperature near the transition from the nematic to the isotropic state (from Porter and Johnson,[35] reprinted with permission from *Rheology,* Vol. 4, Academic Press, 1967).

$$\eta_a = \tfrac{1}{2}\alpha_4 \ ; \ \eta_b = \tfrac{1}{2}(\alpha_3 + \alpha_4 + \alpha_6) \ ; \ \eta_c = \tfrac{1}{2}(-\alpha_2 + \alpha_4 + \alpha_5) \qquad \text{(10-101)}$$

For MBBA, the Miesowicz viscosities are

$$\eta_a = 41.6 \pm 0.7 \ ; \quad \eta_b = 24.8 \pm 0.3 \ ; \quad \eta_c = 103.5 \pm 1.5$$

in centipoise. For a polymeric liquid crystal, poly(γ-benzyl-glutamate),* Tarantuta, et al.[33] measured a large anisotropy: $\eta_B = 0.16$ poise, and $\eta_c = 35$ poise.

* Poly(γ-benzyl-glutamate) or PBG is a racemic mixture of two optically active enantiomers, poly(γ-benzyl-L-glutamate), PBLG, and poly (γ-benzyl-D-glutamate), PBDG. Each enantiomer is a polypeptide that forms rigid helices in solution and, if concentrated enough, orders into a *cholesteric* liquid crystal. In the racemic mixture, the cholesteric character of the enantiomers is negated, and the mix is considered nematic. The cholesteric character of the individual enantiomers seems to have little influence on their rheology,[34] however, and theories for nematics are often used to interpret measurements made on the single enantiomer PBLG.

The magnitudes of the Miesowicz viscosities are always found to obey $\eta_b < \eta_a < \eta_c$. Figure 10-19 shows the Miesowicz viscosities for PAA as a function of temperature in both nematic and isotropic ranges.

The director is held fixed in the flow field when the viscous torque per unit volume is much less than the magnetic torque per unit volume; i.e.,

$$\mu \dot{\gamma} << \chi_a H^2 \tag{10-102}$$

where μ is a typical viscosity, $\mu \sim$ 0.01 to 1.0 poise for small-molecule liquid crystals, and $\mu \sim$ 50 poise to 5,000 poise for polymeric liquid crystals. For a strong, 10 kgauss magnetic field, $\chi_a H^2 \approx 10$ dyn/cm^2. Thus the director can easily be locked in position for small-molecule liquid crystals, while for polymeric liquid crystals the director is dislodged by the flow unless the shear rate is less than $10^{-3} - 10^{-1}$ sec^{-1}.

10.6.2.4 Typical Ericksen Coefficients For small molecules, the α_i's generally lie in the range 1—100 cp. α_2 is usually found to be *negative*, corresponding to a steady-state director orientation in simple shear; that is, the director does not tumble.* Furthermore $|\alpha_3| << |\alpha_2|$ and α_3 is negative, so $\lambda = (\alpha_2+\alpha_3)/(\alpha_2-\alpha_3)$ is only slightly greater than unity. For MBBA, $\lambda = 1.033$, and therefore from Equation 10-85, far from the influence of walls or disclinations the director approaches an angle of $\phi \approx 7°$ with respect to the flow direction. In small-molecule liquid crystals, this angle tends to grow as the temperature is raised toward the isotropic transition temperature.

The Ericksen parameters for MBBA are

$$\lambda = 1.033\ ; \quad \mu_1 = 43.8 \text{ cp}; \quad \mu_2 = -33.64 \text{ cp}; \quad \mu = 41.6 \text{ cp} \qquad \text{(MBBA)}$$

For these parameter values, one can predict the following values for stresses and stress ratios in simple homogeneous flow fields in the absence of magnetic or wall effects:

$$\tau_{xy}/\dot{\gamma} \equiv \eta = 26.2 \text{ cp};$$

$$N_1/\tau_{xy} = 0.41\ ; \quad N_2/N_1 = -0.38\ ; \quad \bar{\eta}_u/\eta = 5.63 \qquad \text{(MBBA)}$$

10.6.2.5 Parameters of the Doi Theory The Doi theory reduces to that of Ericksen at shear rates low enough that the stress, in the Doi theory, is linear in the velocity gradient. The Ericksen parameters calculated from the linearized Doi theory are

* α_2 has been found to become positive in nematics at temperatures close to a transition to a smectic - A phase.[36]

$$\lambda = \frac{1}{3S}(2+S)\,; \qquad \mu = \frac{1}{3}(1-S)\bar{\eta}\,;$$

$$\mu_1 = (-S^2 + \frac{1}{3}(2+S))\bar{\eta}\,; \qquad \mu_2 = -\frac{2}{3}(1-S)\bar{\eta} \tag{10-103}$$

where

$$\bar{\eta} \equiv \frac{\nu kT}{2\bar{D}_r} \tag{10-104}$$

The Leslie parameters are[1]

$$\alpha_1 = -2S^2\bar{\eta}\,; \quad \alpha_2 = -S(1+\frac{3S}{2+S})\bar{\eta}\,; \quad \alpha_3 = -S(1-\frac{3S}{2+S})\bar{\eta}$$

$$\alpha_4 = \frac{2}{3}(1-S)\bar{\eta}\,; \qquad \alpha_5 = 2S\bar{\eta}\,; \qquad \alpha_6 = 0 \tag{10-105}$$

The value of λ is greater than unity, which implies that the director does not tumble in a shear flow. This consequence of the Doi theory, however, is an artifact of the preaveraging approximation; when preaveraging is avoided, one finds that λ is slightly less than unity.[1,37] Also, without preaveraging, $\alpha_6 \neq 0$.

The zero shear viscosity of the Doi theory is[1]

$$\eta = \frac{\bar{\eta}}{3}\frac{(1-S)^2(1+2S)(1+3S/2)}{(1+S/2)^2} \tag{10-106}$$

Taking Equation 10-61 for $\bar{D}_r$ into account,

$$\frac{\eta}{\eta_*} = (\frac{\nu}{\nu^*})^3\frac{(1-S)^4(1+S)^2(1+2S)(1+3S/2)}{(1+S/2)^2} \tag{10-107}$$

At low shear rates, the first and second normal stress differences, N_1 and N_2, are[38]

$$\frac{N_1}{\tau_{12}} = \frac{3S}{(1-S)^{1/2}(1+2S)^{1/2}} \begin{cases} = 1.5 & \text{for } \nu = \nu^* \\ \rightarrow \infty & \text{for } \nu >> \nu^* \end{cases}$$

$$\frac{N_2}{N_1} = \frac{-S}{2+3S} = \begin{cases} -1/7 & \text{for } \nu = \nu^* \\ -1/5 & \text{for } \nu >> \nu^* \end{cases} \tag{10-108}$$

The uniaxial extensional viscosity at low rates of extension is

$$\frac{\overline{\eta}_u}{\eta} = \frac{3S(1+S/2)^2}{(1-S)(1+3S/2)} \begin{cases} = 5.36 & \text{for } \nu = \nu^* \\ \to \infty & \text{for } \nu >> \nu^* \end{cases} \tag{10-109}$$

10.6.3 Boundary Effects

In flow through a confined geometry, such as flow through a duct, the director distortional energy, W_d, competes with the shear field to determine the local orientation of the director $\underset{\sim}{n}$. The distortional stress scales as K/h^2, where h is the diameter of the duct, and K is a curvature stress coefficient, typically of magnitude $\sim 10^{-6}$ dyn. The ratio of the shear stress τ_{xy} to the distortional stress, $\tau^d \sim K/h^2$, thus determines the extent to which the director takes on an orientation dictated by the flow.[13,39] Since $\tau_{xy} \propto \dot{\gamma} \sim \eta V/h$, where η is a true shear viscosity and V is the mean velocity in the duct, the stress ratio τ_{xy}/τ^d is

$$\frac{\tau_{xy} h^2}{K} \approx \eta \frac{Vh}{K} \equiv E \tag{10-110}$$

where E is called the Ericksen number. The average director orientation, and hence the average or apparent viscosity of the fluid is a function of V *multiplied by* h. This contrasts with shear thinning isotropic polymer fluids for which wall effects are negligible; for these in flow through ducts the apparent shear viscosity equals the true viscosity and is a function of $\dot{\gamma}$, which is V *divided by* h.

When $E \approx \eta\dot{\gamma}h^2/K >> 1$, the influence of wall orientation on the bulk of the fluid in the duct is expected to become insignificant. Thus for $K \approx 10^{-6}$ dyn and $\dot{\gamma} = 1\ \text{sec}^{-1}$, wall orientation effects are expected to be insignificant in small-molecule nematics ($\eta \sim 0.10$ poise) for ducts of 3,000 microns or larger, and for polymeric nematics ($\eta \sim 100$ poise) in ducts larger than 100 microns. Wall effects have indeed been found to be significant in flows of small-molecule liquid crystals,[40] but are usually negligible in liquid crystal polymers[38,41] (see, however, Asada et al.[42]).

10.7 RHEOLOGY OF LIQUID CRYSTAL POLYMERS

Important simplifying features of typical isotropic polymers, such as time-temperature superposition, the stress-optic law, and relationships between stresses in steady and oscillatory flows, do not apply to liquid crystalline polymers. Consequently, the rheology of liquid crystalline polymers is much more complex than that of isotropic polymeric liquids. Here we survey some of the contrasts between isotropic and anisotropic polymer rheology. For a more complete review, see Wissbrun.[13]

10.7.1 Temperature-Dependent Rheology

The characterization of *isotropic* polymeric is greatly simplified by time-temperature superposability, which allows one to predict the linear and even nonlinear viscoelastic response of a material over a wide range of temperatures from its response at a single temperature, as discussed in Section 2.8.2. The zero-shear viscosity, or any other viscosity of an isotropic polymer melt or concentrated

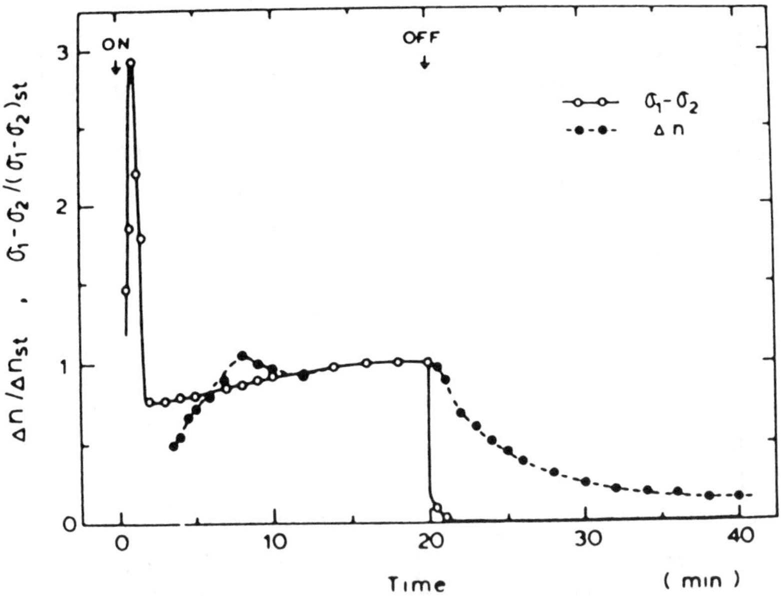

Fig. 10-20 Birefringence, Δn, and principal stress, $\sigma_1-\sigma_2$, after start-up ("on") and cessation ("off") of steady shearing for a liquid crystalline hydroxypropyl cellulose solution. The subscript, st refers to apparent steady state (from Onogi, et al.,[45] reprinted with permission from J. Non-Newt. Fluid Mech.).

solution is proportional to a_T, the temperature-dependent shift factor; a_T monotonically decreases as T increases.

Time-temperature superposability does *not* apply, at least in a simple way, to liquid crystalline polymers. For poly(γ-benzyl-L-glutamate) (PBLG), curves of shear viscosity versus shear rate at various temperatures can sometimes be shifted into superposition, but the shift factor for the steady shear viscosity is different from that for the dynamic viscosity.[41] Furthermore, the transient shear viscosities after cessation of steady shear cannot be made to superpose by shifting. For some thermotropic liquid crystal polymers the viscosity is not even a monotonic function of temperature.[43] As the temperature rises, thermal motion increases, tending to decrease the viscosity; but the nematic character of the polymer may decrease, tending to increase the viscosity. Chemical changes may also occur at the elevated melt temperatures of many thermotropes. Thus competing effects produce a complex temperature dependence in liquid crystalline polymer rheology.[44]

10.7.2 Birefringence

Since the molecules in liquid crystalline polymers have a preferential orientation, these materials are highly birefringent even at equilibrium. Because of its domain-filled texture, the orientation of the principal axes of the birefringence varies from place to place in a quiescent bulk sample, and the sample is on

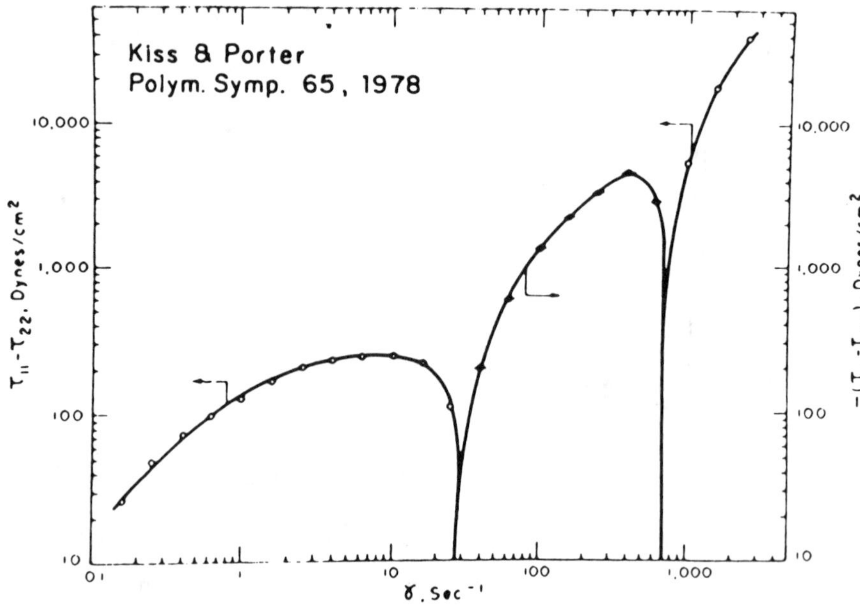

Fig. 10-21 Steady-state first normal stress difference versus shear rate for a liquid crystalline solution of PBG (from Kiss and Porter,[34] reprinted from J. Polym. Sci., Polym. Symp., © 1978 by permission of John Wiley & Sons, Inc.).

average relatively isotropic. When sheared, the domains orient and a large net birefringence results, at least in lyotropics. This birefringence is intimately related to the domain characteristics and is not simply proportional to the stress tensor as is the case for many isotropic polymers. Figure 10-20 shows that the birefringence of liquid crystalline hydroxypropyl cellulose requires many minutes to relax; the stress relaxes in seconds.[45] The time scale for relaxation of the birefringence is comparable to the time required for restoration of the static texture.

10.7.3 Relationship between Linear and Nonlinear Rheology

For *isotropic* fluids, the stresses measured in slow steady shear are directly related to those measured in small-amplitude dynamic oscillatory shearing at low frequency. Thus $\eta(\dot{\gamma}) = \eta^*(\omega) \equiv \sqrt{\eta'^2 + \eta''^2}$, where $\dot{\gamma} = \omega$ is small. Also $N_1(\dot{\gamma}) = 2G'(\omega)$ for small values of $\dot{\gamma} = \omega$. For isotropic fluids the relationship $\eta(\dot{\gamma}) = \eta^*(\omega)$ often holds even at moderate and high frequencies; it is then called the Cox-Merz relationship.

For anisotropic fluids, these relationships should not be expected to hold. Experiments on lyotropic and thermotropic liquid crystalline polymers, reviewed by Wissbrun,[13] show that $\eta^* \approx \eta$ for some polymers, while for others $\eta^* >> \eta$. It is almost always found that $N_1 >> 2G'$.

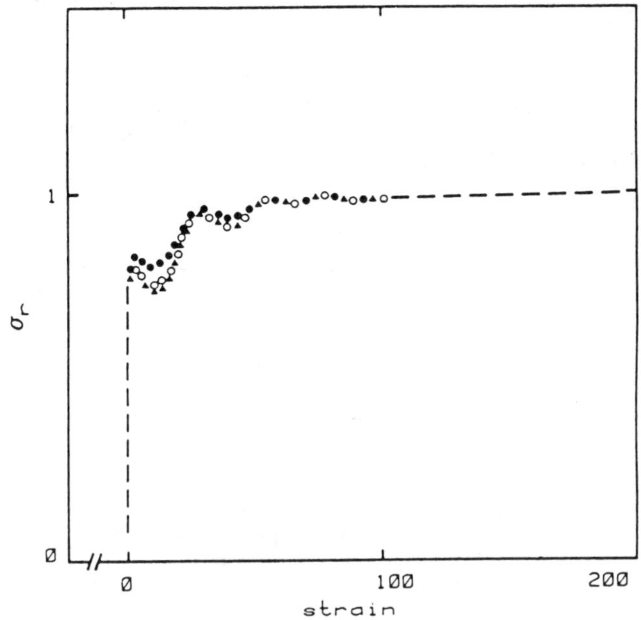

Fig. 10-22 Normalized shear stress after step-up of shear rate from 0.05 sec^{-1} to (•) 0.2 sec^{-1}, (○) 0.5 sec^{-1}, and (Δ) 1.0 sec^{-1} for liquid crystalline PBLG (from Moldenaers[41]).

10.7.4 First Normal Stress Difference

For a nematic fluid, the first normal stress difference is predicted by the Ericksen and Doi theories to be linear in the shear rate at low shear rates, in contrast to a quadratic dependence found in isotropic fluids. All measurements with liquid crystalline polymers have shown $N_1 \propto \dot{\gamma}^m$ with $m < 2$; usually $m \approx 1$, as predicted.[41] The ratio N_1/τ_{xy} has been found to be 0.75 for PBLG.[41] According to the Doi theory, Equation 10-107, this ratio has a minimum value of 1.5, double the experimental value.

Kiss and Porter[34,46] in 1978 reported the highly unusual phenomenon of ***negative first normal stress differences*** in steady-state shearing. Figure 10-21 shows that steady-state positive values of N_1 were obtained for PBG at high and low shear rates with negative values at intermediate shear rates. Moldenaers[41] reported that oscillations are sometimes superposed on the negative values of N_1 and that the curve of shear viscosity versus shear rate is slightly distorted in the region of negative N_1. Kiss and Porter noted that the region of negative N_1 correlated with the appearance in the texture of striations perpendicular to the flow direction. Navard[47] found steady-state negative N_1 values in a solution of HPC in water, and Gotsis and Baird[48] found a negative N_1 in a thermotropic copolyester. Also for thermotropics, Wissbrun[13] has noticed ***transient*** negative N_1 values (with positive values at steady state). The Doi theory predicts only positive values for N_1, both during start-up and at steady state.

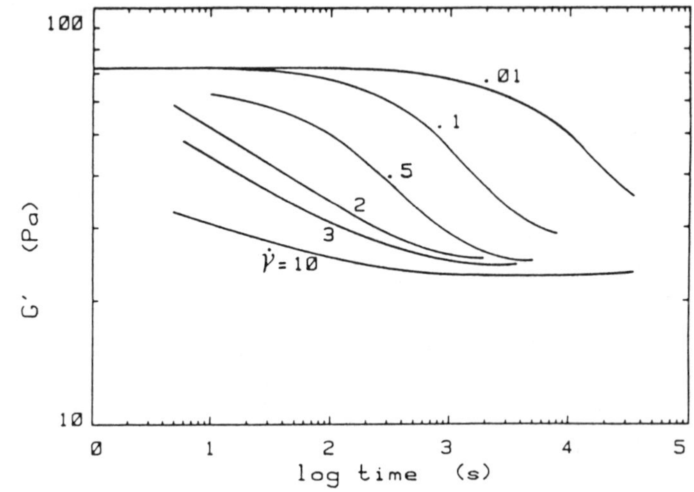

Fig. 10-23 Time-dependence of the storage modulus after cessation of steady shearing at the shear rates shown for liquid crystalline PBLG (from Moldenaers[41]).

10.7.5 Complex Time-Dependent Stresses

Because it ignores texture, the Doi theory is limited in its ability to describe the rheological behavior of liquid crystalline polymers. This is especially true of transient rheological properties. Figure 10-22 shows the shear stress measured by Moldenaers[41] in PBLG after a step-up in shear rate from an initial shear rate of 0.05 sec^{-1} to final rates ranging from 0.2, 0.5, and 1.0 sec^{-1} The stress shows at least three overshoots, and steady state is achieved only after some 50 strain units have been imposed! Note that the peaks in the stress occur at strain values that are independent of the final strain rate. These results may be indicative of gradual changes in structure or texture induced by shearing.

A more definitive mechanical probe of gradual structural changes in PBLG is provided by oscillatory-shear measurements of G' and G'', after cessation of steady shearing at a shear rate $\dot{\gamma}_0$. Mewis and Moldenaers[41,49] found that both of these moduli gradually *decrease* after cessation of steady shearing. Naive intuition would suggest that steady shear ought to disrupt fluid structure; at cessation of shearing, structure would then recover and the modulus would increase, in contrast to experiments. The time required for the modulus to reach its final value decreases as the shear rate $\dot{\gamma}_0$ increases, Figure 10-23. For small $\dot{\gamma}_0$, this time span is of order one hour and is a couple of orders of magnitude longer than the time required for relaxation of the nonoscillatory shear stress after cessation of steady shearing. It is likely that these changes in the moduli are associated with gradual changes in texture, while stress relaxation is influenced predominantly by the more rapid relaxation of molecular orientation. Thus the time required for equilibration of G' and G'' after cessation of shearing may be comparable to that required for relaxation of birefringence. Both of these times are much longer than the time required for mechanical stress relaxation (see Figure 10-20), and both are

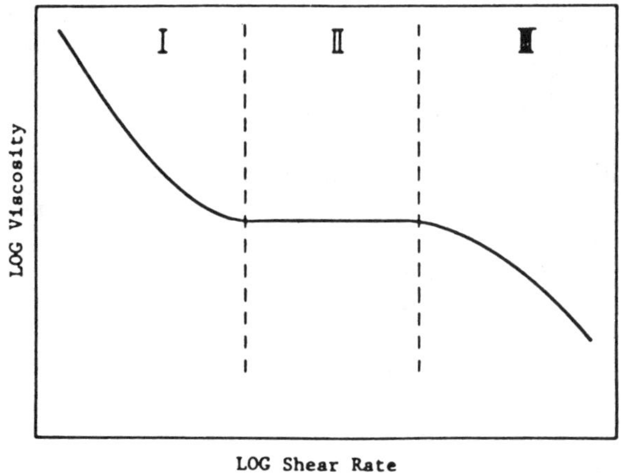

Fig. 10-24 Three regions of flow behavior in liquid crystal polymers, suggested by Onogi and Asada (from Onogi and Asada,[50] reprinted with permission from *Rheology*, Vol. 1, ©1980 Plenum Press).

probably associated with *relaxation of texture*.[41,48]

It is not yet known whether unusual time-dependent behavior like that shown in Figures 10-22—10-23 will also be found in liquid crystal polymers other than PBLG.

10.7.6 Domain Theories

Thus in addition to molecular relaxation, liquid crystalline polymers have much slower *texture relaxation* processes that can affect optical and rheological properties. Some efforts are now being made to account for the effects of texture or domains on orientation and stress in liquid crystal polymers. Onogi and Asada[50,42] noticed that the viscosity-versus-shear-rate curves sometimes have more or less distinct regions that correlate with distinct types of texture, Figure 10-24. There is sometimes a shear thinning region (region I) at low shear rates[51,42] that is associated with a "piled polydomain" texture. As the shear rate increases, the domains decrease in size, and the viscosity may level off to a constant viscosity plateau (region II). At high shear rates, the samples sometimes appear nearly homogeneous, perhaps because domains are too small to be seen. In this high shear-rate region (region III), the viscosity again shows shear thinning. Unfortunately, all three regions have rarely if ever been seen in a single sample, possibly because of the limited range of experimentally accessible shear rates. Region I shear thinning is distinguished from that of region III by the location of the Newtonian plateau of region II in relation to the region of shear thinning.

The Doi theory predicts a low-shear-rate constant-viscosity plateau that gives way to shear thinning at high shear rates, because of saturation of the order parameter. Thus the Doi theory could account for regions II and III,[52] but not region I, which, when it occurs, is thought to originate from the domain texture.

Marrucci[23] has proposed a model to explain shear thinning in region I. The model supposes that, under shear, disclinations are stretched and domains become smaller, producing an increased Frank distortional stress, τ^d. The increased elastic distortional stress is balanced by the viscous stress, τ^v, that is produced by flow that occurs in regions between the domains; thus $\tau^d = \tau^v$. As the shear rate increases, the domains shrink and $\tau^d = \tau^v$ increases. The increase is less than linear in shear rate, because the regions between the domains through which flow occurs enlarge as the domains shrink. In Marrucci's theory, $\tau^d = \tau^v \propto \dot{\gamma}^{1/2}$; hence $\eta \propto \dot{\gamma}^{-1/2}$. The number of domains remains constant in Marrucci's theory even as they decrease in size. This assumption is removed in a modification of the theory by Wissbrun.[53] Based on measurements of relaxation of molecular orientation after alignment in a magnetic field, Moore and Denn[24] suggest that the distortional stress associated with domains is viscoelastic and not merely viscous.

10.8 SUMMARY

In the isotropic semidilute concentration regime, the rheology of rodlike polymers can be described by a simple tube model of Doi and Edwards, if due allowance is made for the effects of polydispersity. The predicted nonlinear behavior is qualitatively like that predicted by the Doi and Edwards theory for entangled flexible polymers. However, solutions of semidilute rods are simpler rheologically and better described by the tube model then are entangled flexible polymers, because the latter can stretch in flow fields and therefore, in addition to reptation, have retraction and Rouse-like modes of relaxation that are not present in solutions of rigid rods.

If the rods are sufficiently concentrated, or if the material is a bulk melt of rigid rodlike molecules, equilibrium anisotropy, namely nematic liquid crystallinity, may be present. The statics and dynamics of small-molecule nematics are well understood, including the effects of flow, magnetic fields, and wall-induced orientational distortions. The Leslie-Ericksen theory, which in the absence of magnetic fields or distortions simplifies to Ericksen's transversely isotropic fluid, provides a thorough description of the statics and dynamics of small-molecule nematics at most flow rates and, in principal, of polymeric nematics at slow flow rates in a defect-free sample.

Familiar laws that apply to isotropic polymeric fluids, such as time-temperature superposability, the stress optic-law, and relationships between linear and nonlinear viscoelasticity, are neither predicted nor observed in nematic fluids. Some new rules, such as a linear dependence of the first normal stress difference on shear rate at small shear rates, are predicted and found to hold approximately.

The linear and nonlinear rheology of polymeric nematics is predicted by the Doi theory, which simply adds an orientation-dependent excluded volume potential of Maier-Saupe form to the usual Brownian dynamics for semidilute rods. The Doi theory, though simple and elegant, is limited in its ability to describe polymer liquid crystal rheology, because it does not account for their defect-filled texture. Defects, which can also be present to a lesser extent in small-molecule liquid crystals, affect both optical and mechanical properties to such a degree that their influence must be accounted for in any realistic rheological description.

A few speculative theories have been proposed and are now being tested and developed.

References

1. M. Doi and S.F. Edwards, *The Theory of Polymer Dynamics*, Oxford Press, New York (1986).
2. Y. Mori, N. Ookubo, R. Hayakawa, and Y. Wada, J. Polym. Sci., *20*, 2111 (1982).
3. M. Doi and S.F. Edwards, J. Chem. Soc., Faraday II, *74*, 560; 918 (1978).
4. Ref. 1, p. 309.
5. N.Y. Kuzuu and M. Doi, Polym. J., *12*, 883 (1980).
6. G. Marrucci and N. Grizzuti, J. Polym. Sci., Polym. Lett Ed., *21*, 83 (1983).
7. A.W. Chow, G.G. Fuller, D.G. Wallace, and J.A. Madri, Macromolecules, *18*, 805 (1985).
8. P.G. de Gennes, *The Physics of Liquid Crystals*, Oxford Press, New York (1974).
9. L. Onsager, Ann. N.Y. Acad. Sci., *51*, 627 (1949).
10. P.J. Flory, Proc. Roy. Soc., *A234*, 73 (1956).
11. P.J. Flory, Advances in Polymer Sci., *59*, 1 (1984).
12. W. Maier and A. Saupe, Z. Naturforsch., *13A*, 564 (1958); *14A*, 882 (1959); *15A*, 287 (1960).
13. K.F. Wissbrun, J. Rheol., *25*, 619 (1981).
14. M. Doi, J. Polym. Sci., Polym. Phys. Ed., *19*, 229 (1981).
15. C.W. Oseen, Trans. Faraday Soc., *29*, 883 (1933).
16. H. Zocher, Trans. Faraday Soc., *29*, 945 (1933).
17. R.C. Frank, Discuss. Faraday Soc., *25* 19 (1958).
18. J.L. Ericksen, Arch. Rat. Mech. Anal., *23*, 266 (1966).
19. M.J. Stephen and J.P. Straley, Rev. Mod. Phys., *46*, 617 (1974).
20. W.F. Brinkman and P.E. Cladis, Physics Today, p 48, May 1982.
21. M.R. Mackley, R. Pinaud, and G. Siekmann, Polymer, *22*, 437 (1981).
22. N.J. Alderman and M.R. Mackley, Faraday Discuss. Chem. Soc., *79*, paper 12 (1985).
23. G. Marrucci, Proceedings of the Ninth Intern. Congress on Rheology, Acapulco, Mexico, Oct. 8—13, 1984.
24. R.C. Moore and M.M. Denn in *The Path to High Modulus Polymers with Stiff and Flexible Chains*, Eds. A.E. Zachariades and R.S. Porter, Marcel Dekker, New York (1987).
25. J.L. Ericksen, Arch. Rat. Mech. Anal., *4*, 231 (1960); Trans. Soc. Rheol., *5*, 23 (1961).

26. E.J. Hinch and L.G. Leal, J. Fluid Mech., *57*, 753 (1973).

27. F.M. Leslie, Quart. J. Appl. Math., *19*, 357 (1966); Arch. Rat. Mech. Anal., *28*, 265 (1968).

28. O. Parodi, J. Phys. (Fr.), *31*, 581 (1970).

29. P.K. Currie, Mol. Cryst. Liq. Cryst., *28*, 335 (1974).

30. Ref. 8, p.189; note however that there are errors in the Miesowicz viscosities.

31. K. Skarp, S.T. Lagerwall, and B. Stebler, Mol. Cryst. Liq. Cryst., *60*, 215 (1980).

32. M. Miesowicz, Nature, *17*, 261 (1935); Bull. Acad. Pol. Sci., *A228*, (1936); Nature, *158*, 27 (1946).

33. V.G. Tarantuta, A.J. Hurd, and R.B. Meyer, Phys. Rev. Lett., *55*, 246 (1985).

34. G. Kiss and R.S. Porter, J. Polym. Sci., Polym. Symp., *65*, 193 (1978).

35. R.S. Porter and J.F. Johnson, *Rheology*, Vol. 4, Ed. F.R. Eirich, Academic Press, New York (1967).

36. P.E. Cladis and S. Torza, Phys. Rev. Lett., *35*, 1283 (1975).

37. N. Kuzuu and M. Doi, J. Phys. Soc. Japan, *52*, 3486 (1983); *53*, 1031 (1984).

38. A. B. Metzner and G.M. Prilutski, J. Rheol., *30*, 661 (1986).

39. P.K. Currie, Rheol. Acta, *16*, 205 (1977).

40. J. Fisher and A.G. Fredrickson, Mol. Cryst. Liq. Cryst., *8*, 267 (1969).

41. P. Moldenaers, Ph.D. Thesis, Katholicke Universiteit Leuven (1987).

42. T. Asada, S. Onogi, and H. Yanase, Polym. Engng. Sci., *24*, 355 (1984).

43. D.A. Simoff and R.S. Porter, Mol. Cryst. Liq. Cryst., *110*, 1 (1984).

44. G. Kiss, J. Rheol., *30*, 585 (1986).

45. Y. Onogi, J.L. White, and J.F. Fellers, J. Non-Newt. Fluid Mech., *7*, 121 (1980).

46. G. Kiss and R.S. Porter, Mol. Cryst. Liq. Cryst., *60*, 267 (1980).

47. P. Navard, presented at the Second Conference of European Rheologists, Prague (1986).

48. A.D. Gotsis and D.G. Baird, J. Rheol., *29*, 538 (1985); Rheol. Acta, *25*, 275 (1986).

49. P. Moldenaers and J. Mewis, J. Rheol., *30*, 567 (1986).

50. S. Onogi and T. Asada, *Rheology*, Vol. 1, Eds. G. Astarita, G. Marrucci, and L. Nicolais, Plenum Press, New York (1980).

51. G.C. Berry, Faraday Discuss. Chem. Soc., April, 1985.

52. D. Doraiswamy and A.B. Metzner, to be published in Rheol. Acta (1987).

53. K. Wissbrun, Faraday Discuss. Chem. Soc., *79*, 161 (1985).

AUTHOR INDEX

SUBJECT INDEX